Molecular, Cellular, and Blood Biomarkers in Diagnosis, Prognosis, Monitoring, and Treatment

Molecular, Cellular, and Blood Biomarkers in Diagnosis, Prognosis, Monitoring, and Treatment

Guest Editors

Jordi Camps
Isabel Fort Gallifa
Xavier Gabaldó Barrios

MDPI

Basel • Beijing • Wuhan • Barcelona • Belgrade • Novi Sad • Cluj • Manchester

Guest Editors
Jordi Camps
United de Recerca Biomèdica
Universitat Rovira i Virgili
Reus
Spain

Isabel Fort Gallifa
Clinical Laboratory of the
Hospital Universitari Joan
XXIII de Tarragona
Universitat Rovira i Virgili
Tarragona
Spain

Xavier Gabaldó Barrios
Clinical Laboratory of the
Hospital Universitari Sant
Joan de Reus
Universitat Rovira i Virgili
Reus
Spain

Editorial Office
MDPI AG
Grosspeteranlage 5
4052 Basel, Switzerland

This is a reprint of the Special Issue, published open access by the journal *Biomolecules* (ISSN 2218-273X), freely accessible at: https://www.mdpi.com/journal/biomolecules/special_issues/1709RA9DHV.

For citation purposes, cite each article independently as indicated on the article page online and as indicated below:

Lastname, A.A.; Lastname, B.B. Article Title. *Journal Name* **Year**, *Volume Number*, Page Range.

ISBN 978-3-7258-7506-1 (Hbk)
ISBN 978-3-7258-7507-8 (PDF)
https://doi.org/10.3390/books978-3-7258-7507-8

Cover image courtesy of Jordi Camps

Contents

Preface vii

Jordi Camps, Isabel Fort-Gallifa and Xavier Gabaldó-Barrios
The Transformative Role of Molecular, Cellular, and Blood Biomarkers in Precision Medicine
Reprinted from: *Biomolecules* **2025**, *15*, 1680, https://doi.org/10.3390/biom15121680 1

Mauricio Murcia-Mejía, Marta Canela-Capdevila, Raquel García-Pablo, Andrea Jiménez-Franco, Juan Manuel Jiménez-Aguilar, Joan Badía, et al.
Combining Metabolomics and Machine Learning to Identify Diagnostic and Prognostic Biomarkers in Patients with Non-Small Cell Lung Cancer Pre- and Post-Radiation Therapy
Reprinted from: *Biomolecules* **2024**, *14*, 898, https://doi.org/10.3390/biom14080898 7

Mateusz Kozłowski, Dominika Borzyszkowska, Natalia Lerch, Agnieszka Turoń-Skrzypińska, Marta Tkacz, Jerzy Lubikowski, et al.
IL-4, IL-7, IL-9, NT, NRP1 May Be Useful Markers in the Diagnosis of Endometrial Cancer
Reprinted from: *Biomolecules* **2024**, *14*, 1095, https://doi.org/10.3390/biom14091095 20

William Middlezong, Victoria Stinnett, Michael Phan, Brian Phan, Laura Morsberger, Melanie Klausner, et al.
Rapid Detection of *PML::RARA* Fusions in Acute Promyelocytic Leukemia: CRISPR/Cas9 Nanopore Sequencing with Adaptive Sampling
Reprinted from: *Biomolecules* **2024**, *14*, 1595, https://doi.org/10.3390/biom14121595 33

Barbara Hegyi, Kristóf György Csikó, Tímea Balatoni, Georgina Fröhlich, Katalin Bőcs, Erika Tóth, et al.
Tumor-Infiltrating Immune Cells and HLA Expression as Potential Biomarkers Predicting Response to PD-1 Inhibitor Therapy in Stage IV Melanoma Patients
Reprinted from: *Biomolecules* **2024**, *14*, 1609, https://doi.org/10.3390/biom14121609 54

Ruomin Xin, Elizabeth Kim, Wei Tse Li, Jessica Wang-Rodriguez and Weg M. Ongkeko
Non-Coding RNAs: lncRNA, piRNA, and snoRNA as Robust Plasma Biomarkers of Alzheimer's Disease
Reprinted from: *Biomolecules* **2025**, *15*, 806, https://doi.org/10.3390/biom15060806 68

Juncheng Wang, Meng Liu, Rukhshan Zahid, Wenjie Zhang, Zecheng Cai, Yan Liang, et al.
Identification of Two Critical Contact Residues in a Pathogenic Epitope from Tetranectin for Monoclonal Antibody Binding and Preparation of Single-Chain Variable Fragments
Reprinted from: *Biomolecules* **2025**, *15*, 1100, https://doi.org/10.3390/biom15081100 82

Carolin J. Sprenzel, Barbara Amann, Cornelia A. Deeg and Roxane L. Degroote
Glycan Signatures on Neutrophils in an Equine Model for Autoimmune Uveitis
Reprinted from: *Biomolecules* **2025**, *15*, 1444, https://doi.org/10.3390/biom15101444 97

Rossella Quarta, Giovanni Martino, Letizia Rosa Romano, Giovanni Lopes, Francesco Fabio Greco, Carmen Anna Maria Spaccarotella, et al.
The Role of Circulating Biomarkers in Patients with Coronary Microvascular Disease
Reprinted from: *Biomolecules* **2025**, *15*, 177, https://doi.org/10.3390/biom15020177 121

Hanna Pawluk, Alina Woźniak, Agnieszka Tafelska-Kaczmarek, Agnieszka Kosinska, Mateusz Pawluk, Krzysztof Sergot, et al.
The Role of IL-6 in Ischemic Stroke
Reprinted from: *Biomolecules* **2025**, *15*, 470, https://doi.org/10.3390/biom15040470 138

Carmela Serpe, Paola De Sanctis, Marina Marini, Silvia Canaider, Provvidenza Maria Abruzzo and Cinzia Zucchini
Human Blood-Derived lncRNAs in Autism Spectrum Disorder
Reprinted from: *Biomolecules* **2025**, *15*, 937, https://doi.org/10.3390/biom15070937 . . . . **169**

Kha Wai Hon and Rakesh Naidu
Unveiling Metabolic Signatures as Potential Biomarkers in Common Cancers: Insights from Lung, Breast, Colorectal, Liver, and Gastric Tumours
Reprinted from: *Biomolecules* **2025**, *15*, 1376, https://doi.org/10.3390/biom15101376 . . . . **195**

Preface

This Reprint brings together the contributions from the Special Issue on molecular, cellular, and blood biomarkers, reflecting both the rapid progress and ongoing challenges in biomarker research. Our motivation was to provide a resource that not only summarizes current findings but also emphasizes the translational potential of these markers for disease diagnosis, prognosis, monitoring, and treatment. By compiling diverse perspectives from experimental, clinical, and translational studies, we hope this Reprint will serve as a practical and inspirational guide for researchers, clinicians, and other professionals seeking to bridge molecular insights with patient care. It is intended for a broad audience interested in understanding how emerging biomarkers can inform both research and clinical decision-making.

Jordi Camps, Isabel Fort Gallifa, and Xavier Gabaldó Barrios
Guest Editors

biomolecules

MDPI

Editorial

The Transformative Role of Molecular, Cellular, and Blood Biomarkers in Precision Medicine

Jordi Camps [1,*], Isabel Fort-Gallifa [2,3] and Xavier Gabaldó-Barrios [4]

1 Unitat de Recerca Biomèdica, Hospital Universitari Sant Joan, Institut d'Investigació Sanitària Pere Virgili, Universitat Rovira i Virgili, Av. Dr. Josep Laporte 2, 43204 Reus, Spain
2 Department of Clinical Laboratory, Hospital Universitari Joan XXIII, Institut d'Investigació Sanitària Pere Virgili, Universitat Rovira i Virgili, C. Dr. Mallafré Guasch 4, 43005 Tarragona, Spain; isabel.fort@urv.cat
3 Oxidative Stress Commission of SemedLab, C. Padilla 323, 08025 Barcelona, Spain
4 Department of Clinical Laboratory and Autoimmunity, Infection and Thrombosis Research Group (GRAIIT), Department of Internal Medicine, Hospital Universitari de Sant Joan, Institut d'Investigació Sanitària Pere Virgili, Universitat Rovira i Virgili, Av. Dr. Josep Laporte 2, 43204 Reus, Spain; xavier.gabaldo@salutsantjoan.cat
* Correspondence: jorge.camps@salutsantjoan.cat; Tel.: +34-977-310-300

1. The Expanding Role of Biomarkers in Modern Medicine

The importance of molecular, cellular, and blood-based biomarkers in modern medicine cannot be overstated. These tools are reshaping how diseases are detected, monitored, and treated, and the contributions gathered in this Special Issue illustrate their growing relevance in the study of oncology, cardiovascular disease, and neurodegeneration [1]. As healthcare moves toward increasingly personalized approaches, biomarkers provide essential insights into disease mechanisms and support the development of more precise and effective therapies.

Early detection remains one of the strongest motivations for biomarker research. Identifying diseases before symptoms appear can dramatically improve survival and quality of life. This is particularly evident in conditions where clinical signs emerge at advanced stages, such as lung cancer. Likewise, the introduction of plasma phosphorylated tau 217 (p-tau217) as a marker for the preclinical stages of Alzheimer's disease demonstrates how accessible biomarkers can expand early diagnostic capabilities. Because amyloid-β pathology develops years before cognitive decline, reliable blood-based indicators, such as p-tau217, have the potential to guide interventions when treatments are most likely to be effective [2,3]. Early detection transforms care from reactive to preventive, enabling timely therapies and targeted follow-up in high-risk individuals.

Equally important is the prognostic value. Understanding how a disease is likely to progress is crucial for guiding treatment decisions and predicting patient outcomes. Prognostic markers can provide valuable insights into the likelihood of recurrence, metastasis, or disease progression, informing therapy choices. In oncology, for instance, tumor-specific markers help clinicians predict how a patient will respond to treatments like immunotherapy or targeted therapy [4]. The ability to anticipate treatment responses is essential for selecting the most effective therapies and minimizing side effects. The relevance of prognostic biomarkers extends far beyond cancer, encompassing other major chronic diseases associated with aging, such as cardiovascular and neurodegenerative disorders [5].

Personalized medicine, often referred to as precision medicine, is another area in which biological indicators are making a profound impact. Instead of assuming uniform responses to therapy, precision medicine accounts for the molecular profile of each patient to select the most appropriate treatment. This strategy increases therapeutic success and

 https://doi.org/10.3390/biom15121680

reduces adverse effects by avoiding ineffective interventions. Classic examples, such as HER2 expression in breast cancer guiding trastuzumab therapy, illustrate how molecular profiles can determine whether a given treatment is truly justified [6].

Beyond guiding therapy selection, biomarkers are also indispensable for monitoring disease progression and treatment efficacy. With the rise of complex cancer therapies such as immunotherapy and targeted therapy, real-time indicators provide crucial information about how patients respond to treatment. For example, liquid biopsies that detect tumor DNA in blood offer a non-invasive means to track tumor dynamics and adjust therapy accordingly [7]. This continuous monitoring is particularly vital in diseases like cancer, where rapid changes in disease status require swift therapeutic decisions.

The utility of these indices extends beyond individual patient care into public health. They can reveal patterns in disease prevalence, risk factors, and population health trends. Screening programs incorporating validated indices help identify at-risk groups who may benefit from early interventions, lifestyle modifications, or prophylactic treatments, ultimately reducing the overall burden of disease [8].

However, despite the promise of biomarkers, their implementation in clinical practice is not without challenges. One significant hurdle is the need for rigorous validation. While the term biomarker is widely used in the scientific community, it is essential to emphasize that not all measurable parameters qualify as such. To be considered a true marker, a parameter must distinctly differentiate between well-defined biological states, such as health versus disease or responsiveness versus resistance to treatment, while minimizing overlap. Achieving such high levels of sensitivity and specificity, especially in complex diseases like cancer and metabolic disorders, remains an ongoing challenge [9].

On the other hand, while blood-based biomarkers offer substantial convenience, their ability to capture the full complexity of disease processes is limited. Complementary approaches, such as imaging or tissue biopsies, are often required to provide a more comprehensive view of disease status. Standardization is another critical issue. Reproducible and comparable results require harmonized protocols across laboratories and healthcare systems. International testing standards are needed to ensure that biomarker-based diagnostics are dependable in real-world settings [10].

Furthermore, integrating these technologies with bioinformatics and artificial intelligence is crucial to enhancing their clinical utility. The vast datasets generated by omic analyses, when coupled with machine learning and deep learning tools, can accelerate the discovery of novel indicators and reveal previously undetected disease signatures. Artificial intelligence-driven algorithms can also assist in interpreting complex data, helping clinicians make faster and more accurate decisions. As these technologies evolve, the potential for discovering new, more precise tools will continue to grow, offering even greater promise for disease management.

Another challenge that often goes underreported is the high cost and the sophisticated analytical methods required for these tools. Many of these advanced diagnostic and prognostic methods rely on complex technologies that are expensive and resource-intensive, making them difficult to implement on a wide scale. The translation of these methods from research settings into routine clinical practice remains a significant barrier, particularly in lower-resource environments. Reducing costs and streamlining these processes will be key to ensuring that these innovations are accessible to all patients, not just those in specialized centers. Moreover, the use of biomarkers in precision medicine raises ethical concerns about privacy, consent, and potential discrimination based on genetic or molecular profiles. These issues need to be carefully considered and addressed to ensure the responsible and equitable use of biomarkers in healthcare.

Looking ahead, multi-biomarker panels are expected to gain prominence. Many diseases, particularly cancer and chronic inflammatory or neurodegenerative conditions, are too heterogeneous to be fully characterized by a single marker. Panels combining complementary indicators provide a more comprehensive view of disease biology and allow for more accurate diagnosis, prognosis, and treatment monitoring. Emerging biomarker families continue to broaden the range of potential applications, opening new opportunities for clinical translation and personalized healthcare [11,12].

In this context, the articles included in this Special Issue provide a representative overview of these advances, illustrating how different types of biomarkers are being applied to specific diseases and offering an up-to-date perspective on their clinical potential.

2. An Overview of Published Articles

2.1. Original Articles

Murcia-Mejía et al. (contribution 1) explored the metabolic profiles of non-small cell lung cancer patients before and after conventionally fractionated radiation therapy (CFRT) or stereotactic body radiation therapy (SBRT). Semi-targeted metabolomics revealed 32 elevated and 18 reduced metabolites in patients compared with healthy controls, with discriminatory metabolites including ethylmalonic acid, taurine, glutamic acid, maltose, glycocolic acid, and d-arabinose. Importantly, the metabolic changes following CFRT and SBRT differed, reflecting the distinct therapeutic impacts of each approach. By combining metabolomics with machine learning, the authors identified signatures capable of perfectly discriminating patients from controls. This study underscores the potential of metabolic biomarkers not only for early diagnosis but also for monitoring therapeutic response, offering a powerful example of precision oncology in practice.

Kozłowski et al. (contribution 2) assessed serum cytokines and related proteins as potential diagnostic biomarkers for endometrial cancer. Several cytokines distinguished endometrial cancer patients from those with noncancerous lesions with promising sensitivity and specificity. Their findings suggest that panels of immune-related biomarkers could enhance early detection and offer a cost-effective, minimally invasive approach for screening, while also providing a foundation for future studies exploring prognostic and predictive utility.

Middlezong et al. (contribution 3) introduced a novel amplification-free nanopore sequencing approach, CRISPR/Cas9-enriched nanopore sequencing with adaptive sampling (CENAS), for rapid detection of PML::RARA fusions in acute promyelocytic leukemia. CENAS enabled fast, low-cost, point-of-care detection of both typical and atypical fusion breakpoints, with concordance to standard diagnostics. This method exemplifies how genomic biomarkers can facilitate immediate treatment initiation, potentially reducing early mortality and providing insights into fusion-associated pathophysiology.

Hegyi et al. (contribution 4) examined tumor HLA class I and II expression and immune cell infiltration in melanoma patients undergoing PD-1 inhibitor therapy. The study demonstrated that higher HLA-II expression, combined with HLA-I/II scores, and increased immune cell infiltration correlated with treatment response and progression-free survival. These results highlight the predictive value of immunological biomarkers in stratifying patients for immunotherapy and guiding personalized treatment decisions.

Xin et al. (contribution 5) profiled circulating non-coding RNAs in Alzheimer's disease (AD), identifying dysregulated lncRNAs, piRNAs, and snoRNAs with potential as diagnostic biomarkers. Using machine learning approaches, the authors highlighted specific ncRNA signatures capable of distinguishing AD patients from controls with high accuracy. This study emphasizes the promise of plasma-based RNA biomarkers and machine learning tools for early, non-invasive detection of neurodegenerative diseases.

Wang et al. (contribution 6) dissected the interaction between a pyroptosis-inducing tetranectin epitope and a monoclonal antibody. By identifying two critical contact residues required for binding, the study refines the molecular understanding of this biomarker and guides the design of synthetic antibodies. Such work exemplifies how structural biomarker studies can inform both diagnostics and therapeutic development in sepsis.

Sprenzel et al. (contribution 7) analyzed neutrophil surface glycosylation in horses with recurrent uveitis, a model for human autoimmune disease. They observed increased O-glycosylation levels in diseased animals and identified Integrin beta-2 and CUB domain-containing protein 1 as potential anchoring proteins. This research highlights glycan modifications as functional biomarkers of immune activation, with potential relevance for early detection and monitoring of autoimmune disorders.

2.2. Review Articles

Quarta et al. (contribution 8) reviewed circulating biomarkers associated with coronary microvascular disease (CMD), including markers of endothelial dysfunction, inflammation, and oxidative stress. They highlighted the utility of non-invasive blood tests to complement invasive assessments, offering insights into pathophysiology, early diagnosis, and risk stratification. Such biomarkers could enhance clinical management of CMD, a frequently underdiagnosed condition.

Pawluk et al. (contribution 9) focused on the pleiotropic cytokine IL-6 in ischemic stroke. Elevated IL-6 levels were found to be implicated in oxidative stress, vascular injury, and leukocyte recruitment. The review underscores the potential of IL-6 as a prognostic biomarker for stroke severity and clinical outcomes, supporting its integration into patient monitoring and treatment planning.

Serpe et al. (contribution 10) summarized the evidence for blood-derived long non-coding RNAs (lncRNAs) as candidate biomarkers in autism spectrum disorder (ASD). Their review highlights lncRNAs linked to disrupted biological pathways in ASD and emphasizes the need for validation in larger, younger cohorts. Integrating multiple datasets and high-throughput approaches may ultimately enable early diagnosis and inform therapeutic strategies.

Hon and Naidu (contribution 11) provided an overview of metabolic reprogramming in the five most lethal cancers, identifying enzymes and transporters involved in glycolysis, amino acid metabolism, and lipid synthesis as potential biomarkers. While only a few markers, such as tumor M2-pyruvate kinase and serum LDH, are currently in clinical use, the review highlights the promise of metabolic biomarkers for early detection, prognosis, and treatment stratification, emphasizing the translational potential of these molecular indicators.

3. Conclusions

The articles in this Special Issue collectively illustrate the breadth and depth of contemporary biomarker research. From metabolomics and glycomics to immunoprofiling, non-coding RNA analysis, and cutting-edge sequencing technologies, these studies highlight how multidisciplinary approaches are driving innovation. They reinforce the crucial role of biomarkers in advancing early diagnosis, refining prognosis, guiding treatment selection, and monitoring disease evolution across diverse medical fields.

As technologies continue to evolve and integrative analytical methods become increasingly sophisticated, the discovery and validation of clinically meaningful biomarkers will accelerate further. Efforts to harmonize methodologies, enhance longitudinal study designs, and incorporate artificial intelligence will be essential to ensure that these advances translate into improved health outcomes. The contributions to this Special Issue reflect an exciting moment in molecular medicine and underscore the promise of biomarkers to reshape the future of personalized healthcare.

Declaration of Generative AI and AI-Assisted Technologies in the Writing Process: During the preparation of this work, the authors utilized CHAT GPT 5.0 (developed by OpenAI, San Francisco, CA, USA) to assist in improving the grammar, syntax, and clarity of the text and Grammarly (from Grammarly Inc., San Francisco, CA, USA) for orthographic corrections. After using these tools, the authors reviewed and edited the content as needed and took full responsibility for the content of the published article. The content, ideas, and scientific conclusions presented in this manuscript are solely the authors' work and have not been generated by AI. The AI tools were utilized exclusively to enhance the readability and presentation of the text.

Conflicts of Interest: The authors declare no conflicts of interest.

List of Contributions:

1. Murcia-Mejía, M.; Canela-Capdevila, M.; García-Pablo, R.; Jiménez-Franco, A.; Jiménez-Aguilar, J.M.; Badía, J.; Benavides-Villarreal, R.; Acosta, J.C.; Arguís, M.; Onoiu, A.I.; et al. Combining metabolomics and machine learning to identify diagnostic and prognostic biomarkers in patients with non-small cell lung cancer pre- and post-radiation therapy. *Biomolecules* **2024**, *14*, 898. https://doi.org/10.3390/biom14080898.
2. Kozłowski, M.; Borzyszkowska, D.; Lerch, N.; Turoń-Skrzypińska, A.; Tkacz, M.; Lubikowski, J.; Tarnowski, M.; Rotter, I.; Cymbaluk-Płoska, A. IL-4, IL-7, IL-9, NT, NRP1 may be useful markers in the diagnosis of endometrial cancer. *Biomolecules* **2024**, *14*, 1095. https://doi.org/10.3390/biom14091095.
3. Middlezong, W.; Stinnett, V.; Phan, M.; Phan, B.; Morsberger, L.; Klausner, M.; Ghabrial, J.; DeMetrick, N.; Zhu, J.; James, T.; et al. Rapid detection of PML::RARA fusions in acute promyelocytic leukemia: CRISPR/Cas9 nanopore sequencing with adaptive sampling. *Biomolecules* **2024**, *14*, 1595. https://doi.org/10.3390/biom14121595.
4. Hegyi, B.; Csikó, K.G.; Balatoni, T.; Fröhlich, G.; Bőcs, K.; Tóth, E.; Mohos, A.; Neumark, A.R.; Menyhárt, C.D.; Ferrone, S.; et al. Tumor-infiltrating immune cells and HLA expression as potential biomarkers predicting response to PD-1 inhibitor therapy in stage IV melanoma patients. *Biomolecules* **2024**, *14*, 1609. https://doi.org/10.3390/biom14121609.
5. Xin, R.; Kim, E.; Li, W.T.; Wang-Rodriguez, J.; Ongkeko, W.M. Non-coding RNAs: lncRNA, piRNA, and snoRNA as robust plasma biomarkers of Alzheimer's disease. *Biomolecules* **2025**, *15*, 806. https://doi.org/10.3390/biom15060806.
6. Wang, J.; Liu, M.; Zahid, R.; Zhang, W.; Cai, Z.; Liang, Y.; Li, D.; Hao, J.; Xu, Y. Identification of two critical contact residues in a pathogenic epitope from tetranectin for monoclonal antibody binding and preparation of single-chain variable fragments. *Biomolecules* **2025**, *15*, 1100. https://doi.org/10.3390/biom15081100.
7. Sprenzel, C.J.; Amann, B.; Deeg, C.A.; Degroote, R.L. Glycan signatures on neutrophils in an equine model for autoimmune uveitis. *Biomolecules* **2025**, *15*, 1444. https://doi.org/10.3390/biom15101444.
8. Quarta, R.; Martino, G.; Romano, L.R.; Lopes, G.; Greco, F.F.; Spaccarotella, C.A.M.; Indolfi, C.; Curcio, A.; Polimeni, A. The role of circulating biomarkers in patients with coronary microvascular disease. *Biomolecules* **2025**, *15*, 177. https://doi.org/10.3390/biom15020177.
9. Pawluk, H.; Woźniak, A.; Tafelska-Kaczmarek, A.; Kosinska, A.; Pawluk, M.; Sergot, K.; Grochowalska, R.; Kołodziejska, R. The role of IL-6 in ischemic stroke. *Biomolecules* **2025**, *15*, 470. https://doi.org/10.3390/biom15040470.
10. Serpe, C.; De Sanctis, P.; Marini, M.; Canaider, S.; Abruzzo, P.M.; Zucchini, C. Human blood-derived lncRNAs in autism spectrum disorder. *Biomolecules* **2025**, *15*, 937. https://doi.org/10.3390/biom15070937.
11. Hon, K.W.; Naidu, R. Unveiling metabolic signatures as potential biomarkers incommon cancers: Insights from lung, breast, colorectal, liver, and gastric tumours. *Biomolecules* **2025**, *15*, 1376. https://doi.org/10.3390/biom15101376.

References

1. Sharma, R.; Ghosh, G.; Abbas, A.; Mahato, K. Biochemical, biophysical, and digital biomarkers in precision healthcare. In *Nano-Bioelectronics for Precision Health Monitoring*; Mahato, K., Chandra, P., Eds.; Springer: Singapore, 2025. [CrossRef]
2. Zafar, S.; Hafeez, A.; Shah, H.; Mutiullah, I.; Ali, A.; Khan, K.; Figueroa-González, G.; Reyes-Hernández, O.D.; Quintas-Granados, L.I.; Peña-Corona, S.I.; et al. Emerging biomarkers for early cancer detection and diagnosis: Challenges, innovations, and clinical perspectives. *Eur. J. Med. Res.* **2025**, *30*, 760. [CrossRef]
3. Salvadó, G.; Janelidze, S.; Bali, D.; Dolado, A.O.; Therriault, J.; Brum, W.S.; Pichet Binette, A.; Stomrud, E.; Mattsson-Carlgren, N.; Palmqvist, S.; et al. Plasma phosphorylated tau 217 to identify preclinical Alzheimer disease. *JAMA Neurol.* **2025**, *82*, 1122–1134. [CrossRef] [PubMed]
4. Zhou, Y.; Tao, L.; Qiu, J.; Xu, J.; Yang, X.; Zhang, Y.; Tian, X.; Guan, X.; Cen, X.; Zhao, Y. Tumor biomarkers for diagnosis, prognosis and targeted therapy. *Signal Transduct. Target. Ther.* **2024**, *9*, 132. [CrossRef] [PubMed]
5. Vo, D.K.; Trinh, K.T.L. Emerging biomarkers in metabolomics: Advancements in precision health and disease diagnosis. *Int. J. Mol. Sci.* **2024**, *25*, 13190. [CrossRef]
6. Zhu, K.; Yang, X.; Tai, H.; Zhong, X.; Luo, T.; Zheng, H. HER2-targeted therapies in cancer: A systematic review. *Biomark. Res.* **2024**, *12*, 16. [CrossRef]
7. Adhit, K.K.; Wanjari, A.; Menon, S.; K, S. Liquid biopsy: An evolving paradigm for non-invasive disease diagnosis and monitoring in medicine. *Cureus* **2023**, *15*, e50176. [CrossRef] [PubMed]
8. Zoccali, C.; Tripepi, G.; Stel, V.; Fu, E.L.; Mallamaci, F.; Dekker, F.; Jager, K.J. Biomarkers in clinical epidemiology studies. *Clin. Kidney J.* **2024**, *17*, sfae130. [CrossRef] [PubMed]
9. FDA-NIH Biomarker Working Group. *BEST (Biomarkers, EndpointS, and Other Tools) Resource [Internet]*; Bookshelf ID: NBK326791; Food and Drug Administration (US): Silver Spring, MD, USA, 2016.
10. Gupta, S.; Venkatesh, A.; Ray, S.; Srivastava, S. Challenges and prospects for biomarker research: A current perspective from the developing world. *Biochim. Biophys. Acta* **2014**, *1844*, 899–908. [CrossRef] [PubMed]
11. Götze, K.; Vrillon, A.; Dumurgier, J.; Indart, S.; Sanchez-Ortiz, M.; Slimi, H.; Raynaud-Simon, A.; Cognat, E.; Martinet, M.; Zetterberg, H.; et al. Plasma neurofilament light chain as prognostic marker of cognitive decline in neurodegenerative diseases, a clinical setting study. *Alzheimers Res. Ther.* **2024**, *16*, 231. [CrossRef] [PubMed]
12. Papa, L.; McKinley, W.I.; Valadka, A.B.; Newman, Z.C.; Nordgren, R.K.; Pramuka, P.E.; Barbosa, C.E.; Brito, A.M.P.; Loss, L.J.; Tinoco-Garcia, L.; et al. Diagnostic performance of GFAP, UCH-L1, and MAP-2 within 30 and 60 minutes of traumatic brain injury. *JAMA Netw. Open* **2024**, *7*, e2431115. [CrossRef] [PubMed]

biomolecules

MDPI

Article

Combining Metabolomics and Machine Learning to Identify Diagnostic and Prognostic Biomarkers in Patients with Non-Small Cell Lung Cancer Pre- and Post-Radiation Therapy

Mauricio Murcia-Mejía [1,2,†], Marta Canela-Capdevila [1,2,†], Raquel García-Pablo [1,2], Andrea Jiménez-Franco [2], Juan Manuel Jiménez-Aguilar [2], Joan Badía [3], Rocío Benavides-Villarreal [1,2], Johana C. Acosta [1,2], Mónica Arguís [1,2], Alina-Iuliana Onoiu [2], Helena Castañé [2], Jordi Camps [2,*], Meritxell Arenas [1,2,*] and Jorge Joven [2]

1 Department of Radiation Oncology, Hospital Universitari de Sant Joan, Institut d'Investigació Sanitària Pere Virgili, Universitat Rovira i Virgili, 43204 Reus, Spain; mauricio.murcia@salutsantjoan.cat (M.M.-M.); marta.canela@iispv.cat (M.C.-C.); raquel.garcia@iispv.cat (R.G.-P.); americadelrocio.benavides@salutsantjoan.cat (R.B.-V.); johana.acosta@salutsantjoan.cat (J.C.A.); monica.arguis@salutsantjoan.cat (M.A.)

2 Unitat de Recerca Biomèdica, Hospital Universitari de Sant Joan, Institut d'Investigació Sanitària Pere Virgili, Universitat Rovira i Virgili, 43204 Reus, Spain; andrea.jimenez@urv.cat (A.J.-F.); juanmaaguilar106@gmail.com (J.M.J.-A.); alinaiuliana.onoiu@urv.cat (A.-I.O.); helena.castane@iispv.cat (H.C.); jorge.joven@salutsantjoan.cat (J.J.)

3 Statistical Support Platform, Hospital Universitari de Sant Joan, Institut d'Investigació Sanitària Pere Virgili, Universitat Rovira i Virgili, 43204 Reus, Spain; joan.badia@iispv.cat

* Correspondence: jorge.camps@salutsantjoan.cat (J.C.); meritxell.arenas@urv.cat (M.A.); Tel.: +34-977310300 (J.C. & M.A.)

† These authors contributed equally to this work.

Abstract: Lung cancer is the leading cause of cancer-related deaths globally, with non-small cell lung cancer (NSCLC) accounting for over 85% of cases and poor prognosis in advanced stages. This study explored shifts in circulating metabolite levels in NSCLC patients versus healthy controls and examined the effects of conventionally fractionated radiation therapy (CFRT) and stereotactic body radiation therapy (SBRT). We enrolled 91 NSCLC patients (38 CFRT and 53 SBRT) and 40 healthy controls. Plasma metabolite levels were assessed using semi-targeted metabolomics, revealing 32 elevated and 18 reduced metabolites in patients. Key discriminatory metabolites included ethylmalonic acid, maltose, 3-phosphoglyceric acid, taurine, glutamic acid, glycocolic acid, and d-arabinose, with a combined Receiver Operating Characteristics curve indicating perfect discrimination between patients and controls. CFRT and SBRT affected different metabolites, but both changes suggested a partial normalization of energy and amino acid metabolism pathways. In conclusion, metabolomics identified distinct metabolic signatures in NSCLC patients with potential as diagnostic biomarkers. The differing metabolic responses to CFRT and SBRT reflect their unique therapeutic impacts, underscoring the utility of this technique in enhancing NSCLC diagnosis and treatment monitoring.

Keywords: biomarkers; lung cancer; metabolomics; radiation therapy; stereotactic body radiation therapy

1. Introduction

Lung cancer stands as the predominant contributor to cancer-related fatalities worldwide, accounting for approximately 12% of all cases of cancer, with an annual incidence exceeding two million patients [1]. Non-small cell lung cancer (NSCLC) encompasses over 85% of lung cancer cases and is categorized into three principal histological subtypes: adenocarcinoma, squamous cell carcinoma, and large cell carcinoma. The five-year survival rate for early stage, operable NSCLC averages around 70%; however, it declines to 10–13% in advanced stages of the disease [2–4]. Presently, diagnosis heavily relies on symptomatology, often leading to late-stage detection and dismal prognoses. If the diagnosis could be shifted to early stages, the overall morbidity for this disease could profoundly decrease. Screening for lung cancer utilizing low-dose computed tomography has shown efficacy in mortality reduction [5]. However, the prevalence of individuals with indeterminate nodules, exorbitant costs, and resource constraints makes it necessary to investigate the discovery of more straightforward early diagnosis methods, ideally utilizing non-invasive or minimally invasive modalities, such as blood biomarkers [6], supplemented by clinical, epidemiological, imaging, and lifestyle data. This approach holds particular promise for individuals at high risk of lung cancer, as they may harbor subclinical disease for extended periods before symptomatic manifestation.

Recent research has underscored the pivotal role of investigating metabolic alterations in cancer development [7]. Cancer cell pathogenesis entails oncogene activation, apoptosis evasion, heightened replicative capacity, angiogenesis, and metabolic dysregulation, marked by alterations in signaling cascades, protein expression, and biochemical adaptation [8]. Predominant metabolic shifts in cancer cells encompass increased glucose consumption or aerobic glycolysis (the Warburg effect), augmented glutaminolysis, and heightened amino acid and lipid biosynthesis pathways [9–11]. Metabolomics emerges as a promising avenue in systems biology, aiming for the comprehensive interrogation of low-molecular-weight metabolites in biological specimens [12]. It is a potent tool for deciphering the biological pathways implicated in disease initiation and progression, furnishing invaluable insights into the molecular underpinnings of pathological processes [13]. Quantitative metabolomic profiling of plasma has been proposed as a candidate approach for lung cancer diagnosis and prognosis [14,15].

NSCLC patients typically undergo a treatment regimen comprising surgical resection, radiotherapy, chemotherapy, immunotherapy, and targeted therapy. In this study, we will refer to Conventionally Fractionated Radiation Therapy (CFRT) as encompassing two treatment schedules: normofractionated radiation therapy, which delivers 2 Gy per session on 30–33 consecutive fractions, and hypofractionated radiation therapy, which administers 4 Gy per session on 15 consecutive fractions. We will compare these treatments with another radiotherapy technique, stereotactic body radiation therapy (SBRT), also known as stereotactic ablative radiotherapy. SBRT has been introduced to improve treatment precision and reduce overall treatment time by delivering a narrow beam of high-dose radiation per fraction (>5 Gy to a specific target in fewer alternating fractions (between three and eight), ensuring accurate delivery to tumor tissue while sparing surrounding healthy tissue [16]. This treatment is commonly used for managing early stage tumors (less than 5 cm and with negative lymph nodes) in patients who are inoperable due to comorbidities, impaired respiratory function, or those who refuse surgery. In contrast, patients with inoperable locally advanced tumors (larger than 5 cm and with involved lymph nodes) are often treated with CFRT [17].

Therefore, this investigation aims to elucidate the shifts in circulating metabolite levels in NSCLC patients compared to healthy volunteers and delineate the effects induced by CFRT and SBRT. The overarching objective is identifying and proposing potential biomarkers for early diagnosis and prognosis of the disease.

2. Materials and Methods

We enrolled a cohort of 91 patients diagnosed with NSCLC who underwent either CFRT (38 patients) or SBRT (53 patients), collectively representing 99 treated lesions. All participants were treated at the Department of Radiation Oncology of the Hospital Universitari de Sant Joan de Reus between February 2013 and October 2022. Eligible patients exhibited a Karnofsky Performance Status Index > 70 and were classified as 0 or 1 on the Eastern Cooperative Oncology Group (ECOG) scale.

The radiotherapy protocols included normofractionated radiotherapy (total dose 54–70 Gy delivered at 2 Gy/day, five days/week), hypofractionated radiotherapy (total dose 60 Gy delivered at 4 Gy/day, five days/week), or risk-adapted SBRT (total dose 54–60 Gy, three alternated days/week) utilizing Volumetric Modulated Arc Therapy (VMAT) via the Varian RapidArc® system (Varian Medical Systems, Palo Alto, CA, USA). Additionally, fourteen NSCLC patients received concurrent treatment with cisplatin (50 mg/m^2) and etoposide (50 mg/m^2) intravenously every three weeks.

Assessment of acute toxicity throughout the course of treatment was conducted weekly, employing the criteria established by the Radiation Therapy Oncology Group (RTOG) and the European Organization for Research and Treatment of Cancer (EORTC) [18].

Fasting blood samples were obtained at baseline and one-month post treatment. EDTA-plasma aliquots were promptly stored at −80 °C until metabolomic analyses were conducted. As a control group, we employed plasma samples from 40 healthy volunteers who had no clinical or analytical evidence of infectious disease, renal insufficiency, liver disease, neoplasia, or neurological disorders. These individuals were recruited from a population-based study conducted within our local region [19].

Semi-targeted metabolomics assessed the plasma levels of 74 metabolites involved in carbohydrates, amino acids, lipids, cofactors, vitamins, nucleotide pathways, energy metabolism, and xenobiotic biodegradation [20,21]. Plasma samples (50 µL) were mixed with an 8:2 (v/v) methanol: water solution containing internal standards, vortexed, and centrifuged. The supernatants (200 µL) were evaporated in a SpeedVac vacuum concentrator (Thermo Fisher Scientific, Waltham, MA, USA) and reconstituted with methoxyamine before silylation. Chromatographic separation was achieved using a 7890A gas chromatograph paired with a 7200-quadruple time-of-flight mass spectrometer equipped with an electron impact source (Agilent Technologies, Santa Clara, CA, USA). Moreover, the system was fitted with a 7693 autosampler module and a J&W Scientific HP-5MS column (30 ms 0.25 mm, 0.25 µm) from Agilent Technologies. Compound identification and semi-quantification were determined using Relative Units (RU) calculated based on compound area/internal standard area ratios, with ion selection guided by electron impact spectra and primary specific ions recorded in the Fiehn-pct-2013 spectral library.

Group comparisons were conducted using appropriate statistical tests based on variable types. Quantitative variables were analyzed by the Mann–Whitney U test (independent variables) or the Wilcoxon signed rank test (dependent variables), and results are shown as medians [interquartile ranges]. Categorical variables were evaluated by the χ^2 square test, and results are shown as number of cases (percentages). Significance was determined at $p < 0.05$. All analyses and plots were executed using RStudio version 4.3.1. Patient clinical characteristics were summarized using the TableOne package version 0.13.2. A logarithmic transformation was applied to metabolomic data. Orthogonal Partial Least Squares Discriminant Analysis (OPLS-DA) and Variable Importance in Projection (VIP) score analyses were performed using the ropls package version 1.32.0. Using the caret package version 6.0-94, a 5-fold cross-validated Support Vector Machine (SVM) multivariate model was trained on 70% of the data. The model was then tested on the remaining 30% of the data. Receiver Operating Characteristics (ROC) curves were generated using the pROC package version 1.18.5. Volcano plots and boxplots were created with the ggplot2 package version 3.5.1. All analyses utilized the latest package versions available on CRAN as of 28 April 2024.

3. Results

3.1. Clinical Characteristics of Study Participants and Their Tumors

While most participants were men, notable differences emerged between NSCLC patients and the control group. These variances encompassed age, comorbidities such as diabetes mellitus, arterial hypertension, and dyslipidemia, as well as lifestyle factors like smoking and alcohol consumption. Patients treated with SBRT had a higher proportion of women and a lower prevalence of smokers than those receiving CFRT. The predominant tumor types among NSCLC patients were adenocarcinoma and squamous cell carcinoma, with most lesions located in the right upper lobe, right lower lobe, or left upper lobe (Table 1).

Table 1. Main characteristics of lung cancer patients, their tumors, and the control group.

	Control Group	All NSCLC Patients	CFRT	SBRT	*p*-Value
Patient characteristics	*n* = 40	*n* = 91	*n* = 38	*n* = 53	
Age (years)	69.50 [65.0–74.0]	73.00 [66.5–79.0]	71.50 [66.8–77.8]	74.00 [67.0–79.0]	*
Women	12 (30.0)	39 (42.9)	8 (21.1)	31 (58.5)	#
Smoking habit	9 (22.5)	69 (76.7)	34 (91.9)	35 (66.0)	* #
Alcohol habit	20 (50.0)	20 (22.0)	11 (28.9)	9 (17.0)	*
Diabetes mellitus	3 (7.5)	27 (30.0)	11 (29.7)	16 (30.2)	*
Arterial hypertension	13 (32.5)	56 (62.2)	21 (56.8)	35 (66.0)	*
Dyslipidemia	6 (15.0)	43 (47.3)	19 (50.0)	24 (45.3)	*
Non-cancer pulmonary disease	NA	58 (63.7)	31 (81.6)	27 (50.9)	NS
Cardiovascular disease	NA	34 (37.4)	20 (52.6)	14 (26.4)	#
Cancer characteristics		*n* = 99	*n* = 38	*n* = 61	
Histology					NS
Adenocarcinoma	NA	44 (44.4)	15 (39.5)	29 (47.5)	
Squamous cell carcinoma	NA	42 (42.4)	21 (55.3)	21 (34.4)	
Others	NA	8 (8.1)	2 (5.3)	6 (9.8)	
Not determined	NA	5 (5.1)	-	5 (8.2)	
Stage					#
Ia	NA	60 (60.6)	13 (34.2)	47 (77.0)	
Ib	NA	15 (15.2)	6 (15.8)	9 (14.8)	
IIa	NA	1 (1.0)	1 (2.6)	0 (0.0)	
IIb	NA	8 (8.1)	3 (7.9)	5 (8.2)	
IIIa	NA	5 (5.1)	5 (13.2)	0 (0.0)	
IIIb	NA	8 (8.1)	8 (21.1)	0 (0.0)	
IIIc	NA	1 (1.0)	1 (2.6)	0 (0.0)	
IV	NA	1 (1.0)	1 (2.6)	0 (0.0)	
Tumor location					NS
RUL	NA	27 (27.6)	10 (26.3)	17 (28.3)	
RML	NA	5 (5.1)	2 (5.3)	3 (5.0)	
RLL	NA	21 (21.4)	11 (28.9)	10 (16.7)	
LUL	NA	27 (27.6)	11 (28.9)	16 (26.7)	
LLL	NA	18 (18.4)	4 (10.5)	14 (23.3)	

* $p < 0.05$, cancer vs. control group; # $p < 0.05$, SBRT vs. CFRT. Age is shown as medians [interquartile ranges], and differences were analyzed by the Mann–Whitney U test. Qualitative variables are shown as n (%), and differences were analyzed by the X^2 test. CFRT: Conventionally Fractionated Radiation Therapy; LLL: left lower lobule; LUL: left upper lobule; NA: not applicable; NS: not significant; NSCLC: non-small cell lung cancer; RLL: right lower lobule; RML: right middle lobule; RUL: right upper lobule; SBRT: stereotactic body radiation therapy.

3.2. *Baseline Metabolite Levels Effectively Differentiate NSCLC Patients from the Control Group*

A comparison of metabolite concentrations revealed 32 significantly higher and 18 lower plasma metabolite levels in NSCLC patients compared to the control group (Figure 1A and Supplementary Table S1).

OPLS-DA successfully distinguished the metabolic profiles of healthy individuals and NSCLC patients. Specific metabolites such as ethylmalonic acid, maltose, glycerol, 3-phosphoglyceric acid, taurine, glutamic acid, glycolic acid, and d-arabinose showed higher discriminatory capacity in segregating the groups, as indicated by their VIP scores exceeding 1.5 points (Figure 1B). Among these, ethylmalonic acid, glycerol, and glycolic acid plasma concentrations were lower in NSCLC patients. In contrast, maltose, 3-phosphoglyceric acid, taurine, glutamic acid, and d-arabinose were higher. These findings suggest potential metabolic alterations associated with NSCLC, particularly in amino acid, carbohydrate, and lipid metabolism (Figure 1C).

For metabolites with VIP scores exceeding 1.5 points, we conducted univariate ROC curves, all of which yielded an area under the curve (AUC) higher than 0.8, indicating a high level of accuracy. Furthermore, the Youden index was computed for each curve to establish the optimal cutoff point for distinguishing NSCLC patients from healthy individuals (Figure 2A,B).

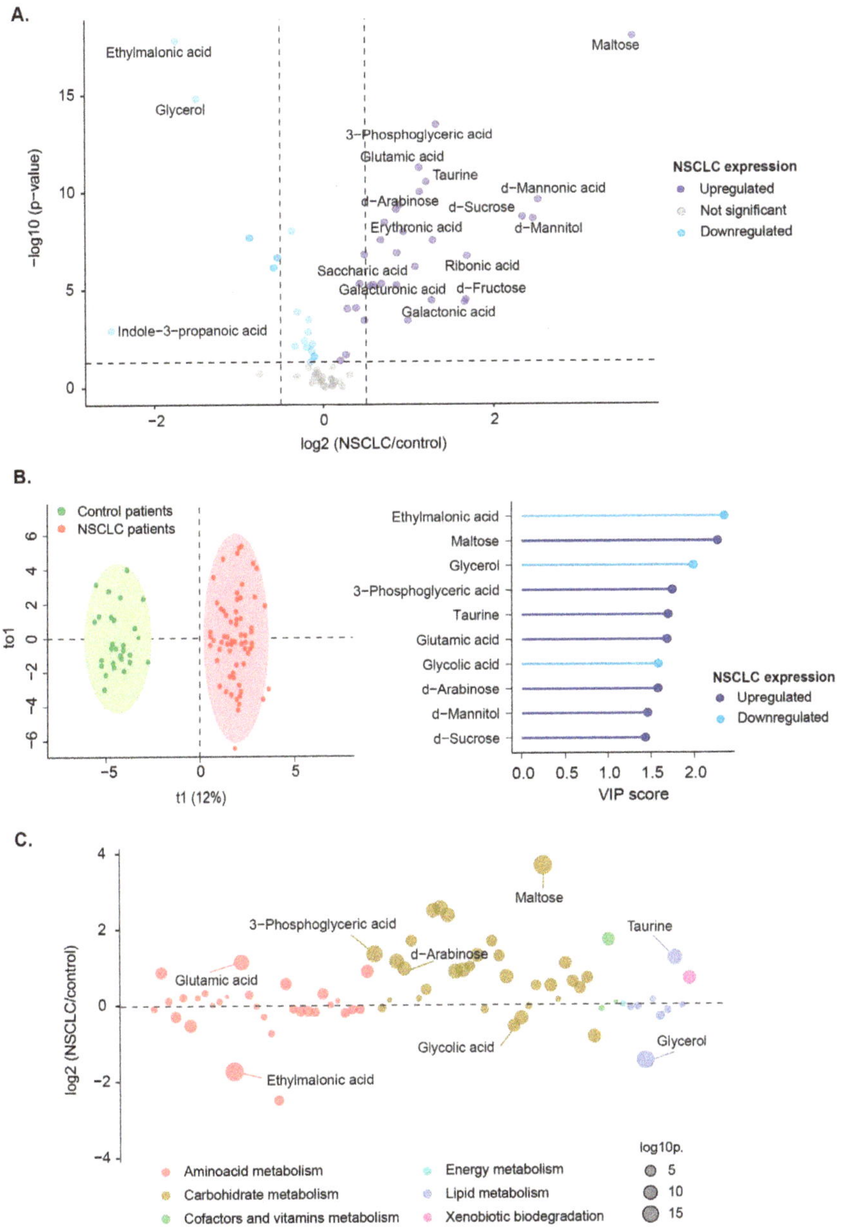

Figure 1. Comparative analysis of plasma metabolite profiles in non-small cell lung cancer (NSCLC) patients and the control group. (**A**) Differential expression of plasma metabolites revealed significantly upregulated and downregulated metabolites in patients compared to the control group. (**B**) Orthogonal Partial Least Squares Discriminant Analysis showed a clear distinction between the metabolic profiles of NSCLC patients and healthy individuals, with eight metabolites exhibiting a Variable Importance in Projection (VIP) score greater than 1.5. (**C**) The most significant metabolites distinguishing NSCLC patients from the control group were associated with amino acid, carbohydrate, and lipid metabolism.

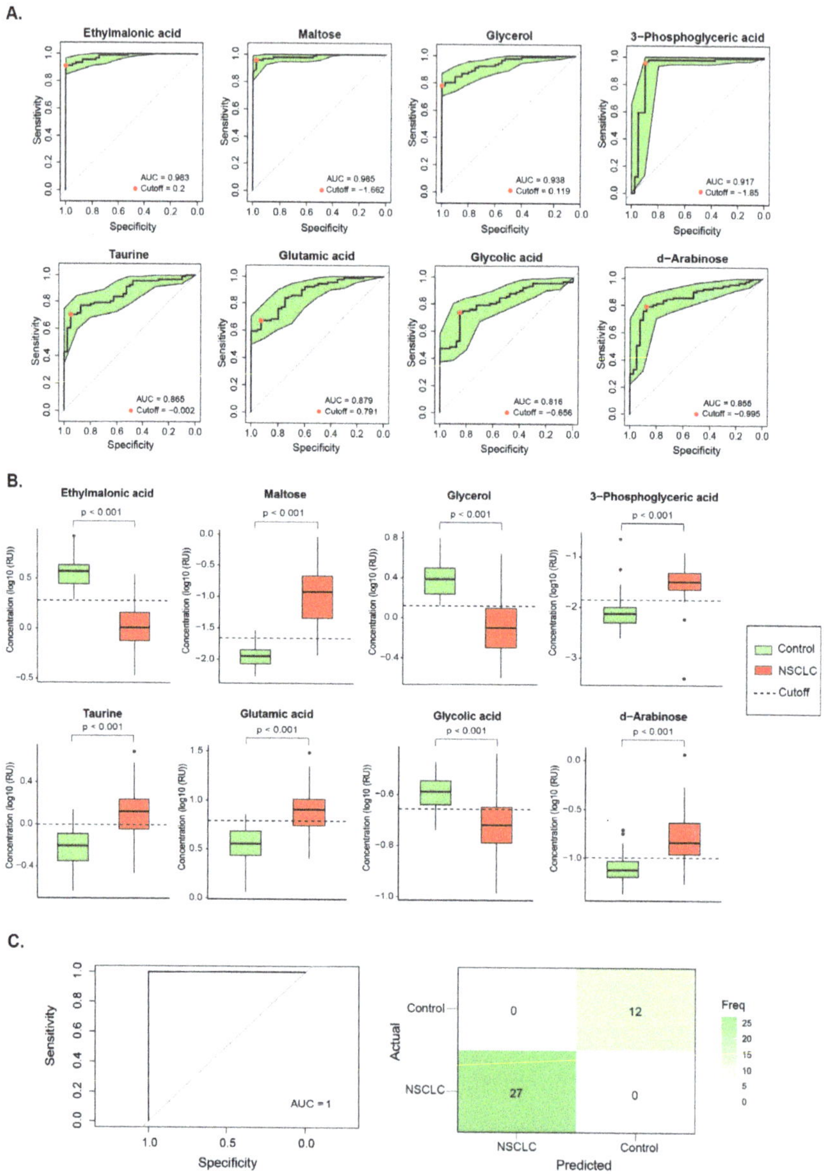

Figure 2. Plasma concentrations of ethylmalonic acid, maltose, glycerol, 3-phospogyceric acid, taurine, glutamic acid, glycolic acid, and d-arabinose had a high discriminating power to classify non-small cell lung cancer (NSCLC) patients from the control group. (**A**) Individual Receiver Operating Characteristics (ROC) curves showed an area under the curve (AUC) higher than 0.8. Red dots represent the optimal cutoff value determined by the Youden index for patient classification. (**B**) The Mann–Whitney U test revealed significant differences in plasma metabolite concentrations between NSCLC patients and healthy individuals. (**C**) The Support Vector Machine multivariate model showcased outstanding performance, with an ROC curve achieving an AUC of 1 and an unequivocal confusion matrix. Results and cutoffs are shown as log10 of Relative Units (RU).

Finally, a SVM multivariate model was trained and tested based on those metabolites with a VIP score higher than 1.5 points. The ROC curve yielded an AUC of 0.999 for the training dataset (Figure S1) and 1 for the testing dataset (Figure 2C), indicating perfect discrimination between classes.

3.3. SBRT and CFRT Induce Distinct Changes in the Plasma Metabolome

The two radiation therapy modalities produced distinct changes in the metabolic profiles of NSCLC patients. SBRT led to decreased plasma concentrations of maltose, 3-phosphoglyceric acid, taurine, d-arabinose, sedoheptulose, malic acid, d-xylitol, and d-threitol, indicating a trend toward normalization. However, it also decreased levels of lactic acid and ornithine while increasing levels of α-tocopherol, further diverging from the profiles of healthy volunteers (Figure 3A and Supplementary Table S2).

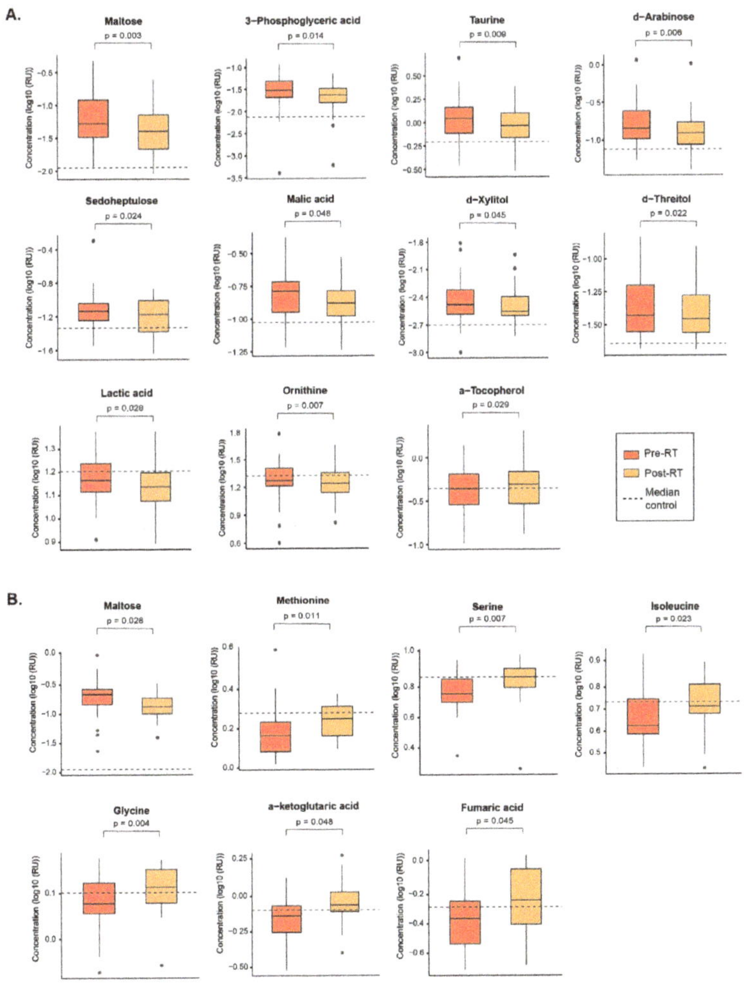

Figure 3. This figure illustrates metabolites that exhibited significant changes after treatment compared to pre-RT values with either (**A**) stereotactic body radiation therapy or (**B**) Conventional Fractionated Radiotherapy. Statistical significance was determined using the Wilcoxon signed-rank test. The dashed line indicates the median metabolite value for healthy volunteers. Results are presented as the log10 of Relative Units (RU).

In contrast, CFRT was associated with a lower concentration of maltose and higher concentrations of methionine, serine, isoleucine, glycine, α-ketoglutaric acid, and fumaric acid. These changes indicate a partial return to normal levels for these metabolites (Figure 3B and Supplementary Table S3).

It is noteworthy that maltose was the only metabolite exhibiting similar changes after both SBRT and CFRT.

4. Discussion

Our findings reveal a distinctive metabolic profile in individuals afflicted with NSCLC compared to healthy controls. The metabolites showing the most pronounced changes included maltose, 3-phosphoglyceric acid, taurine, glutamic acid, and d-arabinose, all of which had higher concentrations in NSCLC. Conversely, ethylmalonic acid, glycerol, and glycolic acid concentrations were decreased. The collective ROC curve generated from these metabolites effectively distinguished between patients and healthy controls, providing a clear and accurate classification.

Maltose, a disaccharide composed of two glucose molecules, arises from the breakdown of complex carbohydrates by the action of maltase. The higher plasma maltose concentrations found in NSCLC patients could be an adaptive response to the altered metabolic environment of the tumor, aiming to satisfy its energetic and biosynthetic needs [22].

3-phosphoglyceric acid is an intermediary metabolite in the glycolytic pathway that plays a crucial role in adenosine triphosphate generation. Under normal conditions, it converts into pyruvate, entering the mitochondria for complete oxidation in the tricarboxylic acid cycle. However, the persistence of heightened glycolytic activity in cancer cells, even in the presence of oxygen, is a hallmark of the Warburg effect [23]. Tumor metabolism prioritizes the generation of intermediates that support macromolecular biosynthesis, thereby providing a proliferative advantage. This metabolic shift likely leads to increased production of 3-phosphoglyceric acid; consequently, higher plasma concentrations in NSCLC patients can be found.

Taurine, also known as 2-aminoethanesulfonic acid, exhibits regulatory effects on various cellular processes, including the modulation of N-acetyl galactosaminyl transferase-2 expression and the downregulation of matrix metalloproteinase-2, thereby impeding invasion and metastasis [24]. Additionally, taurine demonstrates inhibitory effects on proliferation and pro-apoptotic properties in A549 lung cancer cells, with its efficacy being contingent upon dosage and exposure duration [25]. The increase in plasma concentrations of this amino acid in patients with NSCLC could be explained by a positive feedback mechanism attempting to counteract tumor development.

Glutamic acid assumes a pivotal role in metabolic pathways, with its involvement commencing at the onset of glutaminolysis, where glutamine undergoes conversion to glutamic acid catalyzed by glutaminase within the mitochondria. Studies have implicated glutamic acid as a discerning inflammatory marker in pulmonary pathologies [26]. Furthermore, in line with our results, consistent elevations in glutamic acid levels have been documented in patients with NSCLC, juxtaposed with a reduction in glutamine levels [27–29]. In addition, a significant increase in glutamic acid levels within lung cancer tissues compared to adjacent para-carcinomatous tissues has also been observed [30], suggesting a potential association between intratumoral glucogenesis and this amino acid.

The cell utilizes glycerol to synthesize intermediate metabolites essential for producing vital cellular components such as triglycerides and phospholipids. The observed lower concentrations of this metabolite in NSCLC patients may be attributed to its consumption by tumor tissue, where the accelerated synthesis of proteins, nucleic acids, and membranes occurs.

Insufficient data are available to fully interpret the underlying reasons behind the observed increase in d-arabinose concentrations and decreased glycolic acid and ethylmalonic plasma concentrations in NSCLC patients. D-arabinose is a five-carbon monosaccharide primarily found in specific plant sources, yet it is not commonly a part of human dietary intake. In humans, the endogenous synthesis of d-arabinose occurs in negligible quantities. Glycolic acid, a simple organic acid, is derived from the metabolism of lactic acid catalyzed by lactate dehydrogenase. However, its role in central metabolic pathways is relatively limited compared to other compounds, and it does not play a pivotal role in any significant metabolic pathways [31]. The decline in glycolic acid levels of LC patients might be attributed to its utilization as an energy source. On the other hand, ethylmalonic acid is a branched-chain fatty acid that has garnered attention in investigating neurological diseases associated with inborn metabolic disorders. Interestingly, patients with inborn errors of metabolism are predisposed to developing cancer [32]. However, the potential role of ethylmalonic acid in the pathophysiology of cancer onset and development has not been reported to date.

The observed metabolic profile in individuals with NSCLC indicates potential metabolic adaptations associated with cancer progression. Elevated concentrations of maltose, 3-phosphoglyceric acid, taurine, glutamic acid, and d-arabinose suggest increased glycolytic activity, potential regulatory effects on cellular processes, and alterations in glutamine metabolism, which are consistent with metabolic reprogramming observed in cancer cells. Conversely, decreased levels of ethylmalonic acid, glycerol, and glycolic acid may reflect altered utilization of these metabolites in cancer cells for biosynthesis and energy production.

In recent years, some studies have explored metabolic changes in patients with lung cancer, aiming to utilize these alterations for improved disease diagnosis and prognosis. These investigations have identified potential biomarkers to distinguish between patients and healthy individuals and differentiate among various types and stages of the disease. For instance, Qi et al. [33] identified palmitic acid, heptadecanoic acid, 4-oxoproline, tridecanoic acid, and ornithine as potential candidates for lung cancer screening, achieving a fair discriminative ability with AUC values ranging from 0.82 to 0.86. Similarly, Zhang et al. [34] highlighted the predictive potential of phenylalanine, phosphoethanolamine, xanthosine, dehydroepiandrosterone, glucose, fructose, and phenylacetic, hydroxyoxoglutaric, allocolic, aminopentanoic, and uric acids in assessing the stage and progression of lung cancer, exhibiting moderate discrimination with AUC values between 0.70 and 0.80. More promising results were reported by Zhang et al. [35], with the identification of lysophosphatidylcholine 20:3, phosphatidylcholine 40:6, citric acid, hydroxybutyric acid, and fumaric acid showcasing a diagnostic accuracy of 0.90 for early stage NSCLC. Furthermore, a previous study from our group [36] demonstrated that glutamic acid determination alone could diagnose lung cancer with an accuracy of 0.90. Remarkably, the findings presented in our present article underscore the exceptional discriminatory capacity of our identified panel, achieving an AUC of 1.0 and demonstrating perfect discrimination between cases and controls. This development holds great promise for the future of NSCLC diagnosis.

The present study introduces a novel aspect by identifying the metabolic changes induced by CFRT and SBRT in patients with NSCLC. Both modalities tend to normalize the concentrations of some metabolites while altering others, with the affected metabolites differing between treatments. The differences in metabolic effects between CFRT and SBRT reflect the distinct biological responses to these therapies. Understanding these alterations requires considering several factors, including the mechanisms of action, the biological pathways involved, and the intensity and precision of the radiation delivered.

CFRT brought about a partial normalization of plasma levels of maltose, methionine, serine, and isoleucine while elevating the concentrations of glycine, fumaric, and α-ketoglutaric acids beyond the median levels observed in healthy controls. Partial normalization of circulating levels of maltose and amino acids such as methionine, serine, and isoleucine after CFRT may reflect a decrease in carbohydrate metabolism and protein synthesis due to tumor mass reduction [36,37]. This observation suggests that CFRT might help restore normal cellular functions or reduce tumor-related disruptions in these pathways. On the other hand, increases in fumaric acid and α-ketoglutaric acid may indicate heightened oxidative stress and metabolic activity in response to radiation. These metabolites are tricarboxylic acid cycle components, suggesting enhanced mitochondrial activity or a stress response [38,39].

In contrast, SBRT significantly reduced the plasma concentrations of maltose3-phosphoglyceric acid, taurine, d-arabinose, sedoheptulose, malic acid, d-xylitol, and d-threitol toward normalization. Furthermore, SBRT induced decreases in lactic acid and ornithine concentrations, as well as an increase in α-tocopherol, thereby further distinguishing the metabolite profiles of LC patients from those of healthy controls.

The reduction of circulating levels of maltose, 3-phosphoglyceric acid, taurine, sedoheptulose, malic acid, and ornithine produced by SBRT could indicate a restoration of critical metabolic pathways such as the tricarboxylic acid cycle, the pentose cycle, and amino acid metabolism [40–43]. Changes in the levels of molecules generally foreign to the human body, such as d-arabinose, d-xylitol, and d-threitol, are challenging to interpret. However, one could speculate that they indicate alterations in the microbiome [8,44]. The decrease in lactic acid and ornithine may suggest a reduction in aerobic glycolysis (Warburg effect) and altered amino acid metabolism. Indeed, lactic acid is a byproduct of glycolysis, commonly elevated in tumors due to rapid cell proliferation [45]. A decrease in lactic acid suggests more efficient targeting of tumor cells and possibly better oxygenation of tissues post-treatment. Furthermore, an increase in α-tocopherol suggests an enhancement of antioxidant defenses. α-Tocopherol is a potent antioxidant, and its increase might be a protective response to mitigate radiation-induced oxidative damage [46].

The metabolic effects of CFRT and SBRT may differ significantly due to the distinct characteristics of each treatment and the cancer stage they target. CFRT administers lower doses of radiation per fraction over an extended period, causing sustained but less intense metabolic stress. This approach is typically applied to inoperable locally advanced tumors larger than 5 cm with involved lymph nodes. As a result, it may lead to gradual metabolic normalization and a steady stress response. In contrast, SBRT targets early stage tumors smaller than 5 cm with negative lymph nodes, delivering higher doses in fewer fractions, which results in acute and intense metabolic stress [47]. Moreover, CFRT affects a broader area, including surrounding healthy tissue, leading to extensive metabolic alterations as both tumor and normal cells respond to the radiation. Conversely, the high precision of SBRT targets the tumor more accurately, sparing healthy tissue and focusing the metabolic changes on tumor cell death, thereby reducing systemic metabolic disruption. In addition, SBRT activates the immune system more intensely than CFRT, potentially underlying the abscopal effect, which refers to effects observed at sites distant from the radiation administration [48–50]. This phenomenon presents a new avenue of research on the impact of these therapies on the immunometabolism of patients with lung cancer [51].

This study has its limitations. The sample size and demographics of the population studied may not fully represent the broader population affected by NSCLC, potentially limiting the generalizability of the findings. Variations in cancer stage among patients treated with SBRT or CFRT complicate attributing the observed metabolic changes solely to the type of radiotherapy administered. The cross-sectional design also hinders the establishment of causality or assessment of temporal relationships between biomarker levels and disease progression. Furthermore, interpreting metabolic changes in tumor tissue based on the measurement of circulating metabolites remains challenging. Addressing these limitations through more extensive multicenter, prospective studies, leveraging the vast amount of information provided by multi-omics approaches and artificial intelligence, could enable radiotherapy dose adjustments based on biomarkers related to tumor characteristics and radioresistance, thereby significantly improving the accuracy of NSCLC diagnosis and prognosis.

5. Conclusions

Our study identified significant metabolic alterations in NSCLC patients compared to healthy controls, highlighting specific metabolites which exhibited potential as biomarkers for early diagnosis. Both CFRT and SBRT induced distinct changes in the plasma metabolome, reflecting different metabolic responses to these treatments. Our findings underscore the potential of metabolomic profiling in improving lung cancer diagnosis, monitoring therapeutic responses, and guiding personalized treatment strategies.

Supplementary Materials: The following supporting information can be downloaded at: https://www.mdpi.com/article/10.3390/biom14080898/s1, Table S1: Plasma metabolite profiles of the control group and non-small cell lung cancer (NSCLC) patients pre-radiation therapy; Figure S1: Receiver Operating Characteristic (ROC) curve and confusion matrix generated using the training set of the data and a 5-fold cross validated Support Vector Machine model based on the 8 selected metabolites. The model distinguishes between non-small cell lung cancer patients (NSCLC) and the control group; Table S2: Plasma metabolite profiles of non-small cell lung cancer (NSCLC) patients pre- and post-SBRT); Table S3: Plasma metabolite profiles of non-small cell lung cancer (NSCLC) patients pre- and post- CFRT).

Author Contributions: Conceptualization, M.M.-M., M.C.-C., J.C. and M.A. (Meritxell Arenas); methodology, M.M.-M., M.C.-C., R.G.-P. and M.A. (Meritxell Arenas); software, M.C.-C., A.J.-F. and J.B.; validation, M.M.-M., M.C.-C., J.C. and M.A. (Meritxell Arenas); formal analysis, M.C.-C. and J.B.; investigation, M.M.-M., R.G.-P., A.J.-F., J.M.J.-A., R.B.-V., J.C.A., M.A. (Mónica Arguís), A.-I.O. and H.C.; resources, M.A. (Meritxell Arenas) and J.J.; data curation, M.M.-M., M.C.-C. and R.G.-P.; writing—original draft preparation, M.M.-M., M.C.-C. and J.C.; writing—review and editing, M.M.-M., M.C.-C., R.G.-P., A.J.-F., J.C. and M.A. (Meritxell Arenas); visualization, M.M.-M., M.C.-C., J.C. and M.A. (Meritxell Arenas); supervision, J.C., M.A. (Meritxell Arenas) and J.J.; project administration, J.C. and M.A. (Meritxell Arenas); funding acquisition, M.A. (Meritxell Arenas) and J.J. All authors have read and agreed to the published version of the manuscript.

Funding: This study has been funded by the FUNDACIÓN CIENTÍFICA DE LA ASOCIACIÓN ESPAÑOLA CONTRA EL CÁNCER (Madrid, Spain) under the project "PRYES222747", the FUNDACIÓ ONCOLÒGICA DR. AMADEU PELEGRÍ "AODAP/2023" (Salou, Spain), the SOCIETAT CATALANO-BALEAR D'ONCOLOGIA "OncoEsMou 2021" (Barcelona, Spain), and the INSTITUTO DE SALUD CARLOS III (Madrid, Spain) through the project "PI22/00770" and co-funded by the European Union.

Institutional Review Board Statement: This study was performed according to the declaration of Helsinki and was approved by the Comitè d'Ètica i Investigació en Medicaments (Institutional Review Board) of the HOSPITAL UNIVERSITARI DE SANT JOAN DE REUS and INSTITUT D'INVESTIGACIÓ SANITÀRIA PERE VIRGILI (project code: 14/2017, 26 January 2017).

Informed Consent Statement: Written informed consent was obtained from all subjects involved in the study.

Data Availability Statement: The raw data supporting the conclusions of this article will be made available by the authors on request.

Conflicts of Interest: The authors declare no conflicts of interest. The funders had no role in the design of the study; in the collection, analyses, or interpretation of data; in the writing of the manuscript; or in the decision to publish the results.

References

1. Sung, H.; Ferlay, J.; Siegel, R.L.; Laversanne, M.; Soerjomataram, I.; Jemal, A.; Bray, F. Global cancer statistics 2020: GLOBOCAN estimates of incidence and mortality worldwide for 36 cancers in 185 countries. *CA Cancer J. Clin.* **2021**, *71*, 209–249. [CrossRef] [PubMed]
2. Woodard, G.A.; Jones, K.D.; Jablons, D.M. Lung cancer staging and prognosis. *Cancer Treat. Res.* **2016**, *170*, 47–75. [CrossRef] [PubMed]
3. Goldstraw, P.; Chansky, K.; Crowley, J.; Rami-Porta, R.; Asamura, H.; Eberhardt, W.E.; Nicholson, A.G.; Groome, P.; Mitchell, A.; Bolejack, V.; et al. The IASLC Lung Cancer Staging Project: Proposals for Revision of the TNM Stage Groupings in the Forthcoming (Eighth) Edition of the TNM Classification for Lung Cancer. *J. Thorac. Oncol.* **2016**, *11*, 39–51. [CrossRef] [PubMed]
4. Rodríguez De Dios, N.; Navarro-Martin, A.; Cigarral, C.; Chicas-Sett, R.; García, R.; Garcia, V.; Gonzalez, J.A.; Gonzalo, S.; Murcia-Mejía, M.; Robaina, R.; et al. GOECP/SEOR radiotheraphy guidelines for non-small-cell lung cancer. *World J. Clin. Oncol.* **2022**, *13*, 237–266. [CrossRef] [PubMed]

5. National Lung Screening Trial Research Team; Aberle, D.R.; Adams, A.M.; Berg, C.D.; Black, W.C.; Clapp, J.D.; Fagerstrom, R.M.; Gareen, I.F.; Gatsonis, C.; Marcus, P.M.; et al. Reduced lung-cancer mortality with low-dose computed tomographic screening. *N. Engl. J. Med.* **2011**, *365*, 395–409. [CrossRef] [PubMed]
6. Jantus-Lewintre, E.; Usó, M.; Sanmartín, E.; Camps, C. Update on biomarkers for the detection of lung cancer. *Lung Cancer* **2012**, *3*, 21–29. [CrossRef]
7. Kannampuzha, S.; Mukherjee, A.G.; Wanjari, U.R.; Gopalakrishnan, A.V.; Murali, R.; Namachivayam, A.; Renu, K.; Dey, A.; Vellingiri, B.; Madhyastha, H.; et al. A systematic role of metabolomics, metabolic pathways, and chemical metabolism in lung cancer. *Vaccines* **2023**, *11*, 381. [CrossRef] [PubMed]
8. Hanahan, D. Hallmarks of cancer: New dimensions. *Cancer Discov.* **2022**, *12*, 31–46. [CrossRef]
9. Yu, L.; Li, K.; Zhang, X. Next-generation metabolomics in lung cancer diagnosis, treatment and precision medicine: Mini review. *Oncotarget* **2017**, *8*, 115774–115786. [CrossRef]
10. Puchades-Carrasco, L.; Jantus-Lewintre, E.; Pérez-Rambla, C.; García-García, F.; Lucas, R.; Calabuig, S.; Blasco, A.; Dopazo, J.; Camps, C.; Pineda-Lucena, A. Serum metabolomic profiling facilitates the non-invasive identification of metabolic biomarkers associated with the onset and progression of non-small cell lung cancer. *Oncotarget* **2016**, *7*, 12904–12916. [CrossRef]
11. Majem, B.; Nadal, E.; Muñoz-Pinedo, C. Exploiting metabolic vulnerabilities of non small cell lung carcinoma. *Semin. Cell. Dev. Biol.* **2020**, *98*, 54–62. [CrossRef] [PubMed]
12. Holmes, E.; Wilson, I.D.; Nicholson, J.K. Metabolic phenotyping in health and disease. *Cell* **2008**, *134*, 714–717. [CrossRef] [PubMed]
13. Puchades-Carrasco, L.; Pineda-Lucena, A. Metabolomics in pharmaceutical research and development. *Curr. Opin. Biotechnol.* **2015**, *35*, 73–77. [CrossRef] [PubMed]
14. Seijo, L.M.; Peled, N.; Ajona, D.; Boeri, M.; Field, J.K.; Sozzi, G.; Pio, R.; Zulueta, J.J.; Spira, A.; Massion, P.P.; et al. Biomarkers in lung cancer screening: Achievements, promises, and challenges. *J. Thorac. Oncol.* **2019**, *14*, 343–357. [CrossRef] [PubMed]
15. Vanhove, K.; Derveaux, E.; Mesotten, L.; Thomeer, M.; Criel, M.; Mariën, H.; Adriaensens, P. Unraveling the rewired metabolism in lung cancer using quantitative NMR metabolomics. *Int. J. Mol. Sci.* **2022**, *23*, 5602. [CrossRef] [PubMed]
16. Song, X.; Wei, J.; Sun, R.; Jiang, W.; Chen, Y.; Shao, Y.; Gu, W. Stereotactic body radiation therapy versus conventional radiation therapy in pain relief for bone metastases: A systematic review and meta-analysis. *Int. J. Radiat. Oncol. Biol. Phys.* **2023**, *115*, 909–921. [CrossRef] [PubMed]
17. Wirsdörfer, F.; de Leve, S.; Jendrossek, V. Combining radiotherapy and immunotherapy in lung cancer: Can we expect limitations due to altered normal tissue toxicity? *Int. J. Mol. Sci.* **2018**, *20*, 24. [CrossRef]
18. Cox, J.D.; Stetz, J.; Pajak, T.F. Toxicity criteria of the Radiation Therapy Oncology Group (RTOG) and the European Organization for Research and Treatment of Cancer (EORTC). *Int. J. Radiat. Oncol. Biol. Phys.* **1995**, *31*, 1341–1346. [CrossRef]
19. Fort-Gallifa, I.; García-Heredia, A.; Hernández-Aguilera, A.; Simó, J.M.; Sepúlveda, J.; Martín-Paredero, V.; Camps, J.; Joven, J. Biochemical indices of oxidative stress and inflammation in the evaluation of peripheral artery disease. *Free Radic. Biol. Med.* **2016**, *97*, 568–576. [CrossRef]
20. Costanzo, M.; Caterino, M.; Ruoppolo, M. Targeted metabolomics. In *Metabolomics Perspectives: From Theory to Practical Application*; Troisy, J., Ed.; Academic Press: Cambridge, MA, USA, 2022; pp. 219–236. [CrossRef]
21. Bräkling, S.; Hinterleitner, C.; Cappellin, L.; Vetter, M.; Mayer, I.; Benter, T.; Klee, S.; Kersten, H. Gas chromatography coupled to time-of-flight mass spectrometry using parallel electron and chemical ionization with permeation tube facilitated reagent ion control for material emission analysis. *Rapid Commun. Mass Spectrom.* **2023**, *37*, e9461. [CrossRef]
22. Hofman, D.L.; van Buul, V.J.; Brouns, F.J. Nutrition, health, and regulatory aspects of digestible maltodextrins. *Crit. Rev. Food Sci. Nutr.* **2016**, *56*, 2091–2100. [CrossRef]
23. Liberti, M.V.; Locasale, J.W. The Warburg effect: How does it benefit cancer cells? *Trends Biochem. Sci.* **2016**, *41*, 211–218. [CrossRef] [PubMed]
24. Xie, Y.; Meng, W.Y.; Li, R.Z.; Wang, Y.W.; Qian, X.; Chan, C.; Yu, Z.F.; Fan, X.X.; Pan, H.D.; Xie, C.; et al. Early lung cancer diagnostic biomarker discovery by machine learning methods. *Transl. Oncol.* **2021**, *14*, 100907. [CrossRef]
25. Tu, S.; Zhang, X.; Wan, H.; Xia, Y.; Liu, Z.; Yang, X.; Wan, F. Effect of taurine on cell proliferation and apoptosis human lung cancer A549 cells. *Oncol. Lett.* **2018**, *15*, 5473–5480. [CrossRef]
26. Wang, L.; Peng, W.; Wu, T.; Deng, P.; Zhao, Y.L. Increased glutamine anabolism sensitizes non-small cell lung cancer to gefitinib treatment. *Cell. Death Discov.* **2018**, *4*, 24. [CrossRef] [PubMed]
27. Vanhove, K.; Giesen, P.; Owokotomo, O.E.; Mesotten, L.; Louis, E.; Shkedy, Z.; Thomeer, M.; Adriaensens, P. The plasma glutamate concentration as a complementary tool to differentiate benign PET-positive lung lesions from lung cancer. *BMC Cancer* **2018**, *18*, 868. [CrossRef] [PubMed]
28. Fahrmann, J.F.; Kim, K.; DeFelice, B.C.; Taylor, S.L.; Gandara, D.R.; Yoneda, K.Y.; Cooke, D.T.; Fiehn, O.; Kelly, K.; Miyamoto, S. Investigation of metabolomic blood biomarkers for detection of adenocarcinoma lung cancer. *Cancer Epidemiol. Biomark. Prev.* **2015**, *24*, 1716–1723. [CrossRef] [PubMed]
29. Wikoff, W.; Grapov, D.; Fahrmann, J.; DeFelice, B.; Rom, W.; Pass, H.; Kim, K.; Nguyen, U.; Taylor, S.L.; Gandara, D.R.; et al. Metabolomic markers of altered nucleotide metabolism in early stage adenocarcinoma. *Cancer Prev. Res.* **2015**, *8*, 410–418. [CrossRef]

30. Zhao, Q.; Cao, Y.; Wang, Y.; Hu, C.; Hu, A.; Ruan, L.; Bo, Q.; Liu, Q.; Chen, W.; Tao, F.; et al. Plasma and tissue free amino acid profiles and their concentration correlation in patients with lung cancer. *Asia Pac. J. Clin Nutr.* **2014**, *23*, 429–436. [CrossRef]
31. Lachaux, C.; Frazao, C.J.R.; Krauβer, F.; Morin, N.; Walther, T.; François, J.M. A new synthetic pathway for the bioproduction of glycolic acid from lignocellulosic sugars aimed at maximal carbon conservation. *Front. Bioeng. Biotechnol.* **2019**, *7*, 359. [CrossRef]
32. Erez, A.; Shchelochkov, O.A.; Plon, S.E.; Scaglia, F.; Lee, B. Insights into the pathogenesis and treatment of cancer from inborn errors of metabolism. *Am. J. Hum. Genet.* **2011**, *88*, 402–421. [CrossRef]
33. Qi, S.A.; Wu, Q.; Chen, Z.; Zhang, W.; Zhou, Y.; Mao, K.; Li, J.; Li, Y.; Chen, J.; Huang, Y.; et al. High-resolution metabolomic biomarkers for lung cancer diagnosis and prognosis. *Sci. Rep.* **2021**, *11*, 11805. [CrossRef]
34. Zhang, Y.; Cheng, Y.; Qin, L.; Liu, Y.; Huang, S.; Dai, L.; Tao, J.; Pan, J.; Su, C.; Zhang, Y. Plasma metabolomics for the assessment of the progression of non-small cell lung cancer. *Int. J. Biol. Markers* **2023**, *38*, 37–45. [CrossRef]
35. Zhang, L.; Zheng, J.; Ahmed, R.; Huang, G.; Reid, J.; Mandal, R.; Maksymuik, A.; Sitar, D.S.; Tappia, P.S.; Ramjiawan, B.; et al. A high-performing plasma metabolite panel for early-stage lung cancer detection. *Cancers* **2020**, *12*, 622. [CrossRef] [PubMed]
36. Rodríguez-Tomàs, E.; Arguís, M.; Arenas, M.; Fernández-Arroyo, S.; Murcia, M.; Sabater, S.; Torres, L.; Baiges-Gayà, G.; Hernández-Aguilera, A.; Camps, J.; et al. Alterations in plasma concentrations of energy-balance-related metabolites in patients with lung, or head & neck, cancers: Effects of radiotherapy. *J. Proteomics* **2020**, *213*, 103605. [CrossRef] [PubMed]
37. Arenas, M.; Rodríguez, E.; García-Heredia, A.; Fernández-Arroyo, S.; Sabater, S.; Robaina, R.; Gascón, M.; Rodríguez-Pla, M.; Cabré, N.; Luciano-Mateo, F.; et al. Metabolite normalization with local radiotherapy following breast tumor resection. *PLoS ONE* **2018**, *13*, e0207474. [CrossRef] [PubMed]
38. Abla, H.; Sollazzo, M.; Gasparre, G.; Iommarini, L.; Porcelli, A.M. The multifaceted contribution of α-ketoglutarate to tumor progression: An opportunity to exploit? *Semin. Cell. Dev. Biol.* **2020**, *98*, 26–33. [CrossRef] [PubMed]
39. Yang, M.; Soga, T.; Pollard, P.J.; Adam, J. The emerging role of fumarate as an oncometabolite. *Front. Oncol.* **2012**, *2*, 85. [CrossRef]
40. Stincone, A.; Prigione, A.; Cramer, T.; Wamelink, M.M.; Campbell, K.; Cheung, E.; Olin-Sandoval, V.; Grüning, N.M.; Krüger, A.; Tauqeer Alam, M.; et al. The return of metabolism: Biochemistry and physiology of the pentose phosphate pathway. *Biol. Rev. Camb. Philos. Soc.* **2015**, *90*, 927–963. [CrossRef]
41. Kurhaluk, N. Tricarboxylic acid cycle intermediates and individual ageing. *Biomolecules* **2024**, *14*, 260. [CrossRef]
42. Tomar, M.S.; Kumar, A.; Shrivastava, A. Mitochondrial metabolism as a dynamic regulatory hub to malignant transformation and anti-cancer drug resistance. *Biochem. Biophys. Res. Commun.* **2024**, *694*, 149382. [CrossRef] [PubMed]
43. Cruzat, V.; Macedo Rogero, M.; Noel Keane, K.; Curi, R.; Newsholme, P. Glutamine: Metabolism and immune function, supplementation and clinical translation. *Nutrients* **2018**, *10*, 1564. [CrossRef] [PubMed]
44. Xiang, S.; Ge, Y.; Zhang, Y.; Bao, X.; Su, X.; Shi, L.; Xia, Y.; Han, H.; Ying, J.; Lai, S.; et al. L-arabinose exerts probiotic functions by improving gut microbiota and metabolism in vivo and in vitro. *J. Funct. Foods* **2024**, *113*, 106047. [CrossRef]
45. Li, Z.; Cui, J. Targeting the lactic acid metabolic pathway for antitumor therapy. *Mol. Ther. Oncolytics* **2023**, *31*, 100740. [CrossRef] [PubMed]
46. Jomova, K.; Alomar, S.Y.; Alwasel, S.H.; Nepovimova, E.; Kuca, K.; Valko, M. Several lines of antioxidant defense against oxidative stress: Antioxidant enzymes, nanomaterials with multiple enzyme-mimicking activities, and low-molecular-weight antioxidants. *Arch. Toxicol.* **2024**, *98*, 1323–1367. [CrossRef] [PubMed]
47. Kinj, R.; Bourhis, J. How stereotactic radiotherapy changed the landscape in cancer care. *Cancers* **2023**, *15*, 1734. [CrossRef]
48. Demaria, S.; Ng, B.; Devitt, M.L.; Babb, J.S.; Kawashima, N.; Liebes, L.; Formenti, S.C. Ionizing radiation inhibition of distant untreated tumors (abscopal effect) is immune mediated. *Int. J. Radiat. Oncol. Biol. Phys.* **2004**, *58*, 862–870. [CrossRef]
49. Gkika, E.; Firat, E.; Adebahr, S.; Graf, E.; Popp, I.; Radicioni, G.; Lo, S.S.; Nestle, U.; Nicolay, N.H.; Niedermann, G.; et al. Systemic immune modulation by stereotactic radiotherapy in early-stage lung cancer. *NPJ Precis. Oncol.* **2023**, *7*, 24. [CrossRef]
50. Nelson, B.E.; Adashek, J.J.; Lin, S.H.; Subbiah, V. The abscopal effect in patients with cancer receiving immunotherapy. *Med* **2023**, *4*, 233–244. [CrossRef]
51. Swamy, K. Stereotactic body radiotherapy immunological planning—A review with a proposed theoretical model. *Front. Oncol.* **2022**, *12*, 729250. [CrossRef]

biomolecules

MDPI

Article

IL-4, IL-7, IL-9, NT, NRP1 May Be Useful Markers in the Diagnosis of Endometrial Cancer

Mateusz Kozłowski [1,*], Dominika Borzyszkowska [1], Natalia Lerch [1], Agnieszka Turoń-Skrzypińska [2], Marta Tkacz [3], Jerzy Lubikowski [1], Maciej Tarnowski [3], Iwona Rotter [2] and Aneta Cymbaluk-Płoska [1]

1 Department of Reconstructive Surgery and Gynecological Oncology, Pomeranian Medical University in Szczecin, al. Powstańców Wielkopolskich 72, 70-111 Szczecin, Poland
2 Department of Medical Rehabilitation and Clinical Physiotherapy, Pomeranian Medical University in Szczecin, Żołnierska 48, 71-210 Szczecin, Poland
3 Department of Physiology in Health Sciences, Faculty of Health Sciences, Pomeranian Medical University in Szczecin, Żołnierska 54, 70-210 Szczecin, Poland
* Correspondence: mateusz.kozlowski@pum.edu.pl

Abstract: The search for novel endometrial cancer diagnostic biomarkers is pertinent. The purpose of this study was to determine if IL-4, IL-7, IL-9, IL-10, NT, TSP-2, and NRP1 could be used as novel, helpful markers for the detection of endometrial cancer. Ninety-three women diagnosed with endometrial cancer (EC) and sixty-six patients with noncancerous endometrial lesions (NCEL) were included in this study. ELISA was used to measure the concentrations of the proteins tested. Median serum levels of IL-4, IL-7, IL-9, NT, and NRP1 were significantly higher in the EC group compared with NCEL. The cut-off level of IL-4 was set at 802.26 pg/mL with a sensitivity of 83.87% and a specificity of 50% (AUC = 0.7, p = 0.000023). The cut-off level of IL-7 was set at 133.63 ng/L with a sensitivity of 96.77% and a specificity of 75.76% (AUC = 0.91, p < 0.000001). The cut-off level of IL-9 was set at 228.79 pg/mL with a sensitivity of 69.89% and a specificity of 81.82% (AUC = 0.8, p < 0.000001). The cut-off level of NT was set at 275.43 pmol/L with a sensitivity of 94.62% and a specificity of 59.09% (AUC = 0.83, p < 0.000001). The cut-off level of NRP1 was set at 30.37 ng/mL with a sensitivity of 81.72% and a specificity of 57.58% (AUC = 0.71, p = 0.000004). This study suggests the clinical utility of IL-4, IL-7, IL-9, NT, and NRP1 in the diagnosis of endometrial cancer. Nevertheless, these biomarkers may also have prognostic or predictive value, which should be tested in future studies.

Keywords: endometrial cancer; diagnosis; biomarker; IL-4; IL-7; IL-9; IL-10; neurotensin; thrombospondin-2; neuropilin 1

1. Introduction

Endometrial cancer is the most common gynecological malignancy and the fourth most prevalent malignancy in women [1,2]. The incidence of endometrial cancer is steadily increasing worldwide [3]. In 2020, it was observed that 417,367 women globally developed endometrial cancer, with Poland having 9869 cases, representing the highest global incidence rate [4]. The risk of endometrial cancer increases with age. Most cases occur in postmenopausal women, typically between the sixth and seventh decades of life [5,6]. Approximately 90% of women with endometrial cancer experience vaginal bleeding. The presence of postmenopausal bleeding prompts patients to undergo diagnostic procedures, including transvaginal ultrasound, endometrial biopsy, hysteroscopy, and often curettage of the uterine cavity [7]. To date, there is currently no evidence supporting population-based screening for endometrial cancer in asymptomatic women at average risk [8,9]. Even though symptoms of endometrial cancer appear at an early stage of the disease, identifying risk factors and biomarkers useful in diagnosing this malignancy is crucial for early diagnosis. So far, numerous risk factors for endometrial cancer have been documented, with obesity being a major contributor. Women with a BMI > 30 have a significantly higher mortality rate compared with those with a normal body weight [7,10]. Other risk factors include exposure to estrogens, diabetes, hypertension, tamoxifen therapy, early menarche, late menopause, and low parity [5]. Many of these risk factors have strong proinflammatory effects. Increasingly, a strong correlation is being observed between inflammation and cancer [11].

While existing diagnostic procedures have excellent accuracy, there are difficulties due to their intrusive nature and accompanying pain [7,12]. The search for a new, noninvasive, and affordable diagnostic instrument is essential to reduce needless curettage and create new therapeutic approaches as well as preventive strategies. Peripheral blood samples have shown promise as prospective biomarker repositories because they are easily accessible and generally noninvasive. The sensitivity and specificity of currently available serum tumor markers, such as HE4 and CA125, are still unclear, despite their involvement in the diagnosis of EC [13–15]. The examination of new biomarkers can contribute to a better understanding of the underlying mechanism of endometrial cancer's development and progression. The proteins examined in this study are involved in key processes related to tumor formation, such as inflammatory responses [16], angiogenesis [17,18], cell proliferation [19,20], and apoptosis inhibition [21,22]. Given the scarcity of data on these proteins in endometrial cancer, this study aims to evaluate the diagnostic potential of preoperative serum concentrations of IL-4, IL-7, IL-9, IL-10, NT, TSP-2, and NRP1.

Interleukin-4 (IL-4) is an immunomodulatory cytokine that promotes the differentiation of T lymphocytes towards the Th2 phenotype, which in turn supports humoral response and IgE antibody production [23,24]. Due to its antiangiogenic and immunosuppressive properties, it can influence cancer processes [13]. In endometrial cancer, IL-4's role as a potential biomarker remains to be fully understood. Some studies associate higher levels of IL-4 with more advanced cancers, suggesting its potential as a prognostic marker [13,25,26]. On the other hand, IL-4 has been linked to reduced cell proliferation in colon, kidney, and breast cancers, highlighting its complex role in tumor biology [13,27]. Interleukin-7 (IL-7) is crucial for T and B cell development [28]. It demonstrates antitumor activity by enhancing tumor eradication [21]. However, IL-7's protumor effects, including the inhibition of apoptosis and promotion of neoplastic growth via *BCL* gene regulation, have also been noted [21,22]. The limited research on IL-7 in endometrial cancer suggests a need to explore its role as a potential diagnostic marker. Interleukin-9 (IL-9) is another physiologically secreted cytokine that impacts hematopoiesis, mast cells, smooth muscle cells, and respiratory epithelial cells. It plays a role in various cancers, including lung, breast, and thyroid cancer, by promoting cell growth [19]. Studies on the impact of IL-9 on endometrial cancer have highlighted the significant infiltration of IL9+ cells and overexpression of IL9R tumor tissues [29]. These findings suggest that IL-9 may have a crucial diagnostic role in identifying endometrial cancer and a prognostic role in evaluating disease severity and progression. Interleukin-10 (IL-10) is known for its ability to regulate immune responses, including the inhibition of T lymphocyte activity [20]. This suppression can potentially lead to the proliferation of cancer cells [30]. In endometrial cancer, IL-10 has been shown to promote epithelial–mesenchymal transition (EMT) and vascular mimicry, which are processes associated with tumor progression, suggesting a possible role in assessing tumor aggressiveness and patient outcomes [31].

Neurotensin (NT) is a regulatory protein involved in gastrointestinal motility, lipid metabolism, and neuromodulation [32–34]. NT increases the expression of HER2, HER3, and EGFR, indicating its protumor activity [32,34,35]. Research on neurotensin has shown its protumorigenic effects on hormone-dependent tumors, and it has been found that the NT complex and its receptor 1 contribute to cancer progression [35,36]. This suggests the utility of NT as a predictor of metastatic potential in endometrial cancer. Thrombospondin-2 (TSP-2) is a glycoprotein involved in angiogenesis, extracellular matrix formation, and the regulation of proliferation [18,37]. Higher expression of TSP-2 has been observed in endometrial cancer, particularly in the cervical space and lymphatic vessels, suggesting its potential as a prognostic marker [18]. Neuropilin 1 (NRP1) is a transmembrane protein that mediates cytokine signaling and is associated with increased VEGF expression, promoting angiogenesis and tumor growth [38,39]. Additionally, NRP1's role as a co-receptor for TGF-β1 may facilitate processes linked to tumor progression, such as cell proliferation and metastasis [38,40]. Given the above, NRP1 could not only serve as a diagnostic marker but also as a predictor of metastatic potential in endometrial cancer.

The aim of this study was to determine the clinical significance of the proteins studied. First, we compared the serum concentrations of the proteins in patients with noncancerous endometrial lesions (NCEL) and endometrial cancer (EC). Finally, we evaluated the significance of IL-4, IL-7, IL-9, IL-10, NT, TSP-2, and NRP1 in the diagnosis of endometrial cancer. Due to the lack of appropriate and current research, this study will expand the knowledge regarding these proteins in endometrial cancer.

2. Materials and Methods

2.1. Study Design

This study was a retrospective noninterventional one. Patients having abrasions for abnormal uterine bleeding were included in the study. Oncologic surgery was then appropriate for patients who had been diagnosed with endometrial cancer based on histopathology. There was a control group without endometrial cancer and a research group with endometrial cancer. Subgroups within the study group were created according to the histological subtype, grading, and clinical stage (staging). Both the study and control groups had their serum levels of the proteins under investigation measured.

This study was conducted in accordance with the Declaration of Helsinki and approved by the Bioethics Committee of the Pomeranian Medical University in Szczecin (protocol code KB-0012/77/12 on 13 October 2012).

2.2. Participation in the Study

Those receiving treatment for perimenopausal uterine hemorrhage were included in this study. Lack of patient consent, inadequate patient data, treatment history for another malignancy, pelvic inflammatory illness, histological diagnosis of uterine malignancy other than cancer, imbalanced chronic disorders, and autoimmune diseases were the exclusion criteria for this study. In the end, 159 women met the study's eligibility requirements. Based on their weight and height, the patients' body mass index (BMI) was determined at the start of the trial. The formula used to determine the BMI was BMI = weight [kg]/height^2 [m^2]. The patients were classified into two subgroups based on the results: those with normal weight (BMI 18.5–24.9) and those who were overweight or obese (BMI > 30). Every patient also had their blood pressure measured. We separated the patients into groups with (>140/90) and without hypertension based on the results. Furthermore, we evaluated the occurrence of type 2 diabetes (DM2) based on the medical history of the patients. The power of the Mann–Whitney U test was at a very good level in most of the cases analyzed. The exception is the evaluation of the concordance of distributions for BMI in the groups of patients studied and in terms of BMI in the group of patients with a BMI below 25. The group characteristics are shown in Tables 1 and 2.

2.3. Laboratory Analysis

The preoperative serum samples were aliquoted and frozen to a temperature of −80 °C. IL-4, IL-7, IL-9, IL-10, NT, TSP-2, NRP1 levels were determined by an enzyme-linked immunosorbent assay (Shanghai Sunredbio Technology Co., Ltd., Shanghai, China, Cat No. 201-12-0093C, 201-12-1949C, 201-12-0089C, SRB-T-83474, 201-12-1319, 201-12-2386, 201-12-4380, respectively) according to the manufacturer's protocols [41,42]. The specific steps were as follows: the standard of IL-4, IL-7, IL-9, IL-10, NT, TSP-2, NRP1 was diluted to the corresponding concentration series, then the samples, antibodies, and 50 µL of Streptavidin-HRP were added to each well and incubated for 60 min at 37 °C. After washing, the chromogenic agent was used, and the samples were placed in the dark and incubated for

10 min at 37 °C with gently mixed. In the next step, the stop solution was added, and the optical density (OD) under 450 nm wavelength was measured using a microplate reader (Infinite F50, Tecan, Männedorf, Switzerland) for up to 15 min. According to standard concentrations and corresponding OD values, the standard curve linear regression equation was calculated and corresponding sample concentrations were reported in pg/mL, ng/L, pmol/L, µg/mL, and ng/mL.

Table 1. Characteristics of patients including endometrial cancer characteristics.

Characteristics	Number of Patients (%)
Endometrial cancer	
Yes	93 (58)
No	66 (42)
Endometrial Cancer	
Endometrioid endometrial carcinoma	82 (88)
Nonendometrioid endometrial carcinoma	11 (12)
Clinical staging	
FIGO I and II	88 (95)
FIGO III and IV	5 (5)
Histopathological grading	
Grade 1	37 (40)
Grade 2	34 (37)
Grade 3	22 (24)

Table 2. Characteristics of patients including clinical data.

Clinic Demographic Characteristics	Total Cohort (n = 159)	Endometrial Cancer (n = 93)	NCEL (n = 66)	p-Value	Statistical Power
Median (IQR)					
Age (years old)	55 (43–67)	51 (43–63)	59.5 (43–71)	0.1106	0.995
BMI (kg/m^2)	27.9 (24.2–31.7)	28.6 (25.4–31.7)	27 (23–31)	0.0397	0.1671
Number (%)					
Age (years old)					
<65	114 (72)	74 (80)	40 (61)	0.0967	0.999
≥65	45 (28)	19 (20)	26 (39)	0.2889	0.991
BMI					
<25	43 (27)	22 (24)	21 (32)	0.5922	0.0947
≥25	116 (73)	71 (76)	45 (68)	0.1176	0.999
Menopausal status					
Premenopausal	52 (33)	34 (37)	18 (27)	0.0387	0.999
Postmenopausal	107 (67)	59 (63)	48 (73)	0.3035	0.999
Hypertension					
Yes	82 (52)	49 (53)	33 (50)	0.3600	0.999
No	77 (48)	44 (47)	33 (50)	0.5807	0.999
Diabetes mellitus					
Yes	73 (46)	42 (45)	31 (47)	0.5577	0.999
No	86 (54)	51 (55)	35 (53)	0.3123	0.999

2.4. Statistical Calculations

Statistica 13.3.0 was used to conduct the statistical calculation. The gathered research material was assessed using the nonparametric Mann–Whitney U test of significance, correlation analysis, and statistical description. Additionally, a ROC-curve-based diagnostic test was employed. The asymmetry coefficient, minimum and maximum values, arithmetic mean, and median were used to perform the statistical description in the first stage. The normality of the distribution plot and the Shapiro–Wilk test were used to evaluate the distribution of variables. The values of descriptive parameters characterizing the distributions of the variables were also determined. A nonparametric significance test called the Mann–Whitney U test was employed to compare the distributions because the other variables were stated on a nominal scale and the distributions of the quantitative variables were not normal. When examining correlations between quantitative variables, Spearman's rank-order correlation coefficient (test of significance for Spearman's rank-order correlation coefficient) was utilized as a nonparametric measure. This paper also evaluated the power of the significance tests used. The power of the tests was checked to detect the existence of an effect in the form of correlation or difference between mean values. The power analysis was carried out a posteriori, that is, after the data were collected. The following information was used to detect the power of the test: sample size, significance level ($\alpha = 0.05$), and effect size. A test power of at least 0.8 (80%) is sought, meaning that the test has an 80% chance of detecting an effect if it actually exists. Conversely, variables representing protein concentrations and variables expressed on a nominal scale were correlated using Pearson's C contingency coefficient, which is based on the χ^2 statistic (chi square). To differentiate individuals with the feature from those without, a diagnostic test based on the ROC curve—more specifically, the area under the ROC curve—was employed. This was performed using DeLong's nonparametric technique.

3. Results

3.1. Serum Concentrations of the Studied Proteins in the Study Group and the Control Group

First, we determined the serum levels of the proteins studied in the study group and the control group. We then compared the median serum concentrations of the markers studied. We found significantly higher concentrations of IL-4, IL-7, IL-9, NT, NRP1 in the EC group compared with NCEL. We also found significantly lower concentrations of IL-10 in the EC group compared with the NCEL group. The comparison of concentrations for TSP-2 was not significant ($p = 0.6787$), so this protein was not included in further statistical calculations. The power of the Mann–Whitney U test was at a very good level in most of the cases analyzed. However, low power is shown by the test performed for TSP-2 for the groups of female patients studied. Detailed data are shown in Table 3.

Table 3. Comparison of concentrations of studied proteins between patients with endometrial cancer and noncancerous endometrial lesions.

Characteristics		EC	NCEL	*p*-Value	Statistical Power
IL-4 [pg/mL]	Median	982.36	806.42	0.000023	0.999
	Q1–Q3	855.79–1171.3	679.57–991.25		
IL-7 [ng/L]	Median	190.94	106.79	0.000000	0.999
	Q1–Q3	164.55–232.07	89.9–131.77		
IL-9 [pg/mL]	Median	257.15	198.32	0.000000	0.999
	Q1–Q3	220.92–307.78	163.79–223.46		
IL-10 [pg/mL]	Median	1594.97	3245.68	0.000000	0.999
	Q1–Q3	1257.75–1989.90	2998.79–3889.77		
NT [pmol/L]	Median	367.46	242.33	0.000000	0.999
	Q1–Q3	321.16–461.62	177.79–322.46		
TSP-2 [µg/mL]	Median	88.48	87.6	0.6787	0.0845
	Q1–Q3	71.27–113.41	77.57–109.57		
NRP1 [ng/mL]	Median	40.86	28.74	0.000004	0.999
	Q1–Q3	33.59–48.59	15.68–38.46		

3.2. Correlations between Studied Proteins

In the next step, the correlations of the proteins studied were investigated. A significant positive correlation was found between IL-4 and IL-7 ($r = 0.395$, $p = 0.0000001$), IL-4 and IL-9 ($r = 0.396$, $p = 0.0000001$), IL-4 and NT ($r = 0.394$, $p = 0.0000001$), IL-4 and NRP1 ($r = 0.304$, $p = 0.0001$), IL-7 and IL-9 ($r = 0.628$, $p = 0.0000001$), IL-7 and NT ($r = 0.529$, $p = 0.0000001$), IL-7 and NRP1 ($r = 0.259$, $p = 0.000982$), IL-9 and NT ($r = 0.629$, $p = 0.0000001$), IL-9 and NRP1 ($r = 0.264$, $p = 0.000788$), NT and NRP1 ($r = 0.251$, $p = 0.001391$). A significant negative correlation was found between IL-7 and IL-10 ($r = -0.420$, $p = 0.0000001$), IL-9 and IL-10 (r = −0.229, p = 0.003764), IL-10 and NT (r = −0.348, p = 0.000007), IL-10 and NRP1 (r = −0.232, p = 0.003313). The correlation between IL-10 and IL-4 was nonsignificant (r = −0.149, p = 0.061206). Analyzing the power of the test for Spearman's rank correlation coefficient, a very weak power is noted for the correlation between IL-4 and IL-10. A test power of less than 0.8 was observed when verifying the relationship between IL-4 and IL-10, IL-7 and NRP1, IL-9 AND IL-10, IL-9 and NRP1, IL-10 and NRP1, NT and NRP1. All data are shown in Table 4.

Table 4. The correlations between studied proteins, presented as Spearman's ranges; rs—correlation coefficient.

Variable		IL-4	IL-7	IL-9	IL-10	NT	NRP1
IL-4	rs	1	0.395	0.396	−0.149	0.394	0.304
	p-Value	-	0.0000001	0.0000001	0.061206	0.0000001	0.0001
	Statistical power	-	0.979	0.979	0.299	0.978	0.850
IL-7	rs	0.395	1	0.628	−0.420	0.529	0.259
	p-Value	0.0000001	-	0.0000001	0.0000001	0.0000001	0.000982
	Statistical power	0.979	-	0.999	0.990	0.999	0.715
IL-9	rs	0.396	0.628	1	−0.229	0.629	0.264
	p-Value	0.0000001	0.0000001	-	0.003764	0.0000001	0.000788
	Statistical power	0.979	0.999	-	0.604	0.999	0.732
IL-10	rs	−0.149	−0.420	−0.229	1	−0.348	−0.232
	p-Value	0.061206	0.0000001	0.003764	-	0.000007	0.003313
	Statistical power	0.299	0.990	0.604	-	0.934	0.616
NT	rs	0.394	0.529	0.629	−0.348	1	0.251
	p-Value	0.0000001	0.0000001	0.0000001	0.000007	-	0.001391
	Statistical power	0.978	0.999	0.999	0.934	-	0.687
NRP1	rs	0.304	0.259	0.264	−0.232	0.251	1
	p-Value	0.0001	0.000982	0.000788	0.003313	0.001391	-
	Statistical power	0.850	0.715	0.732	0.616	0.687	-

3.3. Receiver Operating Characteristic (ROC) Curve for Using IL-4, IL-7, IL-9, IL-10, NT, NRP1 Distinguishing between Endometrial Cancer and Noncancerous Endometrial Lesions

The diagnostic utility of the tested proteins between EC and NCEL was investigated using the ROC curve. The cut-off level of IL-4 was set at 802.26 pg/mL with a sensitivity of 83.87% and a specificity of 50% (AUC = 0.7, p = 0.000023). The cut-off level of IL-7 was set at 133.63 ng/L with a sensitivity of 96.77% and a specificity of 75.76% (AUC = 0.91, $p < 0.000001$). The cut-off level of IL-9 was set at 228.79 pg/mL with a sensitivity of 69.89% and a specificity of 81.82% (AUC = 0.8, $p < 0.000001$). The cut-off level of IL-10 was set at 187.4 pg/mL with a sensitivity of 100% and a specificity of 0% (AUC = 0.006, $p < 0.000001$). The cut-off level of NT was set at 275.43 pmol/L with a sensitivity of 94.62% and a specificity of 59.09% (AUC = 0.83, $p < 0.000001$). The cut-off level of NRP1 was set at 30.37 ng/mL with a sensitivity of 81.72% and a specificity of 57.58% (AUC = 0.71, p = 0.000004). The detailed data are shown in Table 5 and Figure 1.

Table 5. Diagnostic values of studied proteins for patients with endometrial cancer.

Marker	AUC (95% CI)	Sensitivity [%]	Specificity [%]	PPV [%]	NPV [%]	*p*-Value	Cut-Off Value
IL-4	0.7 (0.61–0.78)	83.87	50	70.27	68.75	0.000023	802.26 pg/mL
IL-7	0.91 (0.85–0.96)	96.77	75.76	84.91	94.34	<0.000001	133.63 ng/L
IL-9	0.8 (0.73–0.87)	69.89	81.82	84.42	65.85	<0.000001	228.79 pg/mL
IL-10	0.006 (0–0.014)	100	0	58.49	NA	<0.000001	187.4 pg/mL
NT	0.83 (0.76–0.9)	94.62	59.09	76.52	88.64	<0.000001	275.43 pmol/L
NRP1	0.71 (0.63–0.8)	81.72	57.58	73.08	69.09	0.000004	30.37 ng/mL

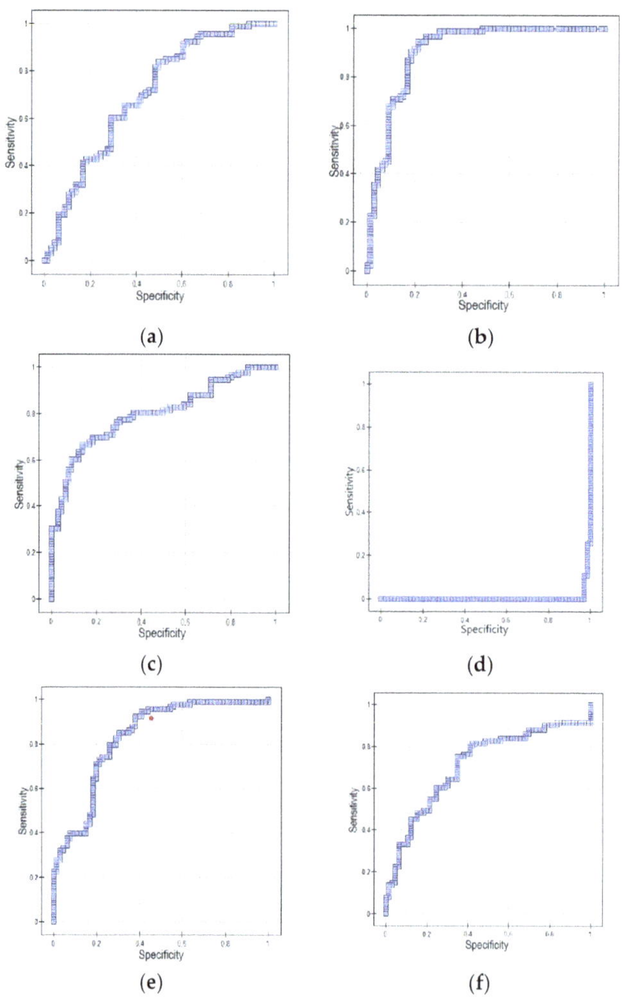

Figure 1. ROC Curve using IL-4 (**a**), IL-7 (**b**), IL-9 (**c**), IL-10 (**d**), NT (**e**), and NRP1 (**f**) distinguishing between EC and NCEL.

4. Discussion

Despite being the most common malignant gynecological tumor in women, endometrial cancer (EC) is not as well-studied as cervical and ovarian cancers, and its incidence continues to rise. To date, no reliable biomarker for the diagnosis of endometrial cancer has been identified, highlighting the importance of continued research in this field. Our study aimed to determine whether IL-4, IL-7, IL-9, IL-10, NT, TSP-2, and NRP1 could be diagnostic markers for endometrial cancer in women. The results obtained show that the median serum concentrations of IL-4, IL-7, IL-9, NT, and NRP1 were higher in women with EC compared with those with NCEL. The evaluation of IL-10's utility revealed an AUC of less than 0.5 for this cytokine, and its specificity was 0%. Moreover, the median serum concentration of IL-10 was higher in the NCEL group compared with those with EC. Therefore, interleukin-10 was excluded as a potential diagnostic marker for EC. In the case of TSP-2, the difference in serum concentration was not significant, with a p-value greater than 0.05, and therefore, it was also disqualified as a diagnostic marker. Although the TSP-2 has been observed to play an important role in other cancers, regulating angiogenesis and extracellular matrix, it lacks significance in our study. This suggests that its role in EC may be less prominent or context-dependent. Variations in study groups, patient demographics, or assay sensitivity could contribute to these findings. Further research is needed to elucidate TSP-2's role in endometrial cancer and determine if there are specific subgroups where it might be relevant.

Interleukin-4 is an immunomodulatory cytokine that plays a dual role in the pathogenesis of tumor development, exhibiting both antiangiogenic and immunosuppressive properties, which may facilitate tumor progression by creating an appropriate immune environment [13,43,44]. IL-4's functional and biological roles have been extensively studied across various cancer types. In breast, kidney, and colorectal cancers, it has been shown to have an inhibitory effect on cell proliferation, while in lung cancer, it has been found to enhance the proliferation of myeloid cells [13,45]. Bermúdez-Morales et al. found high local expression of IL-4 in biopsies from patients with precancerous changes and cervical cancer, simultaneously suggesting the induction of a local immunosuppressive state [46]. Various studies have shown that IL-4 can function as a diagnostic as well as a prognostic marker. Ramspott et al. demonstrated that the presence of IL4I1+ cells was associated with sentinel lymph node invasion and more advanced stages of melanoma [47]. The study conducted by Jincheng Ma et al. aimed to evaluate the role of IL-4 detection in endometrial cancer screening and tumor progression. In their results, they identified low interleukin-4 levels as an independent risk factor for EC in a group of patients with AUB [13]. Although in recent years there has been growing interest in the potential use of Interleukin-4 as a marker in the serum of patients with various types of cancers, its role in endometrial cancer remains controversial. Our study showed that the serum IL-4 concentration in patients with EC was higher compared with patients with NCEL. This result is consistent with the statement that IL-4 may contribute to cancer progression by supporting an immunosuppressive response, promoting the development and invasion of cancer cells. Wieder-Huszla et al. assessing cytokine levels in patients with ovarian cancer and endometrial cancer, found that IL-4 levels were higher in the group of patients with ovarian cancer than in those with endometrial cancer. However, in the group of patients with EC, the IL-4 level was noticeably higher compared with other analyzed cytokines, although the difference was statistically insignificant [24]. All these studies indicate an undeniable role in the pathogenesis of tumor development, but despite promising results, further research is necessary to confirm the role of IL-4 as a marker in endometrial cancer.

Interleukin-7 is a cytokine known for its crucial role in the development and homeostasis of T and B cells. It is also responsible for neoplastic growth and inhibition of apoptosis [21,22]. Some studies attribute IL-7 with the potential to act as an angiogenic enhancer [48]. Although available studies on IL-7 in endometrial cancer are limited, the literature contains data on the assessment of this cytokine in other cancers. In cervical cancer, IL-7 levels were significantly elevated, especially in the invasive stages, indicating its potential involvement in tumor progression [48]. Similarly, in ovarian cancer, increased IL-7 levels have been linked to disease progression and poor clinical outcomes [49,50]. In breast cancer, elevated IL-7 levels have been associated with enhanced tumor growth and poor prognosis [51]. These studies suggest that IL-7 may function not only as a diagnostic marker but also as a prognostic marker. In our study, we observed elevated levels of IL-7 in the serum of patients with endometrial cancer compared with the control group. Chopra et al. also noted an increase in IL-7 levels in the EC group. Moreover, these studies suggest a probable link between elevated IL-7 levels and more advanced stages of the disease, as well as poorer prognosis [52]. Our findings contribute to the growing body of evidence suggesting that IL-7 could be a useful biomarker for the diagnosis of endometrial cancer. Additionally, the association between higher IL-7 levels and advanced disease stages in EC suggests a prognostic role for this cytokine, similar to its role in other cancers.

IL-9 is a cytokine involved in the regulation of inflammatory response and the modulation of immune cell function. It plays a complex role in cancer, exhibiting both protumorigenic and antitumorigenic effects depending on the cancer type [19]. For instance, in lung cancer, IL-9 does not directly increase cancer cell growth but modulates cell proliferation and inhibits apoptosis, thereby supporting tumor survival [19]. In breast and thyroid cancer, IL-9 has been implicated in promoting cell growth and contributing to tumor progression [19]. In endometrial cancer, studies by Tong et al. have highlighted significant infiltration of IL-9+ cells and overexpression of its receptor in tumor tissues, suggesting that IL-9 may facilitate tumor development through its impact on the microenvironment and immune modulation [29]. Zang et al. presented a correlation between IL9+ lymphocyte and patient survival, along with the expression of progesterone receptors in EC [53]. Il-9 role as a diagnostic and prognostic marker varies across different cancers. Research by Habel et al. revealed that IL-9 can be used for early detection of epithelial ovarian cancer [54]. Fu X et al. evaluated the expression of IL-9 and IL-22 in the peripheral blood of patients with endometrial cancer, showing significant differences in cytokine expression between the endometrial cancer group and the control group. They found that the levels of IL-9 and IL-22 were higher in both the EC group and the atypical hyperplasia (AH) group compared with the control group, suggesting their potential role in the pathogenesis and progression of precancerous and cancerous changes. They also observed statistically significant differences in the expression of IL-9 and IL-22 between the EC, AH, and control groups, with the highest levels of these cytokines noted in the EC group and the lowest in the control group, highlighting their value as diagnostic and prognostic biomarkers in endometrial cancer [55]. The increase in IL-9 expression may be associated with the promotion of inflammatory and immunosuppressive processes, which favor tumor development and progression. Our findings are consistent with previous findings, demonstrating a notable increase in IL-9 expression in serum samples from EC patients. This alignment across different sample types reinforces the potential of IL-9 as a diagnostic marker for endometrial cancer. Nonetheless, further research is necessary to fully elucidate the mechanisms by which IL-9 contributes to endometrial cancer pathogenesis and to validate its use as a clinical biomarker.

Neurotensin is known for its regulatory role in the nervous and gastrointestinal system, but its function extends to the modulation of cancer cell behavior. NT plays a significant role in neoplastic progression, including the proliferation of the prostate, colon, pancreas, and lung cancer cells [56]. It also prevents apoptosis, including breast cancer. As a result, elevated NT levels, occurring in tissues, can play a major role in tumor proliferation [57]. The available studies suggest that NT and its receptor NTSR1 may serve not only as diagnostic markers but importantly as prognostic markers. For instance, in breast cancer, elevated levels of NT have been associated with increased tumor aggressiveness and poor survival rates [56,58]. Similarly, in pancreas cancer, increased expression levels of NT and its receptor were found in highly malignant sublines. Further analysis revealed that higher levels were present in more advanced cancer and that high NSTR1 concentrations were correlated with poor prognosis [59]. The research on neurotensin in endometrial cancer is limited. Our study results indicate that in the serum of patients with EC, the concentration of neurotensin was higher compared with the NCEL control group. These findings are consistent with previous reports in the literature suggesting that NT plays a crucial role in the pathogenesis and progression of cancer. Immunohistochemical studies conducted by Agopiantz et al. demonstrated that the expression of NT and NTSR1 is significantly increased in endometrial cancer compared with the control group. Furthermore, a negative correlation has been shown between NTSR1 activity and overall survival, as well as progression-free survival in EC patients, thereby highlighting the role of NT in tumor aggressiveness [35]. NT's elevated serum levels in EC patients, along with its established roles in other cancers, highlight its importance as both a diagnostic and prognostic biomarker. Further research is needed to fully understand the mechanisms of NT in endometrial cancer.

Neuropilin 1 is a transmembrane protein that plays a significant role in the pathogenesis of tumor development. It acts as a co-receptor for various signaling molecules, including vascular endothelial growth factor (VEGF) and transforming growth factor- beta 1(TGF-β1) [38,39]. By binding to VEGF, NRP1 promotes angiogenesis, which is crucial for tumor growth and metastasis across multiple cancer types [60]. Additionally, NPR1's role as a co-receptor for TGF-β1 facilitates cell proliferation and metastasis, further implicating NRP1 in cancer development and spread [61]. For instance, in non-small-cell lung cancer (NSCLC), NRP1 was significantly increased and functioned as a vital promoter of metastasis [62]. In terms of diagnostic and prognostic applications, NRP1 has shown promise across various cancers. As an example, Cao et al. analyzed the expression of NRP1 in gastric cell tissue and it was higher in more advanced stages compared with the group of early stage [63]. Similarly, Oplawski et al. demonstrated that NRP1 expression was elevated in tissue samples from endometrial cancer patients compared with the control group. Furthermore, NRP1 expression was strongly correlated with histological malignancy. In G2 and G3 samples, it was significantly higher than in G1 samples, suggesting that NRP1 could be a useful prognostic marker [40]. These findings are supported by research conducted by Okon et al., where strong NRP-1 expression was also observed in endometrial cancer, along with an association between NRP1 expression and levels of protumorigenic cytokines [38]. In our study, we found a higher level of NRP1 in the serum of EC patients compared with NCEL patients. These results are significant because most available studies on NRP1 in endometrial cancer focus on its expression in tissues rather than serum. Our discovery of elevated NRP1 levels in the serum of EC patients aligns with tissue study results, suggesting that neuropilin-1 may have significance not only at the tissue expression level but also in the serum. Elevated NRP1 levels in serum may reflect its increased expression in tumor tissue and serve as a useful diagnostic marker for detecting cancer and monitoring its progression.

Our study also has limitations. The study presented here is one of the first to present the diagnostic aspect of the proteins studied in endometrial cancer. Future studies are needed to confirm our observations and better understand the role of the proteins we investigated in the diagnosis of endometrial cancer. Existing literature data confirm the utility of the proteins we assessed in diagnosing and monitoring the progression of endometrial cancer, reinforcing the need for continued research in this area. Further research is needed to validate these findings and explore the underlying mechanisms by which these biomarkers contribute to the pathogenesis of endometrial cancer. Moreover, investigating the potential for combining these biomarkers with existing diagnostic modalities, such as imaging or histopathological assessments, could enhance overall diagnostic accuracy. Understanding the biological roles of these biomarkers in the tumor microenvironment could reveal novel insights into their functions and interactions that influence cancer progression. Additionally, the variability in the levels of these biomarkers among different stages and subtypes of endometrial cancer needs to be explored. This would provide information on whether they have stage-specific or type-specific diagnostic values. The role of these biomarkers in predicting treatment responses and patient outcomes also remains an open question. Potential future research could focus on studies assessing how changes in biomarker level correlate with treatment efficacy and disease progression. Addressing these gaps will be crucial in improving the use of these biomarkers for early detection, prognosis, and personalized treatment strategies in endometrial cancer.

5. Conclusions

The initial study presented here demonstrates the clinical utility of IL-4, IL-7, IL-9, NT, and NRP1 in endometrial cancer. IL-4, IL-7, IL-9, NT, and NRP1 may be useful diagnostic markers for endometrial cancer. Perhaps, in the future, these proteins could co-develop a diagnostic protocol to support ultrasound and histological confirmation of endometrial disease. This study showed that IL-10 and TSP-2 should not be used as diagnostic markers in endometrial cancer. In the future, similar studies should be conducted, evaluating combinations of several proteins studied here as diagnostic markers versus single markers. It would also be appropriate to focus on evaluating these proteins as prognostic or predictive markers.

Author Contributions: Conceptualization, A.T.-S. and I.R.; methodology, J.L. and M.T. (Marta Tkacz); validation, M.T. (Maciej Tarnowski) and A.C.-P.; formal analysis, M.T. (Marta Tkacz) and A.T.-S.; resources, J.L.; data curation, M.K., D.B. and N.L.; writing—original draft preparation, M.K., D.B. and N.L.; writing—review and editing, A.C.-P.; visualization, M.K. and D.B.; supervision, A.C.-P. and M.T. (Maciej Tarnowski); project administration, I.R. All authors have read and agreed to the published version of the manuscript.

Funding: This research received no external funding.

Institutional Review Board Statement: This study was conducted in accordance with the Declaration of Helsinki and approved by the Bioethics Committee of the Pomeranian Medical University in Szczecin (protocol code KB-0012/77/12 on 13 October 2012).

Informed Consent Statement: Informed consent was obtained from all subjects involved in the study.

Data Availability Statement: The data presented in this study are available from the corresponding author, M.K., upon reasonable request.

Conflicts of Interest: The authors declare no conflicts of interest.

References

1. Cai, Y.; Wang, B.; Xu, W.; Liu, K.; Gao, Y.; Guo, C.; Chen, J.; Kamal, M.A.; Yuan, C. Endometrial Cancer: Genetic, Metabolic Characteristics, Therapeutic Strategies and Nanomedicine. *Curr. Med. Chem.* **2021**, *28*, 8755–8781. [CrossRef]
2. Braun, M.M.; Overbeek-Wager, E.A.; Grumbo, R.J. Diagnosis and management of endometrial cancer. *Am. Fam. Physician* **2016**, *93*, 468–474. [PubMed]
3. Crosbie, E.J.; Kitson, S.J.; McAlpine, J.N.; Mukhopadhyay, A.; Powell, M.E.; Singh, N. Endometrial cancer. *Lancet* **2022**, *399*, 1412–1428. [CrossRef] [PubMed]
4. Endometrial Cancer Statistics. Available online: https://www.wcrf.org/cancer-trends/endometrial-cancer-statistics/ (accessed on 8 August 2024).
5. Makker, V.; MacKay, H.; Ray-Coquard, I.; Levine, D.A.; Westin, S.N.; Aoki, D.; Oaknin, A. Endometrial cancer. *Nat. Rev. Dis. Primers* **2021**, *7*, 88. [CrossRef]
6. Amant, F.; Moerman, P.; Neven, P.; Timmerman, D.; Van Limbergen, E.; Vergote, I. Endometrial cancer. *Lancet* **2005**, *366*, 491–505. [CrossRef] [PubMed]
7. Clarke, M.A.; Long, B.J.; Del Mar Morillo, A.; Arbyn, M.; Bakkum-Gamez, J.N.; Wentzensen, N. Association of Endometrial Cancer Risk with Postmenopausal Bleeding in Women. *JAMA Intern. Med.* **2018**, *178*, 1210–1222. [CrossRef]
8. Colombo, N.; Creutzberg, C.; Amant, F.; Bosse, T.; González-Martín, A.; Ledermann, J.; Marth, C.; Nout, R.; Querleu, D.; Mirza, M.R.; et al. ESMO-ESGO-ESTRO consensus conference on endometrial cancer: Diagnosis, treatment and follow-up. *Ann. Oncol.* **2016**, *27*, 16–41. [CrossRef]
9. PDQ Screening and Prevention Editorial Board. Endometrial Cancer Screening (PDQ®): Health Professional Version. In *PDQ Cancer Information Summaries*; National Cancer Institute: Bethesda, MD, USA, 2002.
10. Onstad, M.A.; Schmandt, R.E.; Lu, K.H. Addressing the role of obesity in endometrial cancer risk, prevention, and treatment. *J. Clin. Oncol.* **2016**, *34*, 4225–4230. [CrossRef]
11. Cai, J.; Cui, K.; Niu, F.; Jin, T.; Huang, S.; Zhang, Y.; Bao, S. Genetics of IL6 polymorphisms: Case–control study of the risk of endometrial cancer. *Mol. Genet. Genom. Med.* **2019**, *7*, e00600. [CrossRef]
12. Urick, M.E.; Bell, D.W. Clinical actionability of molecular targets in endometrial cancer. *Nat. Rev. Cancer* **2019**, *19*, 510–521. [CrossRef]

13. Ma, J.; Liu, D.; Mao, X.; Huang, L.; Ren, Y.; Xu, X.; Huang, X.; Deng, C.; Shi, F.; Sun, P. Enhanced Diagnostic Efficiency of Endometrial Carcinogenesis and Progression in Women with Abnormal Uterine Bleeding through Peripheral Blood Cytokine Testing: A Multicenter Retrospective Cohort Study. *Int. J. Med. Sci.* **2024**, *21*, 601–611. [CrossRef] [PubMed]
14. Cymbaluk-Płoska, A.; Chudecka-Głaz, A.; Pius-Sadowska, E.; Sompolska-Rzechuła, A.; Machaliński, B.; Surowiec, A.; Menkiszak, J. Clinical importance of serum HE4 and MMP2 levels in endometrial cancer patients. *Onco Targets Ther.* **2017**, *10*, 3169–3175. [CrossRef] [PubMed]
15. Moore, R.G.; Brown, A.K.; Craig Miller, A.; Badgwell, D.; Lu, Z.; Jeffrey Allard, W.; Granai, C.O.; Bast Jr, R.C.; Lu, K. Utility of a novel serum tumor biomarker HE4 in patients with endometrioid adenocarcinoma of the uterus. *Gynecol. Oncol.* **2008**, *110*, 196–201. [CrossRef]
16. Song, X.; Traub, B.; Shi, J.; Kornmann, M. Possible roles of interleukin-4 and-13 and their receptors in gastric and colon cancer. *Int. J. Mol. Sci.* **2021**, *22*, 727. [CrossRef]
17. Tian, Y.; Zhu, P.; Liu, S.; Jin, Z.; Li, D.; Zhao, H.; Zhu, X.; Shu, C.; Yan, D.; Dong, Z. IL-4-polarized BV2 microglia cells promote angiogenesis by secreting exosomes. *Adv. Clin. Exp. Med.* **2019**, *28*, 421–430. [CrossRef]
18. Seki, N.; Kodama, J.; Hashimoto, I.; Hongo, A.; Yoshinouchi, M.; Kudo, T. Thrombospondin-1 and -2 messenger RNA expression in normal and neoplastic endometrial tissues: Correlation with angiogenesis and prognosis. *Int. J. Oncol.* **2001**, *19*, 305–310. [CrossRef]
19. Lee, J.E.; Zhu, Z.; Bai, Q.; Brady, T.J.; Xiao, H.; Wakefield, M.R.; Fang, Y. The Role of Interleukin-9 in Cancer. *Pathol. Oncol. Res.* **2020**, *26*, 2017–2022. [CrossRef] [PubMed]
20. Chang, C.M.; Lam, H.Y.P.; Hsu, H.J.; Jiang, S.J. Interleukin-10: A double-edged sword in breast cancer. *Tzu Chi Med. J.* **2021**, *33*, 203–211. [CrossRef]
21. Wang, C.; Kong, L.; Kim, S.; Lee, S.; Oh, S.; Jo, S.; Jang, I.; Kim, T.-D. The Role of IL-7 and IL-7R in Cancer Pathophysiology and Immunotherapy. *Int. J. Mol. Sci.* **2022**, *23*, 10412. [CrossRef]
22. Lin, J.; Zhu, Z.; Xiao, H.; Wakefield, M.R.; Ding, V.A.; Bai, Q.; Fang, Y. The role of il-7 in immunity and cancer. *Anticancer. Res.* **2017**, *37*, 963–968. [CrossRef]
23. Formentini, A.; Braun, P.; Fricke, H.; Link, K.H.; Henne-Bruns, D.; Kornmann, M. Expression of interleukin-4 and interleukin-13 and their receptors in colorectal cancer. *Int. J. Color. Dis.* **2012**, *27*, 1369–1376. [CrossRef] [PubMed]
24. Wieder-Huszla, S.; Chudecka-Głaz, A.; Gutowska, I.; Karakiewicz, B.; Jurczak, A. Effect of the treatment stage on the serum levels of selected cytokines and antioxidant enzymes in patients with tumors of the reproductive organs. *Eur. Rev. Med. Pharmacol. Sci.* **2023**, *27*, 3117–3133. [CrossRef] [PubMed]
25. Perambakam, S.M.; Srivastava, R.; Peace, D.J. Distinct cytokine patterns exist in peripheral blood mononuclear cell cultures of patients with prostate cancer. *Clin. Immunol.* **2005**, *117*, 94–99. [CrossRef]
26. Wise, G.J.; Marella, V.K.; Talluri, G.; Shirazian, D. Cytokine variations in patients with hormone treated prostate cancer. *J. Urol.* **2000**, *164*, 722–725. [CrossRef] [PubMed]
27. Nagai, S.; Toi, M. Interleukin-4 and breast cancer. *Breast Cancer* **2000**, *7*, 181–186. [CrossRef]
28. Fry, T.J.; Mackall, C.L. Interleukin-7: From bench to clinic. *Blood* **2002**, *99*, 3892–3904. [CrossRef]
29. Tong, H.; Feng, H.; Hu, X.; Wang, M.-F.; Song, Y.-F.; Wen, X.-L.; Li, Y.-R.; Wan, X.-P. Identification of Interleukin-9 Producing Immune Cells in Endometrial Carcinoma and Establishment of a Prognostic Nomogram. *Front. Immunol.* **2020**, *11*, 544248. [CrossRef]
30. Chen, K.Y.; Chien, W.C.; Liao, J.M.; Tsai, C.W.; Chang, W.S.; Su, C.H.; Hsu, S.W.; Wang, H.C.; Bau, D.T. Contribution of interleukin-10 genotype to triple negative breast cancer risk. *Anticancer. Res.* **2021**, *41*, 2451–2457. [CrossRef]
31. Yu, Z.; Zhang, Q.; Wei, S.; Zhang, Y.; Zhou, T.; Zhang, Q.; Shi, R.; Zinovkin, D.; Pranjol, Z.I.; Zhang, J.; et al. $CD146^{+}$CAFs promote progression of endometrial cancer by inducing angiogenesis and vasculogenic mimicry via IL-10/JAK1/STAT3 pathway. *Cell Commun. Signal.* **2024**, *22*, 170. [CrossRef]
32. Nikolaou, S.; Qui, S.; Fiorentino, F.; Simillis, C.; Rasheed, S.; Tekkis, P.; Kontovounisios, C. The role of Neurotensin and its receptors in non-gastrointestinal cancers: A review. *Cell Commun. Signal.* **2020**, *18*, 68. [CrossRef]
33. Grill, S.; Yahiaoui-Doktor, M.; Basrai, M.; Struck, J.; Schulte, J.; Berling-Ernst, A.; Engel, C.; Ullrich, M.; Lammert, J.; Bischoff, S.C.; et al. Precursor fractions of neurotensin and enkephalin might point to molecular mechanisms of cancer risk modulation during a lifestyle-intervention in germline BRCA1/2 gene mutation carriers. *Breast Cancer Res. Treat.* **2021**, *186*, 741–752. [CrossRef] [PubMed]
34. Younes, M.; Wu, Z.; Dupouy, S.; Lupo, A.M.; Mourra, N.; Takahashi, T.; Fléjou, J.F.; Trédaniel, J.; Régnard, J.F.; Damotte, D.; et al. Neurotensin (NTS) and its receptor (NTSR1) causes EGFR, HER2 and HER3 over-expression and their autocrine/paracrine activation in lung tumors, confirming responsiveness to erlotinib. *Oncotarget* **2014**, *5*, 8252–8269. [CrossRef]
35. Agopiantz, M.; Forgez, P.; Casse, J.M.; Lacomme, S.; Charra-Brunaud, C.; Clerc-Urmès, I.; Morel, O.; Bonnet, C.; Guéant, J.L.; Vignaud, J.M.; et al. Expression of neurotensin receptor 1 in endometrial adenocarcinoma is correlated with histological grade and clinical outcome. *Virchows Arch.* **2017**, *471*, 521–530. [CrossRef]
36. Liu, J.; Agopiantz, M.; Poupon, J.; Wu, Z.; Just, P.A.; Borghese, B.; Ségal-Bendirdjian, E.; Gauchotte, G.; Gompel, A.; Forgez, P. neurotensin receptor 1 antagonist SR48692 improves response to carboplatin by enhancing apoptosis and inhibiting drug efflux in ovarian cancer. *Clin. Cancer Res.* **2017**, *23*, 6516–6528. [CrossRef]
37. Liu, J.F.; Lee, C.W.; Tsai, M.H.; Tang, C.H.; Chen, P.C.; Lin, L.W.; Lin, C.Y.; Lu, C.H.; Lin, Y.F.; Yang, S.H.; et al. Thrombospondin 2 promotes tumor metastasis by inducing matrix metalloproteinase-13 production in lung cancer cells. *Biochem. Pharmacol.* **2018**, *155*, 537–546. [CrossRef] [PubMed]

38. Okon, I.S.; Ding, Y.; Coughlan, K.A.; Wang, Q.; Song, P.; Benbrook, D.M.; Zou, M.H. Aberrant NRP-1 expression serves as predicator of metastatic endometrial and lung cancers. *Oncotarget* **2016**, *7*, 7970–7978. [CrossRef]
39. Chaudhary, B.; Khaled, Y.S.; Ammori, B.J.; Elkord, E. Neuropilin 1: Function and therapeutic potential in cancer. *Cancer Immunol. Immunother.* **2014**, *63*, 81–99. [CrossRef]
40. Oplawski, M.; Dziobek, K.; Brabarek, B.; Zmarzły, N.; Dąbruś, D.; Januszyk, P.; Brus, R.; Tomala, B.; Boroń, D. Expression of NRP-1 and NRP-2 in Endometrial Cancer. *Curr. Pharm. Biotechnol.* **2019**, *20*, 254–260. [CrossRef] [PubMed]
41. Rzepakowska, A.; Żurek, M.; Grzybowski, J.; Kotula, I.; Pihowicz, P.; Górnicka, B.; Demkow, U.; Niemczyk, K. Serum and tissue expression of neuropilin 1 in precancerous and malignant vocal fold lesions. *PLoS ONE* **2020**, *15*, e0239550. [CrossRef]
42. Kenawy, M.Z.; Mikhael, N.W.; El-fallah, A.A.; Shehata, R.A. Serum Levels of Insulin Like Growth Factor-1(IGF-1) in Patients with Seborrheic Dermatitis. *Benha J. Appl. Sci.* **2020**, *5*, 229–231. [CrossRef]
43. Wang, H.W.; Joyce, J.A. Alternative activation of tumor-associated macrophages by IL-4: Priming for protumoral functions. *Cell Cycle* **2010**, *9*, 4824–4835. [CrossRef] [PubMed]
44. Baniyash, M. Chronic inflammation, immunosuppression and cancer: New insights and outlook. *Semin. Cancer Biol.* **2006**, *16*, 80–88. [CrossRef]
45. Fu, C.; Jiang, L.; Hao, S.; Liu, Z.; Ding, S.; Zhang, W.; Yang, X.; Li, S. Activation of the IL-4/STAT6 Signaling Pathway Promotes Lung Cancer Progression by Increasing M2 Myeloid Cells. *Front. Immunol.* **2019**, *10*, 2638. [CrossRef]
46. Bermúdez-Morales, V.H.; Peralta-Zaragoza, O.; Alcocer-González, J.M.; Moreno, J.; Madrid-Marina, V. IL-10 expression is regulated by HPV E2 protein in cervical cancer cells. *Mol. Med. Rep.* **2011**, *4*, 369–375. [CrossRef]
47. Ramspott, J.P.; Bekkat, F.; Bod, L.; Favier, M.; Terris, B.; Salomon, A.; Djerroudi, L.; Zaenker, K.S.; Richard, Y.; Molinier-Frenkel, V.; et al. Emerging Role of IL-4–Induced Gene 1 as a Prognostic Biomarker Affecting the Local T-Cell Response in Human Cutaneous Melanoma. *J. Investig. Dermatol.* **2018**, *138*, 2625–2634. [CrossRef] [PubMed]
48. Chopra, V.; Dinh, T.V.; Hannigan, E.V. Circulating serum levels of cytokines and angiogenic factors in patients with cervical cancer. *Cancer Investig.* **1998**, *16*, 152–159. [CrossRef] [PubMed]
49. Lambeck, A.J.A.; Crijns, A.P.G.; Leffers, N.; Sluiter, W.J.; ten Hoor, K.A.; Braid, M.; van der Zee, A.G.J.; Daemen, T.; Nijman, H.W.; Kast, W.M. Serum cytokine profiling as a diagnostic and prognostic tool in ovarian cancer: A potential role for interleukin 7. *Clin. Cancer Res.* **2007**, *13*, 2385–2391. [CrossRef]
50. Xie, X.; Ye, D.; Chen, H.; Lu, W.; Cheng, B.; Zhong, H. Interleukin-7 and suppression of local peritoneal immunity in ovarian carcinoma. *Int. J. Gynecol. Obstet.* **2004**, *85*, 151–158. [CrossRef]
51. Zarogoulidis, P.; Lampaki, S.; Yarmus, L.; Kioumis, I.; Pitsiou, G.; Katsikogiannis, N.; Hohenforst-Schmidt, W.; Li, Q.; Huang, H.; Sakkas, A.; et al. Interleukin-7 and interleukin-15 for cancer. *J. Cancer* **2014**, *5*, 765–773. [CrossRef]
52. Chopra, V.; Ding, T.V.; Hanningan, E.V. Serum levels of interleukins, growth factors and anglogenin in patients with endometrial cancer. *J. Cancer Res. Clin. Oncol.* **1997**, *123*, 167–172. [CrossRef]
53. Zang, Y.; Li, H.; Liu, S.; Zhao, R.; Zhang, K.; Zang, Y.; Wang, Y.; Xue, F. The roles and clinical applications of interleukins in endometrial carcinoma. *Front. Oncol.* **2022**, *12*, 1001693. [CrossRef]
54. Habel, A.; Xu, W.; Ahmed, M.H.; Stayoussef, M.; Bouaziz, H.; Ayadi, M.; Mezlini, A.; Larbi, A.; Yaacoubi-Loueslati, B. Identification of two theranostic biomarker panels for epithelial ovarian cancer. *Cytokine* **2023**, *161*, 156051. [CrossRef] [PubMed]
55. Fu, X.; Song, M.; Zhang, R.; Chen, F. Expression and clinical significance of peripheral blood IL-9 and IL-22 in patients with endometrial cancer. *Minerva Gastroenterol.* **2024**, *70*, 245–247. [CrossRef] [PubMed]
56. Souazé, F.; Dupouy, S.; Viardot-Foucault, V.; Bruyneel, E.; Attoub, S.; Gespach, C.; Gompel, A.; Forgez, P. Expression of neurotensin and NT1 receptor in human breast cancer: A potential role in tumor progression. *Cancer Res.* **2006**, *66*, 6243–6249. [CrossRef] [PubMed]
57. Soma, S.; Gompel, A.; Rostène, W.; Forgez, P. Neurotensin counteracts apoptosis in breast cancer cells. *Biochem. Biophys. Res. Commun.* **2002**, *295*, 482–488. [CrossRef]
58. Morgat, C.; Brouste, V.; Chastel, A.; Vélasco, V.; Macgrogan, G.; Hindié, E. Expression of neurotensin receptor-1 (NTS1) in primary breast tumors, cellular distribution, and association with clinical and biological factors. *Breast Cancer Res. Treat.* **2021**, *190*, 403–413. [CrossRef]
59. Takahashi, K.; Ehata, S.; Miyauchi, K.; Morishita, Y.; Miyazawa, K.; Miyazono, K. Neurotensin receptor 1 signaling promotes pancreatic cancer progression. *Mol. Oncol.* **2021**, *15*, 151–166. [CrossRef]
60. Kofler, N.M.; Simons, M. Angiogenesis versus arteriogenesis: Neuropilin 1 modulation of VEGF signaling. *F1000Prime Rep.* **2015**, *7*, 26. [CrossRef]
61. Glinka, Y.; Stoilova, S.; Mohammed, N.; Prud'homme, G.J. Neuropilin-1 exerts co-receptor function for TGF-beta-1 on the membrane of cancer cells and enhances responses to both latent and active TGF-beta. *Carcinogenesis* **2011**, *32*, 613–621. [CrossRef]
62. Ding, Z.; Du, W.; Lei, Z.; Zhang, Y.; Zhu, J.; Zeng, Y.; Wang, S.; Zheng, Y.; Liu, Z.; Huang, J.A. Neuropilin 1 modulates TGF-β1-induced epithelial-mesenchymal transition in non-small cell lung cancer. *Int. J. Oncol.* **2020**, *56*, 531–543. [CrossRef]
63. Cao, H.; Li, Y.; Huang, L.; Bai, B.; Xu, Z. Clinicopathological Significance of Neuropilin 1 Expression in Gastric Cancer: A Meta-Analysis. *Dis. Markers* **2020**, *2020*, 4763492. [CrossRef] [PubMed]

biomolecules

MDPI

Article

Rapid Detection of *PML::RARA* Fusions in Acute Promyelocytic Leukemia: CRISPR/Cas9 Nanopore Sequencing with Adaptive Sampling

William Middlezong [1], Victoria Stinnett [2], Michael Phan [1], Brian Phan [3], Laura Morsberger [2], Melanie Klausner [2], Jen Ghabrial [2], Natalie DeMetrick [2], Jing Zhu [2], Trisha James [2], Aparna Pallavajjala [2], Christopher D. Gocke [2], Maria R. Baer [4] and Ying S. Zou [2,*]

1 Krieger School of Arts and Sciences, Johns Hopkins University, Baltimore, MD 21218, USA; wmiddle2@jhu.edu (W.M.); mphan4@jhu.edu (M.P.)

2 Department of Pathology, Johns Hopkins University School of Medicine, Baltimore, MD 21287, USA; vlloyd3@jhmi.edu (V.S.); lmorsber@jhmi.edu (L.M.); mhardy22@jhmi.edu (M.K.); jghabri1@jh.edu (J.G.); nescola1@jhu.edu (N.D.); jzhu23@jh.edu (J.Z.); tjames17@jhmi.edu (T.J.); apallav2@jhmi.edu (A.P.); cgocke1@jhmi.edu (C.D.G.)

3 Department of Biology, The College of William and Mary, Williamsburg, VA 23186, USA; bphan3@jh.edu

4 Department of Medicine, University of Maryland Greenebaum Comprehensive Cancer Center, Baltimore, MD 21201, USA; mbaer@umm.edu

* Correspondence: yzou19@jhmi.edu

Abstract: Acute promyelocytic leukemia (APL) accounts for approximately 10–15% of newly diagnosed acute myeloid leukemia cases and presents with coagulopathy and bleeding. Prompt diagnosis and treatment are required to minimize early mortality in APL as initiation of all-trans retinoic acid therapy rapidly reverses coagulopathy. The *PML::RARA* fusion is a hallmark of APL and its rapid identification is essential for rapid initiation of specific treatment to prevent early deaths from coagulopathy and bleeding and optimize patient outcomes. Given limitations and long turnaround time of current gene fusion diagnostic strategies, we have developed a novel amplification-free nanopore sequencing-based approach with low cost, easy setup, and fast turnaround time. We termed the approach CRISPR/Cas9-enriched nanopore sequencing with adaptive sampling (CENAS). Using CENAS, we successfully sequenced breakpoints of typical and atypical *PML::RARA* fusions in APL patients. Compared with the standard-of-care genetic diagnostic tests, CENAS achieved good concordance in detecting *PML::RARA* fusions in this study. CENAS allowed for the identification of sequence information of fusion breakpoints involved in typical and atypical *PML::RARA* fusions and identified additional genes (*ANKFN1* and *JOSD1*) and genomic regions (13q14.13) involving the atypical fusions. To the best of our knowledge, involvements of the *ANKFN1* gene, the *JOSD1* gene, and the 13q14.13 genomic region flanking with the *SIAH3* and *ZC3H13* genes have not been reported in the atypical *PML::RARA* fusions. CENAS has great potential to develop as a point-of-care test enabling immediate, low-cost bedside diagnosis of APL patients with a *PML::RARA* fusion. Given the early death rate in APL patients still reaches 15%, and ~10% of APL patients are resistant to initial therapy or prone to relapse, further sequencing studies of typical and atypical *PML::RARA* fusion might shed light on the pathophysiology of the disease and its responsiveness to treatment. Understanding the involvement of additional genes and positional effects related to the *PML* and *RARA* genes could shed light on their role in APL and may aid in the development of novel targeted therapies.

Keywords: CRISPR/Cas9; nanopore sequencing; *PML::RARA* fusions; acute promyelocytic leukemia; adaptive sampling

1. Introduction

Acute promyelocytic leukemia (APL) accounts for approximately 10–15% of newly diagnosed acute myeloid leukemia cases and presents with coagulopathy and bleeding [1].

Prompt diagnosis and treatment are required to minimize early mortality in APL as initiation of all-trans retinoic acid (ATRA) therapy rapidly reverses coagulopathy [2]. APL is characterized by t(15;17)(q24;q21), a balanced reciprocal translocation between the promyelocytic leukemia (*PML*) gene on chromosome 15 and the retinoic acid receptor alpha (*RARA*) gene on chromosome 17, leading to *PML::RARA* fusion [1]. This *PML::*RARA fusion protein disrupts normal myeloid differentiation and maturation, alters gene expression, deregulates transcriptional control (acting as a transcriptional repressor of *RARA* target genes), and disrupts PML homeostatic function [1]. It plays a critical role in the development of APL. Various breakpoints of *PML::RARA* fusion have been described. Given that *RARA* breakpoints are commonly located in intron 2, typical *PML::RARA* fusion transcripts depend on the *PML* breakpoints located in intron 6, exon 6, or intron 3, also known as bcr1 to bcr3 isoforms, respectively. Although rare, atypical *PML::RARA* fusion transcripts involving other regions of *PML* and *RARA* (such as exons 1 and 3 of the *RARA* gene and exons 4 and 7a–c of the *PML* gene), insertions of *PML::RARA* fusions, three-way translocations, variant complex translocations, and complex rearrangement have also been described [3–8]. The *PML::RARA* fusion is a hallmark of APL and its rapid identification is essential for rapid initiation of specific treatment to prevent early deaths from coagulopathy and bleeding and optimize patient outcomes.

Current gene fusion diagnostic strategies in the clinical setting include conventional karyotyping, fluorescent in situ hybridization (FISH), real-time quantitative reverse transcription polymerase chain reaction (qRT-PCR), and next-generation sequencing (NGS) assays. All these strategies have limitations and turnaround times ranging from days to weeks. While conventional karyotyping usually has a long turnaround time (~1 week), in-house rush/"STAT" FISH is commonly used to detect *PML::RARA* gene fusions within a day. These two standard-of-care cytogenetic methods cannot provide fusion breakpoint sequencing information. qRT-PCR is capable of detecting typical *PML::RARA* transcripts, but it generally fails to detect atypical transcripts involving variant *PML::RARA* fusion breakpoints, insertions of genomic DNA sequences at the junction of the *PML* and *RARA* genes, *PML::RARA* fusions, or partial deletions of the *RARA* or *PML* genes [3–6]. qRT-PCR requires high-quality RNA of adequate length and may exhibit strand bias due to PCR amplification process. Given that atypical *PML::RARA* fusions may be undetectable after standard amplification by routine qRT-PCR, direct sequencing or personalized patient-specific qRT-PCR is usually required to identify atypical breakpoints within the *PML* and *RARA* genes [4,5]. Recently, NGS-based methods have revealed reliable gene fusion data. However, they require high-complexity laboratories, are costly to set up, and are usually performed on batched specimens or sent out to centralized laboratories, leading to a longer turnaround time.

The advances in long-read sequencing allowed us to address these limitations. Nanopore sequencing is one of the long-read sequencing approaches. It functions based on a principle akin to that of a Coulter counter, where the electrical current passing through a small aperture is measured [9]. In nanopore sequencing, the pore is on the nanometer scale in diameter, precisely sized to permit the passage of only a single-stranded DNA molecule at a time. As a DNA molecule translocates through the pore, it partially obstructs the ion flow, leading to fluctuations in the ionic current. These changes are unique to the specific nucleotide bases (adenine, thymine, cytosine, and guanine) passing through the pore, enabling real-time identification of the DNA sequence. This method provides a direct, high-resolution readout of the DNA sequence, leveraging the distinct current disruptions caused by each nucleotide base. This technology offers distinct advantages over traditional sequencing methods, especially in detecting large structural variants (SVs). Long-read nanopore sequencing shows exceptional potential for characterizing chromosomal structural abnormalities, such as reciprocal translocations that result in fusion genes, like *PML::RARA* fusions in APL patients. Given the capability of sequencing stretches of DNA of up to a few kilobases in length, long-read nanopore sequencing enables the identification of breakpoints in structural abnormalities and uncovers genes or genomic regions involved in potential gene fusions or rearrangements caused by these structural changes. It offers a direct insight into large SVs and is capable of identifying and characterizing complex structural rearrangements (e.g., large deletions, duplication/amplification, insertions, translocations, and chromoanagenesis), without the need for additional specialized assays. Unlike other sequencing methods that require the amplification of DNA fragments before sequencing, nanopore sequencing allows for the direct reading of native DNA strands without requiring PCR-based amplification. It eliminates PCR amplification biases and enables the sequencing of complex regions or SVs, which are frequently challenging for traditional methods. In contrast to traditional sequencing methods where sequencing data are generated after extensive laboratory processing and results can take days or even weeks, one of the standout features of nanopore sequencing is its ability to produce data in real time. This real-time capability significantly accelerates the diagnostic process, making it ideal for urgent clinical applications where time is critical, such as detecting a *PML::RARA* fusion in APL patients. By delivering immediate results of *PML::RARA* fusion, nanopore sequencing enables faster decision-making and timely interventions in APL patients. The ability of nanopore sequencing to detect large, complex SVs, its real-time direct DNA reading capability, and its absence of amplification bias make it an invaluable tool for advancing modern genomics and clinical diagnostics.

Furthermore, the nanopore sequencer presents a highly viable and accessible tool for clinical testing. Unlike other sequencing technologies, which require significant capital investment, a considerable physical footprint, and the need for specialized technical expertise, nanopore sequencing provides a more cost-effective and user-friendly alternative. The MinION sequencing device, about the size of a large USB stick, has a low capital cost (around $1000) and is easy to operate (Oxford Nanopore Technologies (ONT), Oxford, UK). These features make the MinION a valuable and increasingly practical tool for clinical diagnostics. Many traditional sequencing methods, particularly those relying on large-scale sequencers, necessitate sophisticated laboratory infrastructure and highly skilled personnel, limiting their availability in low-resource or remote settings. In contrast, nanopore sequencing can be conducted using portable devices including the MinION, and PromethION 2 sequencers, which are compact, lightweight, and designed to operate in less specialized environments. This portability empowers genomic diagnostics in diverse clinical settings, including mobile health facilities and remote or rural healthcare centers, thereby enhancing access to diagnostic tools for structural abnormalities, such as the *PML::RARA* fusion in APL patients, worldwide. Moreover, the affordable nanopore flow cells for MinION and Flongle (~$80, ONT, Oxford, UK) further enhance the accessibility of nanopore sequencing, promising to have a significant impact on the clinical diagnostics field [10,11]. Nanopore technology offers a relatively low cost per sample, making it an attractive option for rapid diagnostics and the detection of structural abnormalities, such as the *PML::RARA* fusion in APL patients, across a wide range of patient populations. As technology evolves, nanopore sequencing has the potential to revolutionize the detection and diagnosis of SVs, providing fast, accurate, and more accessible diagnostic solutions.

Adaptive sampling in nanopore sequencing is an advanced technique to monitor and control individual nanopore in real time and to select specific sequence reads via real-time data analysis during nanopore sequencing [12–16]. The adaptive sampling approach allows us to enrich specific genomic regions and genes of interest via real-time data analysis (base-calling). Based on its enrichment mode, implemented using a simple Browser Extensible Data (BED) or FASTA file contains the target sequence regions or genes of interest. It can also deplete sequence reads from unwanted genomic regions during nanopore sequencing by utilizing real-time base-calling and voltage reversal at specific pores to reject molecules that are not of interest. The adaptive sampling approach has successfully characterized depletion and enrichment performance for DNA nanopore sequencing and complex transcriptome subsets by direct RNA nanopore sequencing [17]. Given real-time data analysis that adaptively targets specific genomic regions or genes of interest (with relevant sequences), adaptive sampling improves the run speed and efficiency of sequencing. This makes it a valuable approach to facilitate the use of nanopore sequencing in research and clinical diagnostic fields.

Clustered, regularly interspaced short palindromic repeats (CRISPR)-associated Cas9 nuclease system, an adaptive bacterial immune system that can cleave DNA, enables programmable targeting at a specific genomic DNA site [18–21]. CRISPR/Cas 9 technology has revolutionized the field of gene editing. It is a powerful genome-engineering tool for a wide range of applications in biomedical research and medicine [22–28]. Given its capability of creating targeted double-strand breaks, CRISPR/Cas9 has been successfully used to investigate gene function, increase our knowledge of human disease etiology through CRISPR-based disease models, and improve the treatment of human diseases through gene therapy and genome editing [25,29–31]. Recently, CRISPR/Cas9 technology has also been employed as a targeted enrichment strategy to focus on specific genomic regions, genes, or chromosomal DNA, utilizing adapter ligation for nanopore sequencing [32,33]. For this strategy, the guide ribonucleic acid (guideRNA) flanks specific genomic regions or genes and is custom designed. After the guideRNA aids in recognizing specific sequences of specific genomic regions or genes, the ends of Cas9 cut sites ligate to adaptors. This strategy allows for sequencing enriched regions and genes of interest to provide high sequencing depth and comprehensive variant detection without the loss of native modifications (such as CpG methylation).

We developed a novel nanopore sequencing-based approach for rapid detection of *PML::RARA* fusions in APL patients. We used CRISPR/Cas9 for targeted enrichment of the regions of interest with no PCR amplification steps before nanopore sequencing [32,33]. During nanopore sequencing, we used adaptive sampling for real-time sequencing [16]. Only nanopore sequencing data associated with the *PML* and *RARA* genes (defined as regions of interest) continued to be sequenced and collected for data analyses. We termed the approach CRISPR/Cas9-enriched nanopore sequencing with adaptive sampling (CENAS). An amplification-free sequencing approach has the advantages of low-cost investment, low cost per sample, and a fast turnaround time. It also offers long reads and real-time sequencing data generation using a palm-sized portable nanopore MinION sequencing system (Oxford Nanopore Technologies, Oxford, UK). Therefore, CENAS could develop as a point-of-care test that enables quick, accessible, and cost-efficient bedside diagnosis of APL patients with a *PML::RARA* fusion.

2. Materials and Methods

2.1. Cell Lines and Patient Specimens

The NB4 cell line was obtained from Accegen (Fairfield, NJ, USA) and the GM12878 cell line from Coriell Institute (Camden, NJ, USA). This study also included eighteen peripheral blood or bone marrow specimens from patients referred to the Johns Hopkins Hospital from 1 January 2022, to 30 June 2024. These patients had routine diagnostic procedures, including morphologic evaluation, flow cytometry, fluorescence in situ hybridization (FISH), conventional chromosome analysis, a real-time qRT-PCR, a gene fusion assay, a targeted next-generation sequencing (NGS) assay, and/or a single-nucleotide polymorphism (SNP) microarray analysis. Disease classification by standard pathology practice and delineated by the World Health Organization was based on clinical, morphologic, immunophenotypic, cytogenetic, and molecular genetic features. There were ten women and eight men, with a median age of 44 years (range, 20 to 65 years). Twelve specimens were APL patients at initial diagnosis and had a *PML::RARA* fusion by standard-of-care genetic diagnostic tests. Six specimens were patients with non-APL acute myeloid leukemia without a *PML::RARA* fusion by standard-of-care genetic diagnostic tests. This study was conducted after approval by the institutional review board (IRB) at the Johns Hopkins School of Medicine in accordance with the Declaration of Helsinki, the 1996 Health Insurance Portability and Accountability Act, and the ethical standards of the Institutional Committee on Human Experimentation (https://ictr.johnshopkins.edu/wp-content/uploads/2014/01/Overall-HSR-Section-FINAL-6-15-2017.pdf, accessed on 28 August 2024).

2.2. Cytogenetics Data: Conventional Chromosome Analysis, Fluorescence In Situ Hybridization (FISH), and SNP Microarray

Conventional G-banded chromosome studies were performed using standard techniques. A minimum of 20 metaphase cells were analyzed from bone marrow or peripheral blood specimens. The abnormal karyotypes were described using the International System for Human Cytogenetic Nomenclature (2020) [34].

FISH was performed on interphase nuclei using disease-specific panels of probes according to the manufacturer's protocol and following the previously published protocol [35]. The FISH panel included a *RARA* (17q) break-apart probe set and a dual color dual fusion *PML/RARA* probe set (Abbott Molecular, Inc., Des Plaines, Cook, IL, USA). For each probe, 200 nuclei were visually evaluated with fluorescence microscopy by two technologists scoring separately using a Zeiss Axioscope system (Carl Zeiss Microscopy, LLC, Oberkochen, Germany), blinded from each other. The analysis was performed using Cytovision software version 7.7. The specimen was considered abnormal if the results exceeded the laboratory-established cutoff for each probe set.

Whole-genome SNP microarray analysis was performed according to the manufacturer's protocol and following the previously published protocol [33]. DNA was extracted from bone marrow or blood specimens by conventional methods (Qiacube). The DNA concentration was assessed using a Qubit fluorometer, using the DNA broad range (BR) reagent (Thermo Fisher Scientific, Waltham, MA, USA). The high-resolution microarray platform utilized was the Illumina Infinium CytoSNP−850 K v1.2 BeadChip containing >850,000 markers (mean spacing, 3.5 kb; Illumina, Inc., San Diego, CA, USA). BeadChips were processed per manufacturer's guidelines and imaged with the Illumina iScan system. Data were analyzed with the CNV Partition 2.4.4.0 algorithm in GenomeStudio version 2010.3 (Illumina) and KaryoStudio version 1.4.3.0 (Illumina). B allele frequency and logR signal intensities were used to examine and identify potential pathogenic regions of genomic imbalance. All analyses were performed using human reference genome assembly hg19 (GRCh37).

2.3. CRISPR/Cas9-Mediated and Adaptive Sampling by Nanopore Sequencing

For the NB4 and GM12878 cell lines, DNA was extracted with MasterPure (VWR, Radnor, PA, USA) following a standard protocol. For patient samples, DNA was extracted with PureGene (Qiagen, Germantown, MD, USA) following a standard protocol. Bone marrow samples typically contain a higher concentration of white blood cells than whole blood samples. After adding the cell lysis solution, we ensured the mixture was thoroughly homogeneous. If the solution remained uneven, we added more of the cell lysis solution and proportionally increased the volumes of the other reagents used in the protocol to maintain the proper balance. Three micrograms of DNA were used to generate the sequencing library. crRNA guides were designed using Custom Alt-R CRISPR-Cas9 guide RNA (https://www.idtdna.com/site/order/designtool/index/ CRISPR_ CUSTOM, accessed on 1 February 2023) and Chopchop (https://chopchop.cbu.uib.no/, accessed on 1 February 2023) with CRISPR-Cas9. crRNA guides were designed to direct Cas9 to cut on introns 2, 3, 4 and exons 3, 4, and 5 of RARA and introns 2, 3, 5, and 7 of PML and were obtained from Integrated DNA Technologies (IDT, Inc., Coralville, IA, USA).

CRISPR/Cas9-guided targeted enrichment and adapter ligation followed the previously published protocol [32,33] with modifications. Briefly, after dephosphorylation of free DNA ends and cleavage and A-tailing of DNA, we added an ethanol wash step before the adapter ligation steps. The ethanol wash step improved pore occupancy by increasing the number of active pores performing sequencing, thereby shortening the sequencing runtime. For the ethanol wash step, we added 43 μL of re-suspended AMPure XP beads (AXP) (Beckman Coulter Life Sciences, Sykesville, MD, USA) to the end-prep reaction and incubated on a HulaMixer sample mixer (Thermo Fisher Scientific, Halethorpe, MD, USA) for 5 min at room temperature. Then, we kept the tube on the magnet, pipetted off the supernatant, and washed the beads two times with 200 μL of freshly prepared 80% ethanol without disturbing the pellet. We removed the ethanol, spun it down, placed the tube back on the magnet, pipetted off any residual ethanol, and allowed it to dry for 1 min. We removed the tube from the magnet and re-suspended the pellet in 43 μL nuclease-free water, incubated it for 4 min at room temperature, kept the pellet on a magnet for 1 min, and removed and retained 43 μL of eluate into a clean 1.5 mL Eppendorf DNA LoBind tube. After the ethanol wash step, we continued to go through the adapter ligation step, library preparation, and nanopore sequencing as previously published [32,33] using ligation sequencing kit V14 (SQK-LSK114, Oxford Nanopore Technologies (ONT), Oxford, UK). Briefly, we washed by adding AXP beads (Agencourt, Brea, CA, USA) at 0.5 × volume, and a long fragment buffer (ONT, Oxford, UK) was used for bead washes. Then, the library was eluted using an elution buffer (SQK- LSK114, ONT, Oxford, UK). The library was loaded on ONT flow cells (FLO-FLG114 R10 flongle flow cells or FLO-MIM114 R10 MinION flow cells) and sequenced on a nanopore sequencer (MinION or GridION sequencer, ONT, Oxford, UK). Since both the GridION and Mk1b with MinION used the same flow cell, and sequencing program and operated at a sequencing temperature of 37 °C or higher, no differences in QC or sequencing results were observed between the devices in this study. The only distinction is that the GridION sequencer can process up to five samples simultaneously, whereas the MinION Mk1b can handle only one sample at a time.

The base-calling algorithm was provided by Oxford Nanopore (ONT, Oxford, UK) and sequences were aligned to human genome builder GRCh37/hg19. The sequencing regions used for adaptive sampling were *PML* at chr15:74287014–74340155 and *RARA* at chr17:38465423–38566941 based on human genome builder GRCh37/hg19. SVs involving *PML* and *RARA* genes were reviewed independently by multiple genetic analysts via the Integrative Genomics Viewer (IGV, Broad Institute, Cambridge, MA, USA). The specimen was considered abnormal (positive for a fusion) if identical fusion breakpoints were revealed in more than two sequencing reads.

2.4. Real-Time Quantitative RT-PCR (qRT-PCR) Assays for PML::RARA Fusions

qRT-PCR assays were performed in a CLIA/CAP-certified molecular diagnostics lab. RNA was extracted from bone marrow or peripheral blood using standard procedures [Qiagen, Germantown, MD, USA]. RNA was reverse transcribed using high-capacity cDNA reverse transcription reagents (ABI), and fusion gene mRNA was detected by real-time PCR using the Qiagen *PML::RARA* Fusion Quant kits on a Taqman 7900 or QuantStudio 6 instrument. A control mRNA (ABL1) was amplified to monitor the quality of the sample. In acute promyelocytic leukemia (APL), the Qiagen Fusion Quant *PML::RARa* bcr 1 (L form), *PML::RARa* bcr2 (V form), and *PML::RARa* bcr3 (S form) assays detected and quantified a t(15;17) translocation.

2.5. RNA-Base Gene Fusion Assay for PML::RARA Fusions

The RNA-based gene fusion assay for *PML::RARA* fusions was performed using a comprehensive, NanoString technology–based targeted fusion panel. A total of 300 ng RNAs extracted from patient blood or bone marrow specimens was used for the assay, following the manufacturer's protocol as previously described [36].

2.6. Next-Generation Sequencing (NGS) Assay

DNA was extracted from blood and bone marrow specimens by conventional methods (Qiacube; Qiagen, Hilden, Germany). DNA concentration was assessed by the Qubit fluorometer, using the DNA broad range (BR) reagent, according to vendor specification (Thermo Fisher Scientific, Halethorpe, MD, USA). The NGS assay has been described in previous studies [35,37]. Briefly, library preparation was performed using Kapa Roche Hyper Prep reagents (Roche Diagnostics, Inc., Wilmington, MA, USA); hybrid capture was executed using 40,670 Integrated DNA Technologies, Inc., probes and reagents, and products were sequenced using NovaSeq 6000 with NovaSeq Rapid Cluster and SBS v2 200-cycle reagents with Illumina paired-end technology (Illumina, Inc., San Diego, CA, USA).

All analyses were performed using human reference genome assembly GRCh37/hg19. SVs were reviewed independently by multiple genetic analysts via the IGV (Broad Institute, Cambridge, MA, USA). Only clusters with at least three reads at each potential SV breakpoint were considered a putative SV junction.

2.7. Statistical Calculators

A comparison of concordance, including sensitivity, specificity, predictive values, and the accuracy, between CENAS and the FISH assay was conducted using MEDCALC statistical (MedCalc's Diagnostic test evaluation calculator at https://www.medcalc.org/, last accessed 28 August 2024). Sensitivity (true positive rate), specificity (true negative rate), and predictive values were calculated from the confusion matrix for each assay. Accuracy was the proportion of correct predictions (both true positives and true negatives) out of all cases. McNemar's test was used to assess the concordance between the two assays. The significance level was set at $p < 0.05$, which was considered statistically significant. No corrections were applied for multiple comparisons.

3. Results

CENAS combined CRISPR/Cas9-targeted enrichment and the adaptive sampling approach to enrich the sequencing reads containing the *PML* and *RARA* genes and reduced sequencing runtime (Figure 1).

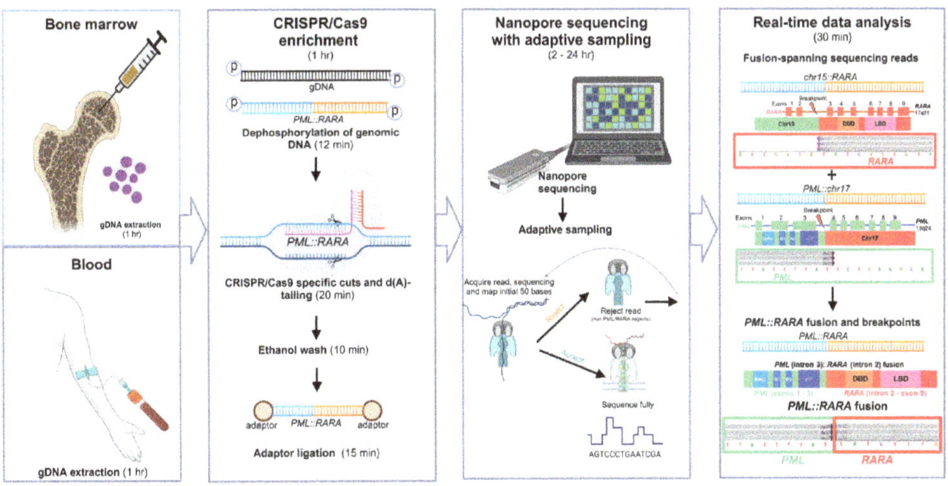

Figure 1. The CRISPR/Cas9-enriched nanopore sequencing with the adaptive sampling (CENAS) approach to reveal *PML::RARA* fusions in APL patients. The entire procedure includes DNA extraction from blood or marrow, CRISPR/Cas9-guided enrichment of targeted *PML* and *RARA* genomic regions, nanopore sequencing with adaptive sampling, and sequencing data analysis to reveal sequences involving *PML::RARA* fusions. hr: hours; min: minutes.

We used CRISPR/Cas9 for targeted enrichment of the regions of interest with no PCR amplification steps before nanopore sequencing. During nanopore sequencing, we used adaptive sampling for real-time sequencing. Only nanopore sequencing data associated with the *PML* and *RARA* genes (defined as regions of interest) continued to be sequenced and collected for data analyses. The CENAS approach optimized sequencing time and efficiency.

3.1. Summary of CENAS-Based Nanopore Sequencing Results

Using CENAS, we successfully sequenced the breakpoints of *PML::RARA* fusion in the APL cell line NB4 (Table 1, Supplemental Figure S1, and Supplemental Table S1).

We performed CENAS on remnant DNA from eighteen blood or bone marrow specimens. Specimens included twelve APL specimens at initial diagnosis with *PML::RARA* fusion by standard-of-care genetic tests and six acute myeloid leukemia specimens without *PML::RARA* fusion (Table 1). Among the twelve APL specimens with *PML::RARA* fusion by dual color dual fusion FISH, eight specimens had the typical *PML::RARA* FISH signal pattern of a classic t(15;17) reciprocal translocation and four had an atypical FISH signal pattern (Cases #11–14 in Table 1). CENAS successfully detected *PML::RARA* fusions and characterized the breakpoint sequences in all twelve APL specimens carrying typical and atypical fusions (Table 1). As expected, CENAS did not detect any *PML::RARA* fusions in the six non-APL specimens (Cases #15–20 in Table 1) and no *PML::RARA* fusion reads were detected in non-APL GM12878 cells. Compared with the FISH test, CENAS achieved good concordance in detecting *PML::RARA* fusions in this study (Table 1).

Table 1. Standard-of-care genetic tests and CENAS-based nanopore sequencing in this study.

Sample ID	FISH (*PML::RARA* Fusion Pattern by Interphase FISH)	Conventional Chromosome Analysis/SNP Microarray	qRT-PCR/the Gene Fusion Assay	Nanopore Sequencing for *PML::RARA* Fusion	
				PML Breakpoint	*RARA* Breakpoint
1. APL cell line (NB4)	Positive (atypical, 3R2G2F)	Complex with t(15;17)(q24;q21) and gain of chr15 and chr17	n.a.	Positive, chr15:74326373	Positive, chr17:38502175
2. Non-APL cell *	Negative	Normal	n.a.	Negative	Negative
3: APL	Positive (typical, 1R1G2F)	46,XX,t(8;16)(q24.1; q13),t(15;17)(q24;q21)	n.a.	Positive, chr15:74317057	Positive, chr17:38500768
4: APL	Positive (typical, 1R1G2F)	46,XX,t(15;17)(q24; q21)	*PML::RARA* S transcripts	Positive, chr15:74315881	Positive, chr17:38499739
5: APL	Positive (typical, 1R1G2F)	46,XY,t(15;17)(q24; q21)	*PML::RARA* S transcripts	Positive, chr15:74317119	Positive, chr17:38496479
6: APL	Positive (typical, 1R1G2F)	46,XX,t(15;17)(q24; q21)	*PML::RARA* S transcripts	Positive, chr15:74316942	Positive, chr17:38487124
7: APL	Positive (typical, 1R1G2F)	46,XY,t(15;17)(q24; q21)	*PML::RARA* S transcripts	Positive, chr15:74316541	Positive, chr17:38489466
8: APL	Positive (typical, 1R1G2F)	46,XY,t(15;17)(q24; q21)	*PML::RARA* L transcripts	Positive, chr15:74325390	Positive, chr17:38503907
9: APL	Positive (typical, 1R1G2F)	46,XX,t(15;17)(q24; q21)	*PML::RARA* S transcripts	Positive, chr15:74316942	Positive, chr17:38487124
10: APL	Positive (typical, 1R1G2F)	46,XY,r(6)(p25q23), t(15;17)(q24;q21)	n.a.	Positive, chr15:74326146	Positive, chr17:38500560
11: APL	Positive (atypical, 1R2G1F)	n.a./normal SNP microarray	*PML::RARA* S transcripts	Positive, chr15:74316787	Positive, chr17:38489189
12: APL	Positive (atypical, 1R2G1F)	46,XY,t(13;15)(q14; q24)/normal SNP microarray	Negative	Positive, chr15:74326463	Positive, chr17:38501782
13: APL	Positive (atypical, 2R2G1F)	47,XY,+ 8,t(15;22; 17)(q24;q13;q21)	*PML::RARA* L transcripts	Positive, chr15:74326356	Positive, chr17:38503146
14: APL	Positive (atypical, 1R1G4F−5F)	47,XX,der(15)t(15;17)(q24;q21),−16,+17,ider (17)(q10)t(15;17)(q24; q21)x2,+22	*PML::RARA* L transcripts	Positive, chr15:74326210	Positive, chr17:38501461
15–20 **: Non-APL	Negative	Normal karyotype	Negative	Negative	Negative

* #2 is a non-APL cell line, GM12878; ** #15 to #20 samples are non-APL, obtained from patients with acute myeloid leukemia; APL: acute promyelocytic leukemia; chr: chromosome; L: the long (bcr1) isoform; n.a.: not available; S: the short (bcr3) isoform. Breakpoint sequences aligned to human genome builder GRCh37/hg19.

The NGS assay was performed on APL cases #3, #6, #8, #9, #10, #12, #13, and non-APL cases #15, #16, #17, 19, and #20. The NGS assay successfully identified the breakpoint sequences associated with *PML::RARA* fusions in the APL specimens, with results that were consistent with those detected by CENAS. The potential fusion breakpoint sequences at the *PML::RARA* junction, identified by the NGS assay, were reviewed using IGV (Supplemental Figure S2). No potential *PML::RARA* fusions were observed in any of the non-APL specimens.

3.2. *CENAS for Atypical and Cryptic PML::RARA Fusions*

Four APL specimens had atypical *PML::RARA fusions* by dual color dual fusion FISH (Cases #11–14 in Table 1). They included three with a FISH signal pattern of only one single fusion due to cryptic insertions, three-way translocations, or complex rearrangements (Cases #11–13 in Table 1) and one with poly-fusions (case #14 in Table 1). CENAS was especially useful in detecting *PML::RARA* fusions due to complex rearrangements.

3.2.1. Case #13: Atypical *PML::RARA* Fusion Due to Three-Way Translocation

Case #13 had a three-way translocation, t(15;22;17) by conventional chromosome analysis and an atypical *PML::RARA* fusion by FISH (Figure 2a). The sequencing-based CENAS approach detected fusions involving three genes (the *PML* gene on chromosome 15q, the *RARA* gene on chromosome 17q, and the *JOSD1* gene on 22q) from three different chromosomes 15, 17, and 22 (Figure 2b). A *PML::RARA* fusion was one of these fusions.

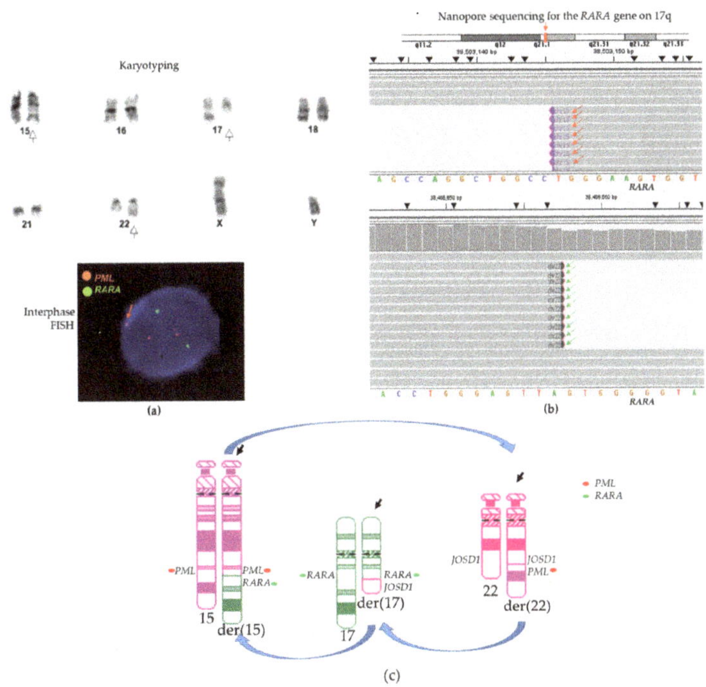

Figure 2. Atypical *PML::RARA* fusions in case #13. (**a**) Partial karyogram shows a t(15;22;17) three-way translocation (white arrows for abnormal derivative chromosomes). Below: Interphase FISH shows an atypical FISH signal pattern (2R2G1F). The red arrows point to a *PML::RARA* fusion. (**b**) CENAS shows sequence reads of a *PML::RARA* fusion involving the *PML* gene on chromosome 15q and the *RARA* gene on 17q (red arrows), and a *JOSD1::RARA* fusion involving the *RARA* gene on 17q and the *JOSD1* gene on chromosome 22 (green arrows). Sequences were aligned to human genome builder GRCh37/hg19. (**c**) Diagram of the t(15;22;17) three-way translocation and atypical *PML::RARA* fusion. Black arrows point to derivative chromosomes involved in this translocation.

3.2.2. Case #12: Cryptic and Atypical *PML::RARA* Fusion

Case #12 had a submicroscopic *PML::RARA* fusion, which was cryptic by conventional chromosome analysis and qRT-PCR. Given a t(13;15) reciprocal translocation by chromosome analysis and an atypical interphase FISH result, we performed metaphase FISH, revealing insertion of a *PML::RARA* fusion into the derivative chromosome 15 formed by the t(13;15) translocation (Figure 3a). With the presence of complex rearrangements involving various breakpoints, qRT-PCR was not sensitive enough to unequivocally detect this *PML::RARA* fusion. CENAS detected this *PML::RARA* fusion and revealed detailed sequencing data involving this case's complex fusion and rearrangement, which also involved the genomic region of 13q14.13 (Figure 3b).

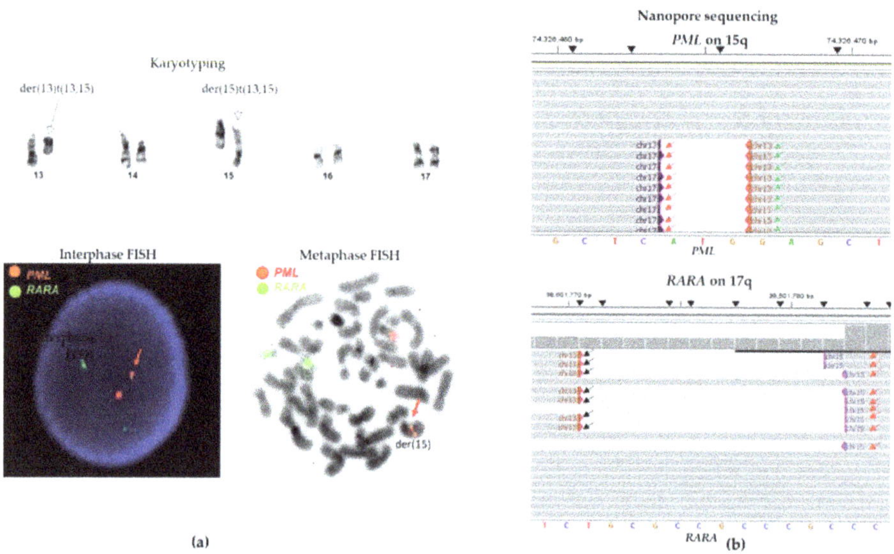

Figure 3. Cryptic *PML::RARA* fusion in case #12. (**a**) Partial karyogram shows a t(13;15) reciprocal translocation (white arrows for abnormal derivative chromosomes) with two normal chromosomes 17. Interphase FISH shows an atypical FISH signal pattern (1R2G1F). The red arrows point to a *PML::RARA* fusion. Metaphase FISH shows an insertional *PML::RARA* fusion into the derivative chromosome 15 formed by t(13;15) translocation, ish der(15)t(13;15)(q14;q24) ins(15;17)(q24;q21q21) (*PML* +, *RARA* +). (**b**) CENAS shows sequence reads of a complex atypical *PML::RARA* fusion involving chromosomes 13, 15, and 17. Red arrows point to a *PML::RARA* fusion/t(15;17) translocation, green arrows point to a fusion of the *PML* gene and 13q14.13 (chr13:46,515,010), and black arrows point to fusions of the *RARA* gene and 13q14.13 (chr13:46,515,891). These data support the presence of complex fusions and rearrangements. Sequences were aligned to human genome builder GRCh37/hg19.

3.2.3. Case #11: Atypical and Likely Insertional *PML::RARA* Fusion

Case #11 had an atypical *PML::RARA* fusion by interphase FISH (Figure 4a) and a normal SNP microarray result without any copy number variants (gains or losses) of the chromosomes 15 and 17, including the *PML* and the *RARA* genes. The sequencing-based CENAS approach detected fusions involving three genes (the *PML* gene on chromosome 15q, the *RARA* gene on chromosome 17q, and the *ANKFN1* gene on 17q22) from two chromosomes 15 and 17 (Figure 4b). Given the presence of a *PML::RARA* fusion and both *PML* and *RARA* genes fused to the *ANKFN1* gene on 17q22, CENAS sequence data suggest a likely insertional *PML::RARA* fusion.

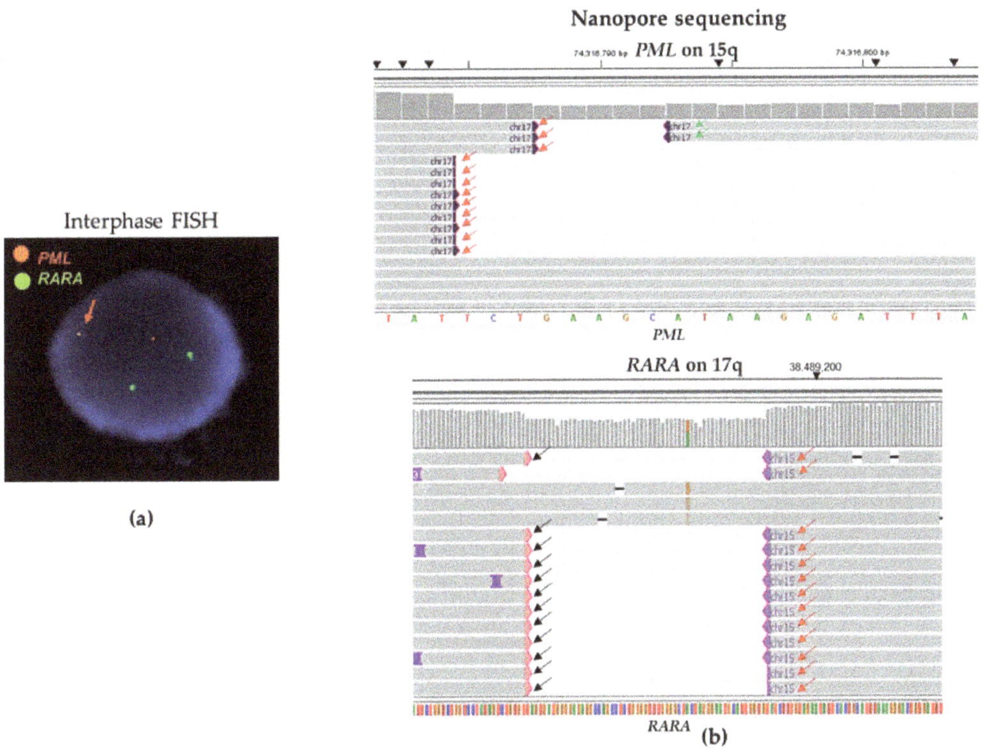

Figure 4. Atypical *PML::RARA* fusion in case #11. (**a**) Interphase FISH shows an atypical FISH signal pattern (1R2G1F). The red arrows point to a *PML::RARA* fusion. (**b**) CENAS reveals an atypical *PML::RARA* fusion (a likely insertional fusion). Red arrows point to a *PML::RARA* fusion, green arrows point to the *PML* gene fused to the *ANKFN1* gene on 17q22, and black arrows point to the *RARA* gene fused to the *ANKFN1* gene on 17q22. These data suggest the presence of a likely insertional *PML::RARA* fusion into a derivative chromosome 17. Sequences were aligned to human genome builder GRCh37/hg19.

3.2.4. Case #14: Multiple *PML::RARA* Fusions by FISH

Case #14 had multiple (four to five) *PML::RARA* fusions due to concurrent isoderivative (ider) chromosome 17q arising from duplication of the ider(17q), with consequent overrepresentation of the reciprocal *RARA::PML* fusion and loss of 17p, including the tumor suppressor gene TP53 (Figure 5a). CENAS revealed *PML::RARA* fusions in this case (Figure 5b).

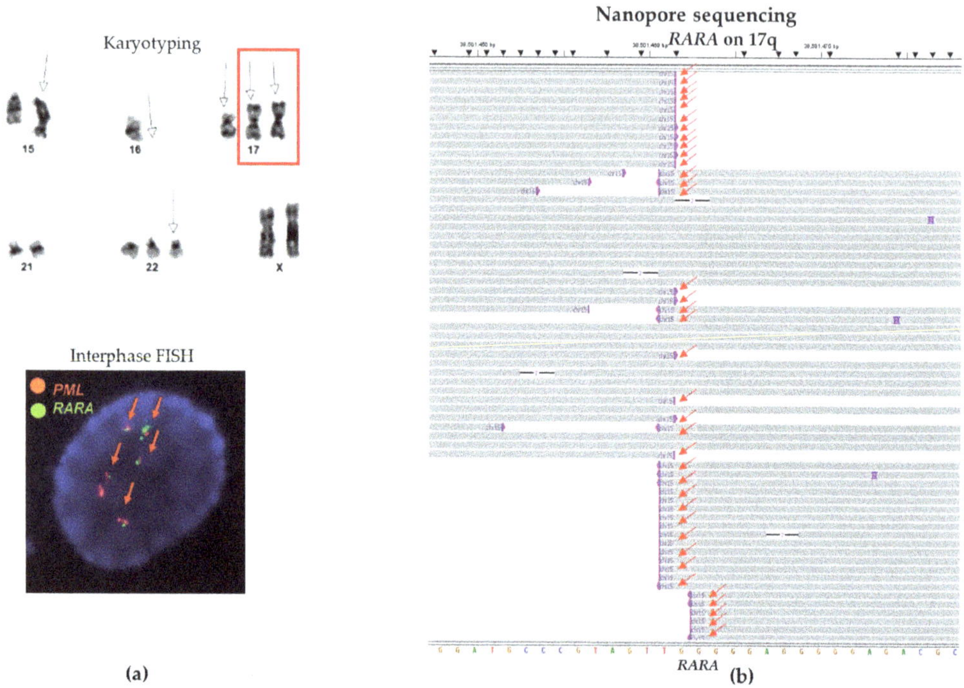

Figure 5. Multiple *PML::RARA* fusions in case #14. (**a**) Partial karyogram shows der(15), −16, +17, ider(17)(q10)t(15;17)x2, and +22 (white arrows for abnormalities). The red box shows isoderivative (ider) chromosome 17q involving t(15;17) translocations on both 17q arms. Interphase FISH reveals multiple fusions. (**b**) CENAS displays a higher number of sequencing fusion reads (indicated by red arrows) compared to non-fusion reads, due to the presence of multiple *RARA::PML* fusions in this case. Sequences were aligned to human genome builder GRCh37/hg19.

4. Discussion

In this study, CENAS successfully detected typical and atypical *PML::RARA* fusions in all APL patients studied. CENAS, a novel amplification-free nanopore sequencing-based approach with low cost and easy setup, has great potential for further development as a point-of-care test to offer an immediate bedside diagnosis of APL patients with a *PML::RARA* fusion, including atypical fusions. While classic/typical balanced t(15;17) rearrangements are found in most (~87%–~92%) APL cases, up to 13% of patients may have atypical and cryptic results by standard-of-care cytogenetic testing [38,39]. As a direct sequencing approach, CENAS has great potential to reveal typical, atypical, and cryptic *PML::RARA* fusions, some of which may be challenging to detect by standard-of-care cytogenetic tests and qRT-PCR. For atypical *PML::RARA* fusions in this study, CENAS not only detected and characterized breakpoints of *PML::RARA* fusions but also revealed additional genes/genomic regions involving additional fusions/rearrangements.

4.1. Pros and Cons of CENAS Compared with Other Emerging Diagnostic Methods

In this study, CENAS nanopore sequencing was performed using the latest R10.4.1 flow cell paired with the updated V14 chemistry. These advancements substantially improve read quality, enabling the generation of high-quality hybrid and long reads while also enhancing base-calling accuracy [40–43]. These improvements result from updates in base-calling algorithms, as well as refinements in library preparation and flow cell chemistry. Additionally, the R10.4.1 flow cell features a dual reader head design, which significantly enhances homopolymer sequencing accuracy [40,44]. Moreover, the V14 chemistry has greatly reduced error rates, achieving an accuracy of approximately 99% with the R10.4.1 flow cell [40,42,43], compared with around 94% with the previous V10 chemistry on the older R9.4.1 flow cell [45]. The combination of the R10.4.1 flow cell and V14 chemistry enhances the clinical applicability of nanopore sequencing for fusion detection. Given the advancements in other emerging diagnostic technologies, CENAS highlights its unique advantages as well as potential disadvantages.

4.1.1. Comparison of CENAS Nanopore Sequencing with Pacific Biosciences' Sequencing

For the past two decades, first- and second-generation short-read sequencing, typically producing reads of around 150 bp, has been the gold standard for genetic profiling. However, these technologies have limitations in detecting and characteriz-ing SVs, as well as in analyzing complex genomic regions, repetitive elements, or vari-ant phasing. Additionally, the PCR amplification used in sequencing templates can in-troduce artifacts and hinder the detection of native base modifications. In contrast, long-read sequencing, or third-generation sequencing, can produce sequence reads crossing kilobases or even megabases, which is extremely functional for detecting SVs. Long-read sequencing is also real-time and direct, PCR-free in both sequencing and li-brary preparation, effectively eliminating PCR-related biases. Single-molecule re-al-time sequencing from Pacific Biosciences and nanopore sequencing from ONT are the two leading technologies for long-read sequencing. Both companies are focused on enhancing the accuracy, throughput, and portability of their long-read sequencing methods while striving to keep costs low. Pacific Biosciences' Revio system, utilizing HiFi reads, is based on the circular consensus sequencing model, enhanced by ad-vanced sequencing chemistry and deep learning algorithms. This combination results in an error rate comparable to that of short-read Illumina sequencing [46]. While na-nopore sequencing generally has higher error rates than Pacific Biosciences' HiFi reads, recent advancements in flow cells, chemistry, and base-calling algorithms have signif-icantly improved its accuracy. These improvements make nanopore sequencing suffi-ciently reliable for many applications, including fusion detection. Both technologies are highly capable of fusion detection, but nanopore sequencing offers distinct advantages in terms of portability and flexibility, making it ideal for real-time, dynamic clinical applications. Nanopore sequencing platforms are typically more cost-effective, partic-ularly for smaller-scale patient populations, such as those with APL. In contrast, Pa-cific Biosciences' systems are less portable, requiring more extensive laboratory infra-structure and a higher initial investment in equipment and consumables, making them better suited for large-scale sequencing projects. The MinION nanopore sequencer provides an affordable entry point for labs and clinicians seeking long-read sequencing for fusion detection without the burden of significant upfront costs. Moreover, na-nopore sequencing's user-friendly workflow facilitates rapid sequencing, which is es-pecially valuable in clinical or point-of-care environments where timely results are critical for patients with conditions like APL.

4.1.2. Comparison of CENAS with Other RNA-Based Gene Fusion Assays

Targeted RNA-based gene fusion assays, such as qRT-PCR, targeted RNA sequencing, and anchored multiplex PCR, are highly effective at detecting low-abundance transcripts [3–6,47,48]. These assays provide detailed and accurate profiles of gene fusions involving specific, predefined target genes. Most of these assays rely on PCR-based enrichment, making RNA integrity and proper sample preparation essential for ensuring accurate and reproducible results [47]. Additionally, most targeted assays are limited to detecting fusions within specific, predefined genes or regions, which means that novel or uncharacterized fusion events may be overlooked. In contrast, whole-transcriptome RNA sequencing (RNA-Seq) does not depend on predefined probes, allowing for the identification of novel genes, isoforms, and previously undiscovered fusion events. Despite this advantage, RNA-Seq generates large amounts of data, requiring significant computational resources and bioinformatics expertise for analysis. It also necessitates high-complexity laboratory setups, making it costly to implement. Typically performed on batched specimens or outsourced to centralized labs, RNA-Seq can result in longer turnaround times. While sequencing costs have decreased over time, RNA-Seq remains expensive, particularly for high-throughput or single-cell applications.

Direct-DNA nanopore sequencing-based CENAS can detect typical, atypical, and cryptic *PML::RARA* fusions, some of which could be difficult to identify using certain targeted RNA-based gene fusion assays. As an amplification-free approach, CENAS offers advantages such as lower cost, simpler setup, and faster turnaround time compared with RNA-Seq. These features position CENAS as a promising tool for point-of-care testing, potentially enabling immediate bedside diagnosis of APL patients with a *PML::RARA* fusion.

4.1.3. Comparison of CENAS with Optical Genome Mapping

In addition to the standard cytogenetic methods, such as conventional karyotyping and FISH, which are routinely used to detect *PML::RARA* fusions in APL patients, optical genome mapping (OGM) is an emerging technology for characterizing genome-wide SVs. OGM provides a detailed view of DNA structure by generating long DNA molecules labeled with fluorescent tags at specific sequence motifs. These labeled molecules are then stretched and imaged, creating a map of the genome's structure, including SVs like *PML::RARA* gene fusions caused by t(15;17) translocations. OGM is particularly useful for characterizing typical, atypical, and cryptic *PML::RARA* fusions in APL patients [49]. However, like conventional karyotyping and FISH, OGM cannot provide sequencing information about fusion breakpoints. Compared with OGM and conventional karyotyping (which takes about a week), CENAS offers a significantly faster turnaround time and is on par with in-house FISH assays.

4.2. Potential Clinical Impacts of Atypical PML::RARA Transcripts

Given different *PML* and *RARA* domains retained in the fusion protein, typical and atypical *PML::RARA* isoforms may be associated with diverse prognoses and treatment responsiveness. However, due to the rarity of t(15;17)-positive APL patients with atypical *PML::RARA* transcripts, the biological and clinical impact of these atypical isoforms is not fully understood. A few patients with atypical *PML::RARA* transcripts have had clinical outcomes similar to patients with typical fusion transcripts [5,50–52]. The V-type (for variable) *PML::RARA* transcripts found in approximately 8% of adult patients with APL are defined molecularly by truncation of *PML* exon 6 and frequent insertion of genetic material from *RARA* intron 2. V-type *PML::RARA* fusion transcripts in APL have shown decreased in vitro responsiveness to ARTA [53] and are reported to display decreased sensitivity to standard ARTA treatments [53,54] and a higher failure rate of standard treatments [55]. Furthermore, atypical *PML::RARA* transcripts, especially *PML* breakpoints at downstream intron 6, are potentially associated with aggressive disease and poor prognosis [4,56]. These transcripts may not respond to standard ATRA treatment [38]. CENAS characterized *PML::RARA* fusion breakpoints at a nucleotide level, which is extremely useful for determining PML and RARA domains retained in typical and atypical *PML::RARA* fusions. Since the early death rate in APL patients still reaches 15%, and up to 10% of APL patients are resistant to initial therapy or prone to relapse, further sequencing studies of typical and atypical *PML::RARA* fusion patients might shed light on the pathophysiology of the disease and its responsiveness to treatment.

4.3. Additional Gene Fusions Involving Atypical PML::RARA Fusions

For two atypical *PML::RARA* fusions in this study, CENAS also revealed additional genes leading to novel gene fusions.

Case # 13 had a three-way translocation, and the *JOSD1* gene on 22q involved fusions of *PML* and *RARA* genes. JOSD1 belongs to a family of Josephin domain-containing deubiquitinating enzymes [57]. The activity of JOSD1 is regulated by monoubiquitination. JOSD1 plays a role in cytoskeletal dynamics, cell motility, and endocytosis [58]. As a deubiquitinase, JOSD1 is reported to be an ideal therapeutic target and a promising diagnostic marker. JOSD1 is a novel regulator of mutant JAK2 [59]. It interacts with and stabilizes JAK2-V617F, which leads to JAK2-V617F protein degradation due to increased ubiquitination levels. Targeting JOSD1 using small molecule inhibition could be a novel targeted therapy for leukemias with mutant JAK2 [59]. Upregulated JOSD1 contributes to the acquisition of chemo-resistance by inhibiting mitochondrial apoptotic signaling in gynecological cancer by stabilizing MCL1 [60,61]. Depletion of JOSD1 leads to the death of gynecological cancer cells [60]. Further studying the potential pathogenesis of novel *JOSD1* fusions with the *PML* and *RARA* genes in case #13 might be useful for targeted therapy and prediction of chemo-resistance and prognosis.

CENAS revealed that the *ANKFN1* gene on 17q involved fusions of *PML* and *RARA* genes in case #11, likely due to an insertional fusion. Besides its roles in vestibular-related functions [62], *ANKFN1* plays pro-tumorigenic and metastatic roles in hepatocellular carcinoma [63]. It is a recurrent HBV integration event of HBV-infected intrahepatic cholangiocarcinoma [64]. It was found to be a pivotal gene corresponding to the diagnosis of early lung cancer [65]. While the *ANKFN1::RARA* and *ANKFN1::PML* gene fusions found in case #11 may lead to disrupting the *ANKFN1* gene, further studies will be valuable to understand the pathogenesis of these novel fusions in this case.

4.4. Complex Rearrangements in a Cryptic PML::RARA Fusion

CENAS revealed a cryptic *PML::RARA* fusion due to a complex fusion/rearrangement (complex insertional) among *PML*, *RARA*, and chromosome 13 in case #12. In addition to a *PML::RARA* fusion, the band of 13q14.13 is the insertion site, which is intergenic and flanks with the *SIAH3* gene (centromeric) and the *ZC3H13* gene (telomeric). These insertional *PML::RARA* fusions led to complex rearrangements, including *PML::*13q14.13 and *RARA::*13q14.13 fusions. While the pathogenesis of these rearrangements is unclear, the positional effect in gene expression is a well-known phenomenon [66–69]. These rearrangements may have impacts on chromatin structure, interactions with regulatory elements (enhancers, silencers, or insulators), transcriptional clustering, epigenetic modifications, and chromosomal positioning [53–56]. These rearrangements may lead to disrupting the normal function of the *PML* and *RARA* genes. Loss and overexpression of PML isoforms can inhibit chromosome-based homology-directed repair and affect the efficiency of homologous recombination [70]. Furthermore, these rearrangements may also have impacts on the *ZC3H13* and *SIAH3* genes located at 13q14.13. The *ZC3H13* gene localizes to nuclear speckles and plays a role in mRNA splicing [71]. Its expression has been used to predict prognosis in acute myeloid leukemia [72,73]. The *SIAH3* gene is a member of the SIAH family and is a RING-containing protein that functions as E3 ubiquitin ligases [74]. Understanding the positional effects related to the *PML*, *RARA*, and 13q14.13-flanking genes (*SIAH3* and *ZC3H13*) could shed light on their role in APL and may aid in predicting the prognosis of the patient.

4.5. Multiple Fusions Due to Two Isoderivative Chromosomes 17q

Isochromosome 17q is a chromosomal abnormality that leads to the gain of the entire long arm of chromosome 17 and the loss of the entire short arm of chromosome 17, including the *TP53* gene. Isochromosome of the long arm of the derivative chromosome 17, originating from the translocation t(15;17), isoderivative chromosome ider(17)(q10)t(15; 17), has been reported in estimated less than 5% of APL cases, ranging from 0.6% to 4.9% [75–78]. The presence of two ider(17)(q10)t(15;17) chromosomes results in extra fusion signals by FISH analysis as found in case #14. Its presence is indicative of additional genetic complexity during the formation and development of APL.

While isochromosome 17q is commonly associated with a worse prognosis in hematologic malignancies [79,80], prognostic significance of ider(17)(q10)t(15;17) in APL remains controversial. Given a low incidence of these ider(17)(q10)t(15;17)-positive APL cases, various prognostic significance has been reported. Few ider(17)(q10)t(15;17)-positive APL cases have been reported to follow an unfavorable course and have short survival and poor prognosis [77,78,81,82]. On the other hand, ider(17)(q10)t(15;17)-positive APL cases have been reported to not cause an unfavorable course of APL in the group of patients treated with ATRA and chemotherapy [83]. Therefore, more studies are required to accurately assess the potential clinical impacts of ider(17)(q10)t(15;17) in APL. Currently, the presence of ider(17)(q10)t(15;17) in APL does not change the therapeutic approach and patients are treated with standard ATRA treatment and chemotherapy.

4.6. Limitation of CENAS Approach

The amplification-free, sequencing-based CENAS approach is designed to rapidly detect *PML::RARA* fusions in APL patients at initial diagnosis, with a high percentage of APL cancer cells (>20%). It is unsuitable for performing a sensitive and specific quantitative evaluation of minimal residual disease (MRD) during follow-up with a low percentage of APL cancer cells.

5. Conclusions

CENAS, an amplification-free nanopore sequencing-based approach with adaptive sampling, is useful for rapidly detecting *PML::RARA* fusions in APL patients. Given its low cost and easy setup, CENAS has the potential for further development as a point-of-care test to offer an immediate bedside diagnosis of APL with a *PML::RARA* fusion. CENAS allowed for the identification of sequence information of fusion breakpoints involved in typical and atypical *PML::RARA* fusions and identified additional genes and genomic regions involving the atypical *PML::RARA* fusions. To the best of our knowledge, involvements of the *ANKFN1* gene, the *JOSD1* gene, and the specific 13q14.13 genomic region have not been reported in the atypical *PML::RARA* fusions. The biological significance of most atypical *PML::RARA* transcripts remains unclear. Thus, there is a need for multi-center studies that combine sequence analysis of atypical *PML::RARA* fusion cases, comprehensive clinical data and genomic data, and uniform treatment strategies. Such studies will shed light on how atypical *PML::RARA* fusions in APL patients are related to treatment response rates, survival, relapse, and overall prognosis. Understanding the involvement of additional genes and positional effects related to the *PML* and *RARA* genes could shed light on their role in APL and may aid in the development of targeted therapies.

Supplementary Materials: The following supporting information can be downloaded at: https://www.mdpi.com/article/10.3390/biom14121595/s1, Figure S1: *PML/RARA* FISH in NB4 and GM12878. Figure S2: *PML::RARA* gene fusion by the NGS sequencing. Table S1: Characteristics of the CENAS runs and enrichment.

Author Contributions: W.M., M.R.B. and Y.S.Z. developed study concept and design; W.M. and Y.S.Z. performed writing; W.M., V.S., M.P., B.P., L.M., M.K., J.G., N.D., J.Z., T.J., A.P. and Y.S.Z. provided acquisition, analysis, and interpretation of data; W.M. and Y.S.Z. performed statistical analysis; M.P., B.P., C.D.G. and A.P. provided technical and materials support. All authors contributed to the review and revision. All authors have read and agreed to the published version of the manuscript.

Funding: This publication was made possible by the Johns Hopkins Institute for Clinical and Translational Research (ICTR), which is funded in part by Grant Number UL1 TR003098 from the National Center for Advancing Translational Sciences (NCATS), a component of the National Institutes of Health (NIH), and NIH Roadmap for Medical Research. Its contents are solely the responsibility of the authors and do not necessarily represent the official view of the Johns Hopkins ICTR, NCATS, or NIH. This study was supported by an ICTR's Accelerated Translational Incubator Pilot (ATIP) grant (Y.S.Z.).

Institutional Review Board Statement: The institutional review board (IRB) of Johns Hopkins Medicine (protocol code JHIRB00388944 on 9 June 2023 and JHIRB00310100 on 10 May 2022) approved this study. The study was conducted according to the guidelines of the Declaration of Helsinki.

Informed Consent Statement: The Johns Hopkins Medicine Institutional Review Board (IRB) approved this HIPAA-compliant study with a waiver of consent. All data (including blood and bone marrow specimens) were provided to us without patient identification.

Data Availability Statement: All requests for primary data and experimental reagents should be addressed to: yzou19@jhmi.edu.

Acknowledgments: We would like to express our sincere gratitude to Alan Scott and David Mohr at the Genetic Resources Core Facility (GRCF) at Johns Hopkins for their invaluable assistance with nanopore sequencing. We sincerely thank all the personnel at the Cytogenomic Laboratories at Johns Hopkins Hospital and Johns Hopkins Genomics for their dedication and involvement in genetic testing. We also appreciate Anne N. Connor at Johns Hopkins University for her valuable editing support.

Conflicts of Interest: The authors declare no conflicts of interest.

Abbreviations

AML	Acute myeloid leukemia
ANKFN1	Ankyrin repeat- and fibronectin type III domain-containing protein 1
APL	Acute promyelocytic leukemia
Cas9	CRISPR-associated Protein 9
CENAS	CRISPR/Cas9-enriched nanopore sequencing with adaptive sampling
CNV	Copy number variants
CRISPR	Clustered regularly interspaced short palindromic repeats
FISH	Fluorescence in situ hybridization
guideRNA	Guide ribonucleic acid ider isoderivative
IDT	Integrated DNA Technologies
IGV	Integrative Genomics Viewer
JOSD1	Josephin domain-containing protein 1
MDS	Myelodysplastic syndrome
NGS	Next-generation sequencing
PML	Promyelocytic leukemia gene
RARA	Retinoic acid receptor alpha gene
SIAH3	SIAH E3 ubiquitin protein ligase family, member 3
SVs	Structural variants
ZC3H13	Zinc finger CCCH domain-containing protein 13

References

1. Grimwade, D.; Lo Coco, F. Acute promyelocytic leukemia: A model for the role of molecular diagnosis and residual disease monitoring in directing treatment approach in acute myeloid leukemia. *Leukemia* **2002**, *16*, 1959–1973. [PubMed]
2. Sanz, M.A.; Fenaux, P.; Tallman, M.S.; Estey, E.H.; Lowenberg, B.; Naoe, T.; Lengfelder, E.; Dohner, H.; Burnett, A.K.; Chen, S.; et al. Management of acute promyelocytic leukemia: Updated recommendations from an expert panel of the European LeukemiaNet. *Blood* **2019**, *133*, 1630–1643. [PubMed]
3. Liquori, A.; Ibanez, M.; Sargas, C.; Sanz, M.A.; Barragan, E.; Cervera, J. Acute Promyelocytic Leukemia: A Constellation of Molecular Events around a Single *PML-RARA* Fusion Gene. *Cancers* **2020**, *12*, 624. [CrossRef] [PubMed]
4. Park, T.S.; Kim, J.S.; Song, J.; Lee, K.A.; Yoon, S.; Suh, B.; Lee, J.; Lee, H.; Kim, J.; Choi, J. Acute promyelocytic leukemia with insertion of *PML* exon 7a and partial deletion of exon 3 of *RARA*: A novel variant transcript related to aggressive course and not detected with real-time polymerase chain reaction analysis. *Cancer Genet. Cytogenet.* **2009**, *188*, 103–107.
5. Iaccarino, L.; Divona, M.; Ottone, T.; Cicconi, L.; Lavorgna, S.; Ciardi, C.; Alfonso, V.; Travaglini, S.; Facchini, L.; Cimino, G.; et al. Identification and monitoring of atypical *PML/RARA* fusion transcripts in acute promyelocytic leukemia. *Genes Chromosomes Cancer* **2019**, *58*, 60–65.
6. Jeziskova, I.; Razga, F.; Gazdova, J.; Doubek, M.; Jurcek, T.; Koristek, Z.; Mayer, J.; Dvoráková, D. A case of a novel *PML/RARA* short fusion transcript with truncated transcription variant 2 of the *RARA* gene. *Mol. Diagn. Ther.* **2010**, *14*, 113–117.
7. Walz, C.; Grimwade, D.; Saussele, S.; Lengfelder, E.; Haferlach, C.; Schnittger, S.; Lafage-Pochitaloff, M.; Hochhaus, A.; Cross, N.C.P.; Reiter, A. Atypical mRNA fusions in *PML-RARA* positive, *RARA-PML* negative acute promyelocytic leukemia. *Genes Chromosomes Cancer* **2010**, *49*, 471–479. [CrossRef]
8. McKinney, C.D.; Golden, W.L.; Gemma, N.W.; Swerdlow, S.H.; Williams, M.E. *RARA* and *PML* gene rearrangements in acute promyelocytic leukemia with complex translocations and atypical features. *Genes Chromosomes Cancer.* **1994**, *9*, 49–56.

9. Bartalucci, N.; Romagnoli, S.; Vannucchi, A.M. A blood drop through the pore: Nanopore sequencing in hematology. *Trends Genet.* **2022**, *38*, 572–586.
10. Avershina, E.; Frye, S.A.; Ali, J.; Taxt, A.M.; Ahmad, R. Ultrafast and Cost-Effective Pathogen Identification and Resistance Gene Detection in a Clinical Setting Using Nanopore Flongle Sequencing. *Front. Microbiol.* **2022**, *13*, 822402. [CrossRef]
11. Gradel, C.; Terrazos Miani, M.A.; Barbani, M.T.; Leib, S.L.; Suter-Riniker, F.; Ramette, A. Rapid and Cost-Efficient Enterovirus Genotyping from Clinical Samples Using Flongle Flow Cells. *Genes* **2019**, *10*, 659. [CrossRef]
12. Loose, M.; Malla, S.; Stout, M. Real-time selective sequencing using nanopore technology. *Nat. Methods* **2016**, *13*, 751–754. [CrossRef] [PubMed]
13. Payne, A.; Holmes, N.; Rakyan, V.; Loose, M. BulkVis: A graphical viewer for Oxford nanopore bulk FAST5 files. *Bioinformatics* **2019**, *35*, 2193–2198. [PubMed]
14. Kovaka, S.; Fan, Y.; Ni, B.; Timp, W.; Schatz, M.C. Targeted nanopore sequencing by real-time mapping of raw electrical signal with UNCALLED. *Nat. Biotechnol.* **2021**, *39*, 431–441. [PubMed]
15. Edwards, H.S.; Krishnakumar, R.; Sinha, A.; Bird, S.W.; Patel, K.D.; Bartsch, M.S. Real-Time Selective Sequencing with RUBRIC: Read Until with Basecall and Reference-Informed Criteria. *Sci. Rep.* **2019**, *9*, 11475.
16. Weilguny, L.; De Maio, N.; Munro, R.; Manser, C.; Birney, E.; Loose, M.; Goldman, N. Dynamic, adaptive sampling during nanopore sequencing using Bayesian experimental design. *Nat. Biotechnol.* **2023**, *41*, 1018–1025.
17. Naarmann-de Vries, I.S.; Gjerga, E.; Gandor, C.L.A.; Dieterich, C. Adaptive sampling for nanopore direct RNA-sequencing. *RNA* **2023**, *29*, 1939–1949.
18. Jinek, M.; Chylinski, K.; Fonfara, I.; Hauer, M.; Doudna, J.A.; Charpentier, E. A programmable dual-RNA-guided DNA endonuclease in adaptive bacterial immunity. *Science* **2012**, *337*, 816–821.
19. Hsu, P.D.; Lander, E.S.; Zhang, F. Development and applications of CRISPR-Cas9 for genome engineering. *Cell* **2014**, *157*, 1262–1278.
20. Bak, R.O.; Dever, D.P.; Porteus, M.H. CRISPR/Cas9 genome editing in human hematopoietic stem cells. *Nat. Protoc.* **2018**, *13*, 358–376.
21. Hendel, A.; Bak, R.O.; Clark, J.T.; Kennedy, A.B.; Ryan, D.E.; Roy, S.; Steinfeld, I.; Lunstad, B.D.; Kaiser, R.J.; Wilkens, A.B.; et al. Chemically modified guide RNAs enhance CRISPR-Cas genome editing in human primary cells. *Nat. Biotechnol.* **2015**, *33*, 985–989. [CrossRef] [PubMed]
22. Dasgupta, I.; Flotte, T.R.; Keeler, A.M. CRISPR/Cas-Dependent and Nuclease-Free *In Vivo* Therapeutic Gene Editing. *Hum. Gene Ther.* **2021**, *32*, 275–293. [CrossRef] [PubMed]
23. Wu, Y.; Zeng, J.; Roscoe, B.P.; Liu, P.; Yao, Q.; Lazzarotto, C.R.; Clement, K.; Cole, M.A.; Luk, K.; Baricordi, C.; et al. Highly efficient therapeutic gene editing of human hematopoietic stem cells. *Nat. Med.* **2019**, *25*, 776–783. [CrossRef] [PubMed]
24. Canver, M.C.; Smith, E.C.; Sher, F.; Pinello, L.; Sanjana, N.E.; Shalem, O.; Chen, D.D.; Schupp, P.G.; Vinjamur, D.S.; Garcia, S.P.; et al. *BCL11A* enhancer dissection by Cas9-mediated in situ saturating mutagenesis. *Nature* **2015**, *527*, 192–197. [CrossRef]
25. Gillmore, J.D.; Gane, E.; Taubel, J.; Kao, J.; Fontana, M.; Maitland, M.L.; Seitzer, J.; O'Connell, D.; Walsh, K.R.; Wood, K.; et al. CRISPR-Cas9 In Vivo Gene Editing for Transthyretin Amyloidosis. *N. Engl. J. Med.* **2021**, *385*, 493–502. [CrossRef]
26. Li, H.; Yang, Y.; Hong, W.; Huang, M.; Wu, M.; Zhao, X. Applications of genome editing technology in the targeted therapy of human diseases: Mechanisms, advances and prospects. *Signal Transduct. Target. Ther.* **2020**, *5*, 1. [CrossRef]
27. Kim, J.S. Genome editing comes of age. *Nat. Protoc.* **2016**, *11*, 1573–1578. [CrossRef]
28. Strong, A.; Musunuru, K. Genome editing in cardiovascular diseases. *Nat. Rev. Cardiol.* **2017**, *14*, 11–20. [CrossRef]
29. Frangoul, H.; Altshuler, D.; Cappellini, M.D.; Chen, Y.S.; Domm, J.; Eustace, B.K.; Foell, J.; de la Fuente, J.; Grupp, S.; Handgretinger, R.; et al. CRISPR-Cas9 Gene Editing for Sickle Cell Disease and beta-Thalassemia. *N. Engl. J. Med.* **2021**, *384*, 252–260. [CrossRef]
30. Sharma, A.; Boelens, J.J.; Cancio, M.; Hankins, J.S.; Bhad, P.; Azizy, M.; Lewandowski, A.; Zhao, X.; Chitnis, S.; Peddinti, R.; et al. CRISPR-Cas9 Editing of the *HBG1* and *HBG2* Promoters to Treat Sickle Cell Disease. *N. Engl. J. Med.* **2023**, *389*, 820–832. [CrossRef]
31. Longhurst, H.J.; Lindsay, K.; Petersen, R.S.; Fijen, L.M.; Gurugama, P.; Maag, D.; Butler, J.S.; Shah, M.Y.; Golden, A.; Xu, Y.; et al. CRISPR-Cas9 In Vivo Gene Editing of *KLKB1* for Hereditary Angioedema. *N. Engl. J. Med.* **2024**, *390*, 432–441. [CrossRef] [PubMed]
32. Gilpatrick, T.; Lee, I.; Graham, J.E.; Raimondeau, E.; Bowen, R.; Heron, A.; Downs, B.; Sukumar, S.; Sedlazeck, F.J.; Timp, W. Targeted nanopore sequencing with Cas9-guided adapter ligation. *Nat. Biotechnol.* **2020**, *38*, 433–438. [CrossRef] [PubMed]
33. Phan, M.; Gomes, M.A.; Stinnett, V.; Morsberger, L.; Hoppman, N.L.; Pearce, K.E.; Smith, K.; Phan, B.; Jiang, L.; Zou, Y.S. An Integrated Approach Including CRISPR/Cas9-Mediated Nanopore Sequencing, Mate Pair Sequencing, and Cytogenomic Methods to Characterize Complex Structural Rearrangements in Acute Myeloid Leukemia. *Biomedicines* **2024**, *12*, 598. [CrossRef] [PubMed]
34. McGowan-Jordan, J.H.R.; Moore, S. (Eds.) *ISCN 2020: An International System for Human Cytogenomic Nomenclature*; Karger: Basel, Switzerland, 2020; ISBN 978-3-318-06867-2.
35. Jiang, L.; Pallavajjala, A.; Huang, J.; Haley, L.; Morsberger, L.; Stinnett, V.; Hardy, M.; Park, R.; Ament, C.; Finch, A.; et al. Clinical Utility of Targeted Next-Generation Sequencing Assay to Detect Copy Number Variants Associated with Myelodysplastic Syndrome in Myeloid Malignancies. *J. Mol. Diagn.* **2021**, *23*, 467–483. [CrossRef]

36. Haley, L.; Parimi, V.; Jiang, L.; Pallavajjala, A.; Hardy, M.; Yonescu, R.; Morsberger, L.; Stinnett, V.; Long, P.; Zou, Y.S.; et al. Diagnostic Utility of Gene Fusion Panel to Detect Gene Fusions in Fresh and Formalin-Fixed, Paraffin-Embedded Cancer Specimens. *J. Mol. Diagn.* **2021**, *23*, 1343–1358. [CrossRef]
37. Pallavajjala, A.; Haley, L.; Stinnett, V.; Adams, E.; Pallavajjala, R.; Huang, J.; Morsberger, L.; Hardy, M.; Long, P.; Gocke, C.D.; et al. Utility of targeted next-generation sequencing assay to detect 1p/19q co-deletion in formalin-fixed paraffin-embedded glioma specimens. *Hum. Pathol.* **2022**, *126*, 63–76. [CrossRef]
38. Grimwade, D.; Biondi, A.; Mozziconacci, M.J.; Hagemeijer, A.; Berger, R.; Neat, M.; Howe, K.; Dastugue, N.; Jansen, J.; Radford-Weiss, I.; et al. Characterization of acute promyelocytic leukemia cases lacking the classic t(15;17): Results of the European Working Party. Groupe Francais de Cytogenetique Hematologique, Groupe de Francais d'Hematologie Cellulaire, UK Cancer Cytogenetics Group and BIOMED 1 European Community-Concerted Action "Molecular Cytogenetic Diagnosis in Haematological Malignancies". *Blood* **2000**, *96*, 1297–1308.
39. Gagnon, M.F.; Berg, H.E.; Meyer, R.G.; Sukov, W.R.; Van Dyke, D.L.; Jenkins, R.B.; Greipp, P.T.; Thorland, E.C.; Hoppman, N.L.; Xu, X.; et al. Typical, atypical and cryptic t(15;17)(q24;q21) (*PML::RARA*) observed in acute promyelocytic leukemia: A retrospective review of 831 patients with concurrent chromosome and *PML::RARA* dual-color dual-fusion FISH studies. *Genes Chromosomes Cancer* **2022**, *61*, 629–634. [CrossRef]
40. Sereika, M.; Kirkegaard, R.H.; Karst, S.M.; Michaelsen, T.Y.; Sorensen, E.A.; Wollenberg, R.D.; Albertsen, M. Oxford Nanopore R10.4 long-read sequencing enables the generation of near-finished bacterial genomes from pure cultures and metagenomes without short-read or reference polishing. *Nat. Methods* **2022**, *19*, 823–826. [CrossRef]
41. Linde, J.; Brangsch, H.; Holzer, M.; Thomas, C.; Elschner, M.C.; Melzer, F.; Tomaso, H. Comparison of Illumina and Oxford Nanopore Technology for genome analysis of *Francisella tularensis, Bacillus anthracis,* and *Brucella suis*. *BMC Genom.* **2023**, *24*, 258. [CrossRef]
42. Sanderson, N.D.; Hopkins, K.M.V.; Colpus, M.; Parker, M.; Lipworth, S.; Crook, D.; Stoesser, N. Evaluation of the accuracy of bacterial genome reconstruction with Oxford Nanopore R10.4.1 long-read-only sequencing. *Microb. Genom.* **2024**, *10*, 001246. [CrossRef] [PubMed]
43. Zhao, W.; Zeng, W.; Pang, B.; Luo, M.; Peng, Y.; Xu, J.; Kan, B.; Li, Z.; Lu, X. Oxford nanopore long-read sequencing enables the generation of complete bacterial and plasmid genomes without short-read sequencing. *Front. Microbiol.* **2023**, *14*, 1179966. [CrossRef] [PubMed]
44. Wick, R.R.; Judd, L.M.; Holt, K.E. Assembling the perfect bacterial genome using Oxford Nanopore and Illumina sequencing. *PLoS Comput. Biol.* **2023**, *19*, e1010905. [CrossRef] [PubMed]
45. Delahaye, C.; Nicolas, J. Sequencing DNA with nanopores: Troubles and biases. *PLoS ONE* **2021**, *16*, e0257521. [CrossRef] [PubMed]
46. Zhang, X.; Liu, C.G.; Yang, S.H.; Wang, X.; Bai, F.W.; Wang, Z. Benchmarking of long-read sequencing, assemblers and polishers for yeast genome. *Brief. Bioinform.* **2022**, *23*, bbac146. [CrossRef]
47. Ramani, N.S.; Patel, K.P.; Routbort, M.J.; Alvarez, H.; Broaddus, R.; Chen, H.; Rashid, A.; Lazar, A.; San Lucas, F.A.; Yao, H.; et al. Factors Impacting Clinically Relevant RNA Fusion Assays Using Next-Generation Sequencing. *Arch. Pathol. Lab. Med.* **2021**, *145*, 1405–1412. [CrossRef]
48. Zheng, Z.; Liebers, M.; Zhelyazkova, B.; Cao, Y.; Panditi, D.; Lynch, K.D.; Chen, J.; Robinson, H.E.; Shim, H.S.; Chmielecki, J.; et al. Anchored multiplex PCR for targeted next-generation sequencing. *Nat. Med.* **2014**, *20*, 1479–1484. [CrossRef]
49. Klausner, M.; Stinnett, V.; Ghabrial, J.; Morsberger, L.; DeMetrick, N.; Long, P.; Zhu, J.; Smith, K.; James, T.; Adams, E.; et al. Optical Genome Mapping Reveals Complex and Cryptic Rearrangement Involving *PML::RARA* Fusion in Acute Promyelocytic Leukemia. *Genes* **2024**, *15*, 1402. [CrossRef]
50. Cao, Y.; Yao, L.; Liu, Y.; Gu, Q.; Dong, W.; Wang, Z.; Wang, F.; Lin, R.; Xie, X.; Cen, J.; et al. An Atypical *PML-RARA* Rearrangement Resulting from Submicroscopic Insertion of the *RARA* Gene at the *PML* Locus with Novel Breakpoints within *PML* Exon 7b and *RARA* Exon 3. *Acta Haematol.* **2019**, *142*, 98–104. [CrossRef]
51. Lauricella, C.; Greco, R.; Mancini, V.; Motta, V.; Ciraolo, A.; De Canal, G.; De Paoli, E.; Paglino, G.; Guido, V.; Bonoldi, E.; et al. Acute Promyelocytic Leukemia with del(6)(p22) and Atypical bcr2 *PML::RARA* Fusion Transcript: A Case Report. *Acta Haematol.* **2023**, *146*, 58–64. [CrossRef]
52. Ismail, S.; Ababneh, N.; Awidi, A. Identification of atypical *PML-RARA* breakpoint in a patient with acute promyelocytic leukemia. *Acta Haematol.* **2007**, *118*, 183–187. [CrossRef] [PubMed]
53. Gallagher, R.E.; Li, Y.P.; Rao, S.; Paietta, E.; Andersen, J.; Etkind, P.; Bennett, J.M.; Tallman, M.S.; Wiernik, P.H. Characterization of acute promyelocytic leukemia cases with *PML-RAR* alpha break/fusion sites in *PML* exon 6: Identification of a subgroup with decreased in vitro responsiveness to all-trans retinoic acid. *Blood* **1995**, *86*, 1540–1547. [CrossRef] [PubMed]
54. Gu, B.W.; Xiong, H.; Zhou, Y.; Chen, B.; Wang, L.; Dong, S.; Yu, Z.; Lu, L.; Zhong, M.; Yin, H.; et al. Variant-type PML-RARα fusion transcript in acute promyelocytic leukemia: Use of a cryptic coding sequence from intron 2 of the *RARα* gene and identification of a new clinical subtype resistant to retinoic acid therapy. *Proc. Natl. Acad. Sci. USA* **2002**, *99*, 7640–7645. [CrossRef] [PubMed]
55. Slack, J.L.; Willman, C.L.; Andersen, J.W.; Li, Y.P.; Viswanatha, D.S.; Bloomfield, C.D.; Tallman, M.S.; Gallagher, R.E. Molecular analysis and clinical outcome of adult APL patients with the type V PML-RARα isoform: Results from intergroup protocol 0129. *Blood* **2000**, *95*, 398–403.

56. Chillon, M.C.; Gonzalez, M.; Garcia-Sanz, R.; Balanzategui, A.; Gonzalez, D.; Lopez-Perez, R.; Mateos, M.V.; Alaejos, I.; Rayón, C.; Arbeteta, J.; et al. Two new 3′ *PML* breakpoints in t(15;17)(q22;q21)-positive acute promyelocytic leukemia. *Genes Chromosomes Cancer* **2000**, *27*, 35–43. [CrossRef]
57. Nomura, N.; Nagase, T.; Miyajima, N.; Sazuka, T.; Tanaka, A.; Sato, S.; Seki, N.; Kawarabayasi, Y.; Ishikawa, K.; Tabata, S. Prediction of the coding sequences of unidentified human genes, I.I. The coding sequences of 40 new genes (KIAA0041-KIAA0080) deduced by analysis of cDNA clones from human cell line KG-1. *DNA Res.* **1994**, *1*, 223–229. [CrossRef]
58. Seki, T.; Gong, L.; Williams, A.J.; Sakai, N.; Todi, S.V.; Paulson, H.L. JosD1, a membrane-targeted deubiquitinating enzyme, is activated by ubiquitination and regulates membrane dynamics, cell motility, and endocytosis. *J. Biol. Chem.* **2013**, *288*, 17145–17155. [CrossRef]
59. Yang, J.; Weisberg, E.L.; Liu, X.; Magin, R.S.; Chan, W.C.; Hu, B.; Schauer, N.J.; Zhang, S.; Ilaria Lamberto, I.; Doherty, L.; et al. Small molecule inhibition of deubiquitinating enzyme JOSD1 as a novel targeted therapy for leukemias with mutant JAK2. *Leukemia* **2022**, *36*, 210–220. [CrossRef]
60. Wu, X.; Luo, Q.; Zhao, P.; Chang, W.; Wang, Y.; Shu, T.; Ding, F.; Li, B.; Liu, Z. JOSD1 inhibits mitochondrial apoptotic signalling to drive acquired chemoresistance in gynaecological cancer by stabilizing MCL1. *Cell Death Differ.* **2020**, *27*, 55–70. [CrossRef]
61. Li, M.; Gao, F.; Li, X.; Gan, Y.; Han, S.; Yu, X.; Liu, H.; Li, W. Stabilization of MCL-1 by E3 ligase TRAF4 confers radioresistance. *Cell Death Dis.* **2022**, *13*, 1053. [CrossRef]
62. Ross, K.D.; Ren, J.; Zhang, R.; Chi, N.C.; Hamilton, B.A. *Ankfn1*-mutant vestibular defects require loss of both ancestral and derived paralogs for penetrance in zebrafish. *G3* **2022**, *12*, jkab446. [CrossRef] [PubMed]
63. Wang, Y.; Zhang, Y.; Mi, J.; Jiang, C.; Wang, Q.; Li, X.; Zhao, M.; Geng, Z.; Song, X.; Li, J. ANKFN1 plays both protumorigenic and metastatic roles in hepatocellular carcinoma. *Oncogene* **2022**, *41*, 3680–3693. [CrossRef] [PubMed]
64. Zhao, L.; Wang, Y.; Tian, T.; Rao, X.; Dong, W.; Zhang, J.; Yang, Y.; Tao, Q.; Peng, F.; Shen, C.; et al. Analysis of viral integration reveals new insights of oncogenic mechanism in HBV-infected intrahepatic cholangiocarcinoma and combined hepatocellular-cholangiocarcinoma. *Hepatol. Int.* **2022**, *16*, 1339–1352. [CrossRef]
65. Qureshi, N.; Chi, J.; Qian, Y.; Huang, Q.; Duan, S. Looking for the Genes Related to Lung Cancer from Nasal Epithelial Cells by Network and Pathway Analysis. *Front. Genet.* **2022**, *13*, 942864. [CrossRef]
66. Quina, A.S.; Parreira, L. Telomere-surrounding regions are transcription-permissive 3D nuclear compartments in human cells. *Exp. Cell Res.* **2005**, *307*, 52–64. [CrossRef]
67. Baur, J.A.; Zou, Y.; Shay, J.W.; Wright, W.E. Telomere position effect in human cells. *Science* **2001**, *292*, 2075–2077. [CrossRef]
68. Spielmann, M.; Lupianez, D.G.; Mundlos, S. Structural variation in the 3D genome. *Nat. Rev. Genet.* **2018**, *19*, 453–467. [CrossRef]
69. Shaul, O. How introns enhance gene expression. *Int. J. Biochem. Cell Biol.* **2017**, *91*, 145–155. [CrossRef]
70. Attwood, K.M.; Salsman, J.; Chung, D.; Mathavarajah, S.; Van Iderstine, C.; Dellaire, G. PML isoform expression and DNA break location relative to PML nuclear bodies impacts the efficiency of homologous recombination. *Biochem. Cell Biol.* **2020**, *98*, 314–326. [CrossRef]
71. Horiuchi, K.; Kawamura, T.; Iwanari, H.; Ohashi, R.; Naito, M.; Kodama, T.; Hamakubo, T. Identification of Wilms' tumor 1-associating protein complex and its role in alternative splicing and the cell cycle. *J. Biol. Chem.* **2013**, *288*, 33292–33302. [CrossRef]
72. Liao, X.; Chen, L.; Liu, J.; Hu, H.; Hou, D.; You, R.; Wang, X.; Huang, H. m^{6}A RNA methylation regulators predict prognosis and indicate characteristics of tumour microenvironment infiltration in acute myeloid leukaemia. *Epigenetics* **2023**, *18*, 2160134. [PubMed]
73. Xu, J.R.; Yang, D.H.; Long, G.F.; Sun, H.; Chen, H.B. [Establishment and validation of prognosis predictive model using m^{6}A RNA methylation regulators in children acute myeloid leukemia]. *Zhonghua Yu Fang. Yi Xue Za Zhi* **2021**, *55*, 983–989. [PubMed]
74. Robbins, C.M.; Tembe, W.A.; Baker, A.; Sinari, S.; Moses, T.Y.; Beckstrom-Sternberg, S.; Beckstrom-Sternberg, J.; Barrett, M.; Long, J.; Chinnaiyan, A.; et al. Copy number and targeted mutational analysis reveals novel somatic events in metastatic prostate tumors. *Genome Res.* **2011**, *21*, 47–55. [PubMed]
75. Gohring, G.; Lange, K.; Atta, J.; Krauter, J.; Holzer, D.; Schlegelberger, B. Cryptic t(15;17) in a patient with AML M3 and a complex karyotype. *Cancer Genet. Cytogenet.* **2007**, *175*, 77–80. [PubMed]
76. De Botton, S.; Chevret, S.; Sanz, M.; Dombret, H.; Thomas, X.; Guerci, A.; Fey, M.; Rayon, C.; Huguet, F.; Sotto, J.J.; et al. Additional chromosomal abnormalities in patients with acute promyelocytic leukaemia (APL) do not confer poor prognosis: Results of APL 93 trial. *Br. J. Haematol.* **2000**, *111*, 801–806.
77. Kim, M.J.; Yoon, H.S.; Cho, S.Y.; Lee, H.J.; Suh, J.T.; Lee, J.; Yoon, H.; Lee, W.; Park, T.S. ider(17)(q10)t(15;17) associated with relapse and poor prognosis in a pediatric patient with acute promyelocytic leukemia. *Cancer Genet. Cytogenet.* **2010**, *201*, 116–121. [CrossRef]
78. Tong, H.; Li, K.; Mei, C.; Wang, H.; Chen, Z.; Jin, J. Arsenic trioxide may improve the prognosis of APL with ider(17)(q10): Report of a rare adult case of acute promyelocytic leukemia with ider(17)(q10)t(15;17) showing poor response to all-trans retinoic acid. *Ann. Hematol.* **2011**, *90*, 1493–1494. [CrossRef]
79. Campo, E.; Cymbalista, F.; Ghia, P.; Jager, U.; Pospisilova, S.; Rosenquist, R.; Schuh, A.; Stilgenbauer, S. *TP53* aberrations in chronic lymphocytic leukemia: An overview of the clinical implications of improved diagnostics. *Haematologica* **2018**, *103*, 1956–1968.
80. Middeke, J.M.; Fang, M.; Cornelissen, J.J.; Mohr, B.; Appelbaum, F.R.; Stadler, M.; Sanz, J.; Baurmann, H.; Bug, G.; Schäfer-Eckart, K.; et al. Outcome of patients with abnl(17p) acute myeloid leukemia after allogeneic hematopoietic stem cell transplantation. *Blood* **2014**, *123*, 2960–2967. [CrossRef]

81. Cervera, J.; Montesinos, P.; Hernandez-Rivas, J.M.; Calasanz, M.J.; Aventin, A.; Ferro, M.T.; Luño, E.; Sánchez, J.; Vellenga, E.; Rayón, C.; et al. Additional chromosome abnormalities in patients with acute promyelocytic leukemia treated with all-trans retinoic acid and chemotherapy. *Haematologica* **2010**, *95*, 424–431.
82. Lee, G.Y.; Christina, S.; Tien, S.L.; Ghafar, A.B.; Hwang, W.; Lim, L.C.; Lim, T.H. Acute promyelocytic leukemia with *PML-RARA* fusion on i(17q) and therapy-related acute myeloid leukemia. *Cancer Genet. Cytogenet.* **2005**, *159*, 129–136. [CrossRef]
83. Manola, K.N.; Karakosta, M.; Sambani, C.; Terzoudi, G.; Pagoni, M.; Gatsa, E.; Papaioannou, M. Isochromosome der(17)(q10)t(15;17) in acute promyelocytic leukemia resulting in an additional copy of the *RARA-PML* fusion gene: Report of 4 cases and review of the literature. *Acta Haematol.* **2010**, *123*, 162–170. [CrossRef]

biomolecules

MDPI

Article

Tumor-Infiltrating Immune Cells and HLA Expression as Potential Biomarkers Predicting Response to PD-1 Inhibitor Therapy in Stage IV Melanoma Patients

Barbara Hegyi [1,2,3,†], Kristóf György Csikó [1,2,3,†], Tímea Balatoni [2,4], Georgina Fröhlich [5,6], Katalin Bőcs [7], Erika Tóth [2,8], Anita Mohos [9,10], Anna Rebeka Neumark [11], Csenge Dorottya Menyhárt [12], Soldano Ferrone [13,‡] and Andrea Ladányi [2,8,*]

1 Department of Chest and Abdominal Tumors and Clinical Pharmacology, National Institute of Oncology, H-1122 Budapest, Hungary; hegyi.barbara@oncol.hu (B.H.); csiko.kristofgyorgy@gmail.com (K.G.C.)
2 National Tumor Biology Laboratory, National Institute of Oncology, H-1122 Budapest, Hungary; balatoni.timea@oncol.hu (T.B.); dr.toth.erika@oncol.hu (E.T.)
3 Doctoral College, Semmelweis University, H-1085 Budapest, Hungary
4 Department of Oncodermatology, National Institute of Oncology, H-1122 Budapest, Hungary
5 Center of Radiotherapy, National Institute of Oncology, H-1122 Budapest, Hungary; frohlich.georgina@oncol.hu
6 Department of Biophysics, Eötvös Loránd University, H-1117 Budapest, Hungary
7 Department of Diagnostic Radiology, National Institute of Oncology, H-1122 Budapest, Hungary; bocs.katalin@oncol.hu
8 Department of Surgical and Molecular Pathology, National Institute of Oncology, H-1122 Budapest, Hungary
9 Department of Pathology and Experimental Cancer Research, Semmelweis University, H-1085 Budapest, Hungary; mohos.anita@gmail.com
10 Department of Dermatology, Venereology and Dermatooncology, Semmelweis University, H-1085 Budapest, Hungary
11 Faculty of Medicine, Semmelweis University, H-1085 Budapest, Hungary
12 Faculty of Science, Eötvös Loránd University, H-1117 Budapest, Hungary
13 Department of Surgery, Massachusetts General Hospital, Harvard Medical School, Boston, MA 02115, USA
* Correspondence: ladanyi.andrea@oncol.hu
† These authors contributed equally to this work.
‡ Professor Ferrone has deceased.

Abstract: PD-1 inhibitors are known to be effective in melanoma; however, a considerable proportion of patients fail to respond to therapy, necessitating the identification of predictive markers. We examined the predictive value of tumor cell HLA class I and II expression and immune cell infiltration in melanoma patients treated with PD-1 inhibitors. Pretreatment surgical samples from 40 stage IV melanoma patients were studied immunohistochemically for melanoma cell expression of HLA class I molecules (using four antibody clones with different specificities), HLA-II, and immune cell infiltration (using a panel of 10 markers). Among the responders, the ratio of patients showing melanoma cell HLA-II expression was higher compared to non-responders ($p = 0.0158$), and similar results were obtained in the case of two anti-HLA-I antibodies. A combined score of HLA-I/II expression also predicted treatment response ($p = 0.0019$). Intratumoral infiltration was stronger in the responders for most immune cell types. Progression-free survival showed an association with HLA-II expression, the combined HLA score, and the density of immune cells expressing CD134 and PD-1, while overall survival was significantly associated only with HLA class II expression. Our findings corroborate previous results indicating the importance of immune cell infiltration and tumor cell HLA-II expression in the efficacy of PD-1 inhibitor treatment in a "real world" patient cohort and suggest the potential predictive role of HLA class I expression.

Keywords: melanoma; immunotherapy; PD-1 inhibitors; biomarker; tumor-infiltrating immune cells; HLA class I and II expression

1. Introduction

Immune checkpoint inhibitors (ICIs), especially antibodies targeting the PD-1/PD-L1 pathway, have transformed cancer treatment, with documented favorable effects in a variety of tumor types, even in advanced stages. However, a considerable proportion of patients (the size of which varies depending on the tumor type) fail to respond to therapy or develop resistance after the initial response. The factors causing either primary or acquired resistance are only partly elucidated, and it is of primary importance to search for biomarkers that could predict the likelihood of therapeutic effects [1–3].

Many potential biomarkers have been proposed that could predict the efficacy of ICI therapies. They include, among others, PD-L1 expression in the case of PD-1/PD-L1 targeted agents, tumor mutational burden (TMB), neoantigen load, microsatellite instability, antigen presentation deficiencies, tumor infiltration by immune cells, and immune-related gene expression in tumors [1–7]. However, few of the above biomarkers have been validated, and most of them were investigated only in one or a few studies. Although malignant melanoma is one of the tumor types with the best response rates to ICIs, more than half of stage IV patients receiving anti-PD-1 therapy still experience treatment resistance [8–10]. There is no approved predictive biomarker for PD-1 blocking therapy in metastatic melanoma patients. PD-L1 expression of tumor cells and/or tumor-infiltrating immune cells, which is a prerequisite of anti-PD-1/PD-L1 treatment for several cancer types, is not a clinically validated biomarker in melanoma; it was found to be associated with clinical response or post-therapy survival in most but not all studies [4,11–14]. Among other components of the tumor immune microenvironment (TIME), infiltration by $CD8^+$ T lymphocytes was consistently found to be predictive of response while results concerning $CD4^+$ T cells are controversial [4,13,15–18]. Other immune cell subset markers (e.g., CD16, CD45RO) and immune checkpoints such as PD-1 and CTLA-4 have also been found to be associated with the treatment outcomes in some studies [4,13,18,19].

Since antigen presentation by MHC class I molecules is the prerequisite of tumor cell recognition by cytotoxic T lymphocytes, considered to be major players in antitumor immune responses, it could be logically expected that MHC-I expression by tumor cells would be required for the efficacy of T cell-based immunotherapies, including ICIs. Human cancer cells frequently downregulate the components of HLA class I antigen processing machinery (APM), which was found to be associated with poor prognosis in several tumor types [20]. The mutation or loss of heterozygosity of β2-microglobulin (β2M) was identified as a mechanism of resistance to immunotherapies, including PD-1 inhibitors [7,21]. However, decreased HLA class I expression is most frequently caused by epigenetic or transcriptional mechanisms [20,22]. The potential association of HLA class I protein expression loss with ICI therapy resistance has been explored only to a limited extent [23–25]. Surprisingly, the studies demonstrated the predictive role of tumor cell HLA class I expression in melanoma patients treated with the CTLA-4 inhibitor ipilimumab [24,25] but not in those receiving PD-1 blocking agents [23,24]. Conversely, HLA class II expression on tumor cells proved to be predictive of anti-PD-1 efficacy but not anti-CTLA-4 effects [23–25].

In our previous studies, we proved the role of HLA class I expression and infiltration by several immune cell types (e.g., $CD4^+$ and $CD8^+$ T lymphocytes, regulatory T cells, B lymphocytes, NK cells, $CD68^+$ macrophages, PD-1^+ and $CD134^+$ lymphocytes) in predicting responses to treatment and/or the survival of melanoma patients receiving ipilimumab therapy [25,26]. The aim of the present study was to explore tumor cell HLA class I and class II expression and tumor-infiltrating immune cell types as potential predictive markers in melanoma patients treated with PD-1 inhibitors.

2. Materials and Methods

2.1. Patients and Tumor Samples

Archived paraffin blocks of pretreatment surgical samples from 40 stage IV melanoma patients receiving anti-PD-1 therapy (nivolumab 240 mg every two weeks or pembrolizumab 200 mg every 3 weeks, start of treatment in 2015–2022) at the National Institute of Oncology were selected. Only patients receiving at least 3 cycles of anti-PD-1 therapy were included. Tumor resection samples of lymph node (n = 70) and skin/subcutaneous metastases (n = 42), obtained within 4 years of starting anti-PD-1 therapy (median 10 months, range 1–48), were selected; the study cohort consisted of 40 patients (1–9 examined lesions per patient). The primary site was cutaneous in 38 cases and unknown in 2 cases. Most patients (n = 28) received the anti-PD-1 treatment as first-line therapy. Response assessment was based on immune-related response criteria (irRC) [27]; patients with complete or partial remission as best overall response were considered responders in the evaluation. Patients' follow-up was performed by CT every 3 months and if it was necessary, MRI or PET/CT was also performed. Follow-up time of surviving patients was median 62.5 months (22–91). Progression-free survival (PFS) and overall survival (OS) were defined as the time from commencing anti-PD-1 treatment to disease progression or death or last follow-up, and death or last follow-up, respectively. Patient characteristics are shown in Table 1.

Table 1. Patient and sample characteristics.

	Responder (CR, PR) n **= 28**	**Non-Responder (SD, PD)** n **= 12**	p **Value**
Age, years: median (range)	62 (27–87)	53 (34–77)	NS [a]
Gender			
female	11	2	
male	17	10	NS [b]
ECOG performance status			
0	27	8	
1	1	4	0.0223 [b]
BRAF mutation status			
wild type	20	10	
mutant	8	2	NS [b]
LDH level			
normal	23	4	
>ULN	5	8	0.0075 [b]
Anti-PD-1 drug			
nivolumab	9	8	
pembrolizumab	19	4	NS [b]
Line of therapy			
1	20	8	
2–5	8	4	NS [b]
PFS, months: median (range)	27 (4–90+)	3 (1–11)	0.0000 [a]
OS, months: median (range)	46 (13–91+)	9 (4–62+)	0.0002 [a]

[a] Mann–Whitney test, [b] Fisher's exact test. CR: complete response, PR: partial response, SD: stable disease, PD: progressive disease, ECOG: Eastern Cooperative Oncology Group, BRAF: B-Raf proto-oncogene, LDH: lactate dehydrogenase, PFS: progression-free survival, OS: overall survival, ULN: upper limit of normal, NS: not significant.

2.2. Immunohistochemical Staining

Immunohistochemical staining of tissue sections of formalin-fixed, paraffin-embedded tumor samples was performed utilizing standard methodology as described earlier [25,26]. Briefly, after deparaffination, sections were treated with 3% H_2O_2 in methanol to block endogenous peroxidases, followed by antigen retrieval by heating at 98 °C for 40 min in citrate buffer (pH 6.0). After incubation with protein blocking solution (Protein Block, Serum-Free, Dako, Glostrup, Denmark) for 10 min at room temperature, primary antibodies were applied overnight at 4 °C. For the detection of immune cell markers, the following monoclonal antibodies were used: CD8 (C8/144B; Dako), CD45R0 (UCHL1; Dako), CD20y (L26; Dako), FOXP3 (236/E7; eBioscience, San Diego, CA, USA), NKp46/NCR1 (195314;

R&D Systems, Minneapolis, MN, USA), CD103 (EPR4166(2); Abcam, Cambridge, UK), CD134 (Ber-ACT35; BioLegend, San Diego, CA, USA), CD137 (BBK-2; Santa Cruz Biotechnology, Dallas, TX, USA), PD-1 (NAT-105; Bio SB, Santa Barbara, CA, USA), PD-L1 (73-10; Abcam). For detecting MHC class I molecules, HCA2 (Origene, Rockville, MD, USA), recognizing β2M-free HLA-A (excluding -A24), -B7301, and -G heavy chains; HC10 (Origene), recognizing β2M-free HLA-A3, -A10, -A28, -A29, -A30, -A31, -A32, -A33, HLA-B (excluding -B5702, -B5804, and -B73) and HLA-C heavy chains; the β2-microglobulin (β2M) specific monoclonal antibody NAMB-1; and the anti-pan HLA class I EMR8-5 (Abcam) were utilized; HLA-DR,DQ,DP was detected using the monoclonal antibody LGII-612.14. Staining was detected using the EnVision+ System HRP Labeled Polymer Anti-mouse reagent (Dako) and the MACH2 Rabbit HRP-Polymer (Biocare Medical, Pacheco, CA, USA) for mouse and rabbit primary antibodies, respectively, followed by staining with 3-amino-9-ethylcarbazole (AEC; Vector Laboratories, Inc., Burlingame, CA, USA) and hematoxylin counterstaining.

2.3. *Evaluation of the Immune Reactions*

Evaluation of immunohistochemical staining was performed independently by two researchers who were blinded to the clinical information, as described earlier [25,26], using light microscope equipped with an eyepiece graticule, and the mean value of their separate counts was used for the analysis. For immune cell evaluation, labeled cells within the metastases were counted in at least 10 randomly chosen fields per section, using the graticule of 10 × 10 squares designating an area of 0.0625 mm^2 at 400× magnification. PD-L1 expression by immune cells was evaluated similarly to the other immune cell markers, while its expression by tumor cells (tumor proportion score, TPS) was also registered. HLA class I staining was scored as 0, 1, and 2 when the percentage of stained melanoma cells was <25%, 25–75%, and >75%, respectively, based on the criteria established by the 12th International Histocompatibility Workshop (1996). Expression on normal cells in the samples (e.g., immune cells, cells of the vasculature) served as positive control. In the case of HLA-DR,DQ,DP, the percentage of the area displaying positive tumor cell staining was determined in the metastases. We also calculated a combined score of HLA class I and class II expression (HLA I/II score) based on the number of anti-HLA-I antibodies showing higher positivity than the cutoff level, combined with HLA-II positivity in a given sample (score of 1 in the case of high labeling with at least 3 anti-HLA-I antibodies, and/or HLA-II expression ≥3%; score of 0 when neither of the above criteria are met). For patients with more than one metastasis available, the average scores were calculated for each marker. Cutoff levels were set up based on the median of the given variable in the whole patient cohort, with modifications for better discriminating power in some cases, and the proportion of patients with a mean cell density/HLA expression score higher than the cutoff level was also determined for all markers.

2.4. *Statistical Analysis*

For the statistical analysis of differences in cell densities and HLA expression levels between different patient groups, we applied the Mann–Whitney U test, while Fisher's exact test was used to compare the proportions of cases in different groups. Correlations were analyzed by Pearson's nonparametric correlation method. Survival analysis was carried out using the Kaplan–Meier method and log-rank test. Differences were considered significant in the case of p values ≤ 0.05. Statistics were calculated using the Statistica software version 12.5 (StatSoft, Tulsa, OK, USA).

3. Results

Of the 40 stage IV melanoma patients involved in this study, 28 exhibited a complete (n = 14) or partial response (n = 14) (Table 1). Among clinical parameters, the ECOG performance status and pretreatment LDH level showed an association with the response. Pretreatment surgical samples of lymph node and cutaneous/subcutaneous melanoma metastases were immunohistochemically analyzed for HLA expression (Figure 1a), evaluating it using scores between 0 and 2 for HLA class I and determining the percentage of positive tumor area in the case of HLA class II. Considering all metastases, tumor cell HLA-I positivity levels determined by the four different antibodies (HCA2, HC10, NAMB-1, EMR8-5) showed a strong positive correlation, with the lowest correlation coefficient (0.6812) in the case of HCA2/HC10 and highest (0.8236) in the case of NAMB-1/EMR8-5 (all p values < 0.0001), while HLA-II staining showed a weaker correlation with HLA-I positivity (correlation coefficients 0.2182–0.2944). HLA class I expression scores were the highest in the case of the EMR8-5 pan-HLA-I antibody (mean ± SD, 1.4 ± 0.7) and the lowest in the case of HC10 (1.1 ± 0.8). Melanoma cells in metastases of 20 patients (50%) did not express HLA class II, while mean tumor cell staining higher than 10% was observed in 11 patients (27.5%). The tumor samples of eight patients showed 3–4% HLA-II staining, mainly in tumor cells near the inflammatory cells at the margin of metastases, consistent with locally induced HLA expression.

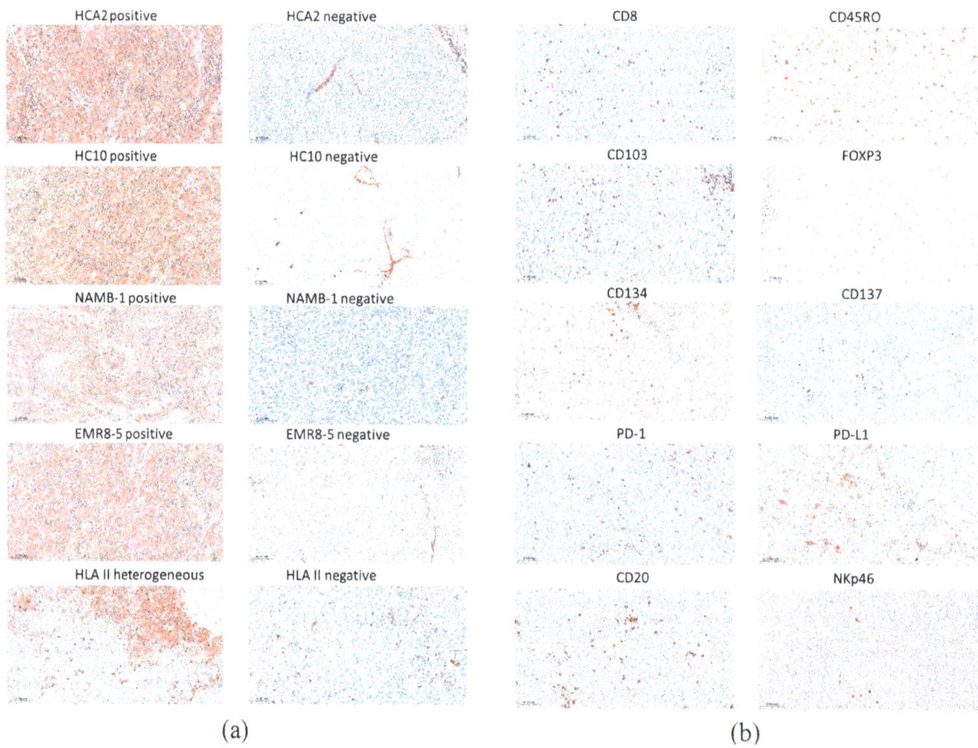

Figure 1. Immunohistochemical staining of melanoma metastases with HLA class I and class II-specific antibodies (**a**) and labeling of immune cell markers (**b**) (3-amino-ethylcarbazole, red). Scale bars: 100 μm.

Among the responders, the proportion of patients showing HLA-II expression in ≥3% of melanoma cells was higher compared to non-responders (17/28 vs. 2/12, p = 0.0158). Similarly, a significant difference was found in the case of two anti-HLA-I antibodies (HC10 and EMR8-5), with fewer cases showing decreased expression in responders (Table 2). A combined score of HLA-I and -II expression was introduced (being positive in the case of high HLA class I and/or HLA class II expression), which proved very effective in predicting treatment response (p = 0.0019) (Table 2).

Table 2. Relationship of treatment response with the proportion of patients with high tumor cell HLA expression in metastases.

	Responder (CR, PR) $n = 28$	Non-Responder (SD, PD) $n = 12$	p Value [a]
HLA class I; Ab clone (cutoff score)			
HCA2 [b] (≥1.3)	16/27 (59%)	5/12 (42%)	0.4877
HC10 (≥0.6)	24/28 (86%)	6/12 (50%)	0.0411
NAMB-1 (≥1.7)	14/28 (50%)	3/12 (25%)	0.1788
EMR8-5 (≥1.4)	20/28 (71%)	4/12 (33%)	0.0367
HLA class II (cutoff: ≥3%)	17/28 (61%)	2/12 (17%)	0.0158
HLA I/II score [c] (cutoff: ≥3)	24/28 (86%)	4/12 (33%)	0.0019

[a] Fisher's exact test. [b] Staining with HCA2 could not be evaluated in one patient because of negativity of normal cells. [c] Combined score of high expression with the anti-HLA-I and/or the anti-HLA-II antibodies. CR: complete response, PR: partial response, SD: stable disease, PD: progressive disease.

We also determined the intratumoral density of immune cell subsets in the melanoma metastases using 10 different markers (CD8, CD45RO, CD20, FOXP3, NKp46, CD103, CD134, CD137, PD-1, PD-L1) (Figure 1b). Among the immune cell types, CD45RO$^+$ T lymphocytes were present in the highest number, followed by CD8$^+$ T lymphocytes, PD-L1$^+$, CD103$^+$, and PD-1$^+$ cells, while NKp46$^+$ NK cells and cells expressing the CD134 and CD137 activation markers were the least numerous (Table 3). Evaluating the infiltration of the labeled cells in relation to immunotherapy response revealed that most of the immune cell subsets were present at higher amounts in the metastases of responding patients compared to non-responders (Table 3, Figure 2). Analyzing the proportion of patients with high mean intratumoral cell density showed higher prevalence in responders in the case of eight of the ten immune cell markers (Table 3). The most significant differences were observed in the case of PD-1 (23/28 vs. 3/12, $p = 0.0010$) and PD-L1 (19/28 vs. 1/12, $p = 0.0012$). Interestingly, PD-L1 expression (≥1%) by tumor cells did not predict treatment response significantly (9/28 of responders, 1/12 of non-responders; $p = 0.2307$), in contrast to its expression by immune cells.

Table S1 presents individual patients' data and a heatmap showing all markers (HLA expression and immune cell density), demonstrating an overall higher HLA expression and immune cell prevalence in responders. A coordinated high expression of at least six of the ten immune cell markers was found in 18 of the 28 responders (64%) and only 1 of the 12 non-responders (8%) ($p = 0.0015$).

Table 3. Relationship of treatment response with the density of immune cells infiltrating metastases.

Immune Cell Marker	Density of Labeled Cells (n/mm^2), Median (Range) Responder (CR, PR) $n = 28$	Density of Labeled Cells (n/mm^2), Median (Range) Non-Responder (SD, PD) $n = 12$	p Value [a]
CD8	96.0 (12.8–540.8)	31.2 (1.6–347.2)	0.0112
CD45RO	416.8 (64.0–769.6)	194.4 (83.2–774.4)	0.1918
CD20	18.4 (0.0–195.2)	7.2 (0.0–27.2)	0.0093
NKp46	3.0 (0.0–16.3)	1.3 (0.0–10.6)	0.3132
FOXP3	24.0 (4.8–100.8)	16.0 (6.4–108.8)	0.1084
CD134	4.9 (0.0–68.0)	2.5 (0.0–10.7)	0.0390
CD137	7.4 (1.1–37.3)	3.8 (0.2–17.0)	0.2992
CD103	76.0 (17.6–246.4)	29.6 (1.6–131.2)	0.0077
PD-1	71.2 (4.8–393.6)	13.6 (0.0–326.4)	0.0011
PD-L1	86.4 (1.6–705.6)	16.8 (0.0–123.2)	0.0046

Table 3. *Cont.*

	Proportion of Cases with High Cell Density (%)		
Immune Cell Marker (Cutoff)	Responder (CR, PR) $n = 28$	Non-Responder (SD, PD) $n = 12$	p Value [b]
CD8 (≥54 cells/mm^2)	20/28 (71%)	3/12 (25%)	0.0130
CD45RO (≥264 cells/mm^2)	18/28 (64%)	3/12 (25%)	0.0378
CD20 (≥28 cells/mm^2)	10/28 (36%)	0/12 (0%)	0.0188
NKp46 (≥3.4 cells/mm^2)	13/28 (46%)	3/12 (25%)	0.2969
FOXP3 (≥20 cells/mm^2)	18/28 (64%)	3/12 (25%)	0.0378
CD134 (≥4.1 cells/mm^2)	17/28 (61%)	2/12 (17%)	0.0158
CD137 (≥6.7 cells/mm^2)	14/28 (50%)	3/12 (25%)	0.1788
CD103 (≥36 cells/mm^2)	21/28 (75%)	4/12 (33%)	0.0297
PD-1 (≥20 cells/mm^2)	23/28 (82%)	3/12 (25%)	0.0010
PD-L1 (≥54 cells/mm^2)	19/28 (68%)	1/12 (8%)	0.0012

[a] Mann–Whitney test, [b] Fisher's exact test. CR: complete response, PR: partial response, SD: stable disease, PD: progressive disease.

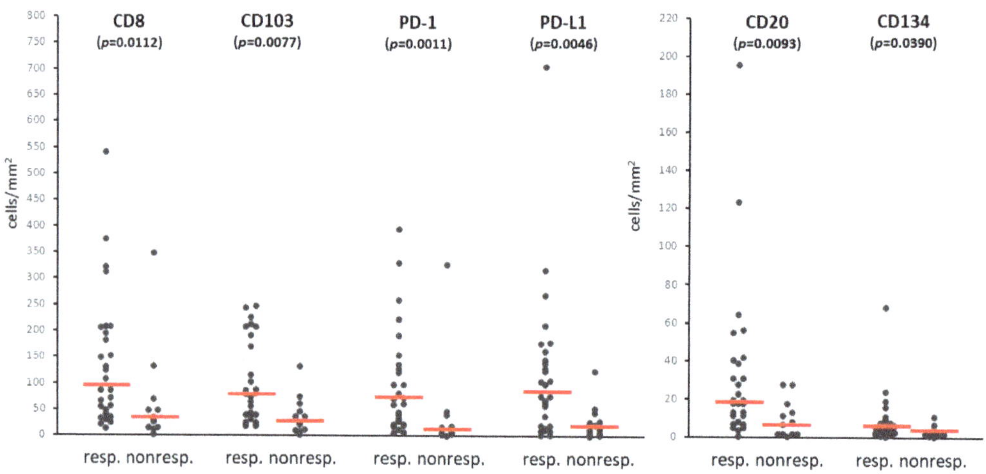

Figure 2. Immune cell infiltration in metastases of responder and non-responder patients treated with PD-1 inhibitors (dots: mean density values in samples from individual patients; horizontal line: median).

Kaplan–Meier survival analysis demonstrated a near-significant ($p = 0.0530$) association of PFS with HLA class I expression detected by the HC10 antibody and a highly significant association ($p = 0.0025$) with HLA class II expression (Figure 3a,b). The combined score of HLA class I and class II expression was also significantly linked to PFS ($p = 0.0166$; Figure 3c). Among immune cell markers, a significant survival correlation was demonstrated in the case of CD134 and PD-1, both showing longer PFS in cases with high immune cell density ($p = 0.0318$ and $p = 0.0230$, respectively; Figure 3d,e). Overall survival was significantly associated only with HLA class II expression ($p = 0.0126$; Figure 3f).

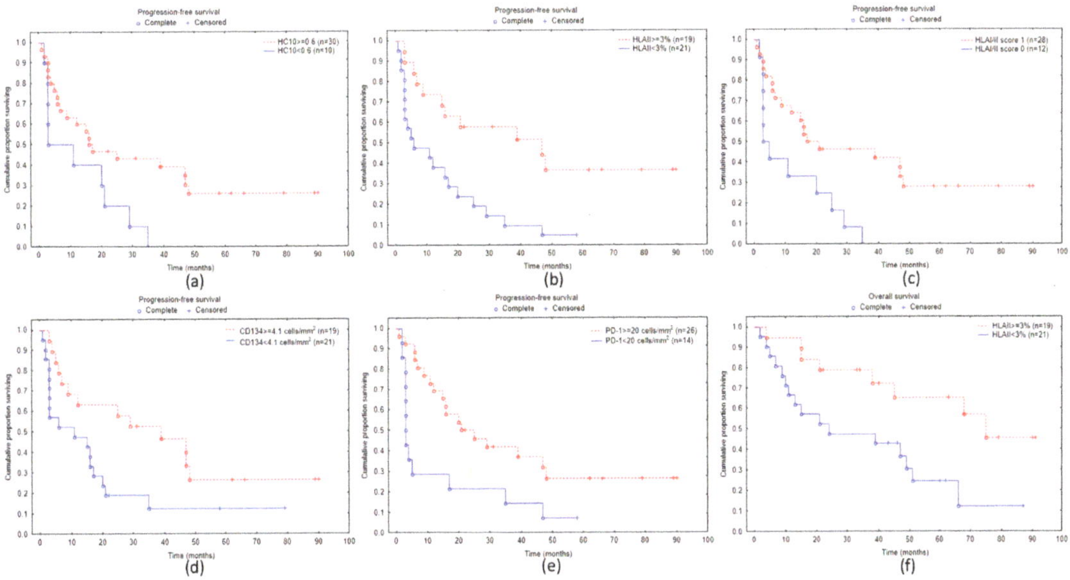

Figure 3. Kaplan–Meier curves of progression-free survival for melanoma patients subdivided according to staining with HLA class I-specific antibody HC10 (**a**), tumor cell HLA-II expression (**b**), HLA I/II score (**c**), density of CD134$^+$ (**d**), PD-1$^+$ cells (**e**), overall survival according to tumor cell HLA class II expression (**f**).

4. Discussion

Although ICI therapy has revolutionized the treatment of many cancer types, the majority of patients fail to achieve durable responses, and continued efforts to unravel the mechanisms of immunotherapy resistance are necessary to allow the optimization of therapeutic strategies for patients.

In our study, we focused on investigating local immunological parameters, i.e., tumor cell HLA expression and immune cell infiltration, as potential biomarkers of the efficacy of anti-PD-1 treatment in metastatic melanoma patients. As antigen presentation by MHC-I molecules on target cells is essential for recognition and killing by CD8$^+$ cytotoxic T lymphocytes, it is a logical assumption that MHC-I expression by tumor cells would be needed for the efficiency of T cell-based immunotherapies such as ICIs. Surprisingly, however, the few earlier studies aiming at demonstrating the predictive role of tumor cell HLA class I protein expression in patients treated with PD-1 inhibitors could not prove an association with therapy outcomes in melanoma [23,24], although it was found to be predictive in the case of CTLA-4 blocking with ipilimumab [24,25]. On the other hand, HLA-II expression by tumor cells predicted anti-PD-1/PD-L1 therapy efficacy in melanoma and breast cancer, while no such association was found with ipilimumab efficacy in melanoma patients [23–25,28].

In our present study on stage IV melanoma patients, a higher level of tumor cell HLA class I expression was detected by two of the four antibodies used: HC10, recognizing mainly HLA-B and -C antigens and the pan-HLA class I EMR8-5; this result is similar to our findings in patients receiving ipilimumab therapy [25]. The discrepancy regarding the predictive role of melanoma cell HLA class I expression between our results and those obtained in previous studies [23,24] may partly be due to methodological differences, for example, using biopsies vs. whole sections; moreover, in one of the previous studies, only HLA-A expression was tested while HLA-B and -C were not [23]. On the other hand, in our cohort, HLA class II expression seemed to have a more significant impact on the treatment outcome, which is in line with the results of previous investigations on melanoma patients receiving anti-PD-1 treatment [23,24] and different from the findings of studies on ipilimumab therapy [24,25]. The different mechanisms of action of anti-CTLA-4 and anti-PD-1 agents may explain the observed dissimilarities in the associations of HLA class I vs. class II tumor cell expression with therapy outcomes; however, the exact mechanisms explaining this divergence are unclear at present.

An additional potential explanation for the stronger association of HLA class II expression with treatment response is that it is highly inducible by cytokines such as IFN-γ, which is mainly produced by T lymphocytes. Consequently, HLA-II expression generally shows a strong correlation with the density of T cells [25,29,30], which is often reported as a factor predicting immunotherapy response, which is also supported by our results; thus, it is possible that the association between HLA-II expression and treatment response results, at least in part, indirectly from the above correlations. In the present study, the associations with either clinical response or survival did not appear to depend on the percentage of HLA-II-positive tumor cells (comparing cases with 3–4% vs. >10% positivity; see Table S1), which may also support the hypothesis of indirect effect. Another alternative mechanism that may be claimed to be in the background of the impact of HLA-II expression in immunotherapy efficiency is the possibility of the activation of helper and cytotoxic $CD4^+$ T lymphocytes by tumor antigens presented by these molecules [31]. MHC-II molecules can bind an even greater diversity of peptides than MHC-I, and a study on 5942 tumors demonstrated that mutated peptides that poorly bound to MHC-II were positively selected during tumorigenesis, which had a stronger effect compared to selective pressure against MHC-I-restricted neoantigens [32,33]. In a clinical trial of personalized neoantigen vaccine plus anti-PD-1 therapy, the successful induction and cytotoxic potential of neoantigen-specific $CD4^+$ and $CD8^+$ T cells was demonstrated [34]. Furthermore, a study analyzing infused TIL products in melanoma patients receiving adoptive cell therapy revealed that neoantigen-specific T cell clones were predominantly $CD4^+$ and displayed cytotoxicity [35].

With regard to the seemingly lower influence of pretreatment tumor cell HLA class I expression (compared to HLA-II expression) on the efficacy of PD-1 blocking agents observed in our and others' studies, it should be mentioned that in our cohort, there were very few patients with completely or almost completely negative tumor cell staining with the four anti-HLA-I antibodies used, and it is possible that a low level of expression is sufficient for T cell recognition is some cases. Furthermore, the ICI treatment was reported to result in immune activation in the TME, for example, enhanced $CD8^+$ T cell infiltration [4,19,36], which could induce HLA class I expression on tumor cells in response to increased lymphocyte cytokine production. The potential changes in TME could not be taken into account in our analysis, which evaluated only pretreatment samples. Nevertheless, we could detect an additional value of determining HLA class I expression in the pretreatment tumor samples since the combined score of HLA-I and HLA-II expression proved a more robust predictor of clinical response than HLA-II expression alone. The same results could be obtained when using a simplified score consisting of the combination of HLA-II expression with HLA-I expression based on labeling by the pan-HLA class I antibody EMR8-5 (instead of on labeling by all four antibodies used), which would provide a biomarker that could be used more easily in routine clinical practice.

Besides tumor cell MHC expression, we also investigated the predictive potential of a panel of immune cell markers and found a significant association with response to anti-PD-1 therapy in the case of eight of the ten immune cell types studied: $CD8^+$ and $CD45RO^+$ T lymphocytes, regulatory T cells ($FOXP3^+$), $CD20^+$ B lymphocytes, and cells expressing activation markers/immune checkpoints CD134, PD-1, PD-L1, and CD103, a marker of tissue-resident T cells. Correlations of the prevalence of immune cell subsets with disease outcomes in metastatic melanoma patients after anti-PD-1 therapy have been consistently reported for $CD8^+$ T cells [4,13,15–18] and found in some studies for other immune cell markers, including CD45RO and FOXP3 [13], PD-1, PD-L1 [4,13], and CD103 [37].

To the best of our knowledge, our study is the first that identified the density of $CD134^+$ cells as a factor predicting the response to treatment with anti-PD-1 agents in metastatic melanoma patients and also showed an association with PFS. CD134 (OX40) is an important costimulatory immune checkpoint molecule, also a T cell activation marker, expressed mainly by $CD4^+$ T cells, also pointing to the potential importance of $CD4^+$ T cell activation upon antigen presentation by HLA-II. It shows transient expression after antigen stimulation, which may be one of the reasons for the relatively low number of $CD134^+$ cells in tumors; furthermore, they generally show a preferential location in peritumoral/stromal areas while, in our study, intratumoral density was evaluated. The prevalence of $CD134^+$ cells was proven to be a prognostic factor in primary melanoma [38] and other cancers [39,40] and a biomarker that could predict the response to ipilimumab therapy [26] in previous studies.

The results regarding $CD20^+$ B lymphocytes are controversial; no correlation with response was found in one study [17], although an association with treatment response was demonstrated in a neoadjuvant trial [41]. The role of B cells in antitumor immune responses is also not clear, with controversial findings regarding their pro- vs. antitumor effects. This discrepancy may be related to the diversity of their functional activities; they can promote tumor growth via multiple mechanisms, e.g., driving chronic inflammation, or immune suppression by regulatory B cells, but may also function as effective antigen-presenting cells promoting antitumor T cell responses [42,43]. Costimulation via the OX40L/OX40 pathway has been described as a mechanism of the B lymphocyte-mediated expansion of $CD4^+$ T cells in a murine model [44]. In this regard, our earlier findings of the colocalization of B cells and $OX40^+$ T lymphocytes in primary melanoma may be of interest, suggesting a possible role of B cells in antigen presentation and costimulation [45].

The densities of most of the studied immune cell types strongly correlated with each other and they frequently showed coordinate presence, which occurred more frequently in the responders. Besides the factors involved in this study, many other elements of the TME may affect the efficacy of immunotherapy, such as macrophages, myeloid-derived suppressor cells, cytokines, and chemokines, among others. According to recent analyses, the application of multiplex immunohistochemistry (IHC)/immunofluorescence techniques yields more accurate predictive markers than studying single factors [18,46]. Multiplex imaging techniques allowing the evaluation of the spatial organization of the tumor microenvironment (TME) are increasingly applied in the search for factors predicting the efficacy of immunotherapy [47,48].

Among the immune cell markers, the most significant association with treatment response was observed in the case of PD-1 and PD-L1; the former showed a correlation with PFS as well. Interestingly, while the intratumoral density of PD-L1-positive immune cells was demonstrated to predict clinical responses to PD-1 blocking therapy in our patient cohort, its expression by melanoma cells did not prove predictive. Evaluating PD-L1 expression by using the CPS (combined positive score) and TPS (tumor proportion score) supported these findings. Previous studies exploring the predictive role of PD-L1 in melanoma patients treated with anti-PD-1 immunotherapy mainly measured its expression by tumor cells, or by tumor and immune cells without distinction [4,11–14], and did not evaluate the impact of immune cell expression separately. In a study applying digital spatial profiling, PD-L1 expression in the $CD68^+$ (macrophage) compartment but not in the tumor compartment proved predictive of immunotherapy effects [49]. In the study by Herbst et al. [50], PD-L1 expression by tumor-infiltrating immune cells, but not by tumor cells showed an association with the response to atezolizumab (anti-PD-L1) treatment in a mixed cohort of different cancer and in the subgroup of NSCLC. Furthermore, according to several studies in mouse models, PD-L1 expression on host myeloid cells, and not on tumor cells, proved essential for PD-L1 pathway-mediated tumor regression [51,52].

The analysis of progression-free survival demonstrated a tendency of better PFS in cases with high HLA class I expression detected by the HC10 antibody. However, significantly longer PFS was demonstrated in cases with high tumor cell expression of HLA class II, high HLA I/II scores, and a high density of $PD\text{-}1^+$ and $CD134^+$ cells. Overall survival, on the other hand, showed a significant association only with HLA-II expression, which was also the factor most significantly influencing PFS. The less prominent associations of the examined factors with OS compared to the clinical response and PFS may be due to the fact that after progression on anti-PD-1 therapy, other treatment modalities were applied in most patients, so their overall survival was not only influenced by the effectivity of PD-1 blocking therapy.

While our study identified several potential biomarkers of responses to PD-1 blocking therapy, we recognize its inherent limitations, mainly caused by its retrospective nature. Moreover, the number of cases included in the analysis was constrained by the availability of sufficient surgical samples and the selection criteria we used in an attempt to decrease patient and sample variability (e.g., including only stage IV melanoma patients, whole sections of surgical samples, and only lymph node and skin/s.c. metastases). On the other hand, we believe that the applied selection criteria enhance the reliability of our results compared to studies performed on samples of unspecified or very heterogeneous locations and studies with a wide range of time intervals between sample acquisition and the implementation of immunotherapy. Also, the use of whole sections vs. small biopsies (or tissue microarrays) could reduce the impact of intratumoral heterogeneity, which is important in the case of immune markers that are often highly heterogeneously distributed within the tissue.

A further limitation of our study is that it included only pretreatment tumor samples, intending to identify potential predictive markers. However, the tumor microenvironment is dynamically changing during progression and therapies, and in order to explore the reasons for acquired resistance to immunotherapy, longitudinal analyses would be beneficial. In our previous investigation comparing HLA class I expression and immune cell infiltration in pre- and post-treatment tumor samples of melanoma patients receiving ipilimumab therapy, we found HLA-I downregulation in the progressing metastases of non-responding patients [53]. Our future plans include using this approach in a longitudinal study of patients in the present cohort treated with PD-1 inhibitors.

5. Conclusions

Summarizing our findings, they corroborate previous results, indicating the importance of tumor infiltration by $CD8^+$ T lymphocytes, $PD\text{-}1^+$ and $PD\text{-}L1^+$ cells, as well as immune cell subsets that are less frequently studied, such as $CD45RO^+$ memory T cells, $FOXP3^+$ regulatory T cells, $CD103^+$ tissue-resident T cells, $CD20^+$ B lymphocytes, in the efficacy of PD-1 blocking therapy. This study also proposes the density of T cells expressing CD134 as a potential new biomarker in a "real world" patient cohort. A coordinated high expression of the immune cell markers was also strongly associated with the treatment response. Furthermore, our results support earlier reports on the predictive value of tumor cell HLA class II expression and suggest the potential additive value of the HLA-I expression level.

Supplementary Materials: The following supporting information can be downloaded at https://www.mdpi.com/article/10.3390/biom14121609/s1, Table S1. Clinicopathological data of individual patients and heatmap of immunohistochemistry results.

Author Contributions: Conceptualization, A.L. and S.F.; investigation, B.H., K.G.C., T.B., K.B., E.T., A.R.N., A.M., C.D.M. and A.L.; formal analysis, T.B., G.F. and A.L.; writing—original draft preparation, A.L.; writing—review and editing, B.H., K.G.C., T.B. and G.F.; visualization, G.F. and A.L.; supervision, A.L.; funding acquisition, A.L. All authors have read and agreed to the published version of the manuscript.

Funding: This research was funded by the National Research, Development and Innovation Office grant NKFI ANN 128524 and the National Laboratories Program (National Tumor Biology Laboratory, 2022-2.1.1-NL-202200010).

Institutional Review Board Statement: The study was conducted in accordance with the Declaration of Helsinki, and approved by the Scientific and Ethical Committee of Medical Research Council, Hungary (12120-1/2019/EKU).

Informed Consent Statement: Informed consent from patients was not required by the board in the case of retrospective studies, where it is not possible to obtain consent from the majority of patients as most patients were deceased at the time of the study. In the case of non-interventional, retrospective studies, patients' informed consent is not required. https://njt.hu/jogszabaly/2009-235-20-22 (accessed on 12 December 2024).

Data Availability Statement: Data supporting the reported results can be found in the Supplementary Materials or provided by the corresponding author upon reasonable request.

Acknowledgments: The authors thank Katalin Derecskei and Tamás Ágoston Udvarhelyi (National Institute of Oncology, Budapest) for their excellent technical assistance.

Conflicts of Interest: The authors declare no conflicts of interest. The funders had no role in the design of the study; in the collection, analyses, or interpretation of data; in the writing of the manuscript; or in the decision to publish the results.

Note on Previous Publication: Part of this work was presented as a poster at EACR 2024, 10–13 June 2024, Rotterdam, The Netherlands [54].

References

1. Topalian, S.L.; Taube, J.M.; Anders, R.A.; Pardoll, D.M. Mechanism-driven biomarkers to guide immune checkpoint blockade in cancer therapy. *Nat. Rev. Cancer* **2016**, *16*, 275–287. [CrossRef]
2. Sharma, P.; Hu-Lieskovan, S.; Wargo, J.A.; Ribas, A. Primary, adaptive, and acquired resistance to cancer immunotherapy. *Cell* **2017**, *168*, 707–723. [CrossRef]
3. Monette, A.; Warren, S.; Barrett, J.C.; Garnett-Benson, C.; Shalper, K.A.; Taube, J.M.; Topp, B.; Snyder, A. Biomarker development for PD-(L)1 axis inhibition: A consensus view from the SITC Biomarkers Committee. *J. Immunother. Cancer* **2024**, *12*, e009427. [CrossRef]
4. Tumeh, P.C.; Harview, C.L.; Yearley, J.H.; Shintaku, I.P.; Taylor, E.J.M.; Robert, L.; Chmielowski, B.; Spasic, M.; Henry, G.; Ciobanu, V.; et al. PD-1 blockade induces responses by inhibiting adaptive immune resistance. *Nature* **2014**, *515*, 568–571. [CrossRef]
5. Taube, J.M.; Klein, A.; Brahmer, J.B.; Xu, H.; Pan, X.; Kim, J.H.; Chen, L.; Pardoll, D.M.; Topalian, S.L.; Anders, R.A. Association of PD-1, PD-1 ligands, and other features of the tumor immune microenvironment with response to anti-PD-1 therapy. *Clin. Cancer Res.* **2014**, *20*, 5064–5074. [CrossRef]
6. Ayers, M.; Lunceford, J.; Nebozhyn, M.; Murphy, E.; Loboda, A.; Kaufman, D.R.; Albright, A.; Cheng, J.D.; Kang, S.P.; Shankaran, V.; et al. IFN-γ-related mRNA profile predicts clinical response to PD-1 blockade. *J. Clin. Investig.* **2017**, *127*, 2930–2940. [CrossRef]
7. Sade-Feldman, M.; Jiao, Y.J.; Chen, J.H.; Rooney, M.S.; Barzily-Rokni, M.; Eliane, J.P.; Bjorgaard, S.L.; Hammond, M.R.; Vitzthum, H.; Blackmon, S.M.; et al. Resistance to checkpoint blockade therapy through inactivation of antigen presentation. *Nat. Commun.* **2017**, *8*, 1136. [CrossRef]
8. Topalian, S.L.; Sznol, M.; McDermott, D.F.; Kluger, H.M.; Carvajal, R.D.; Sharfman, W.H.; Brahmer, J.R.; Lawrence, D.P.; Atkins, M.B.; Powderly, J.D.; et al. Survival, durable tumor remission, and long-term safety in patients with advanced melanoma receiving nivolumab. *J. Clin. Oncol.* **2014**, *32*, 1020–1030. [CrossRef]
9. Larkin, J.; Chiarion-Sileni, V.; Gonzalez, R.; Grob, J.J.; Cowey, C.L.; Lao, C.D.; Schadendorf, D.; Dummer, R.; Smylie, M.; Rutkowski, P.; et al. Combined nivolumab and ipilimumab or monotherapy in untreated melanoma. *N. Engl. J. Med.* **2015**, *373*, 23–34. [CrossRef]
10. Ribas, A.; Hamid, O.; Daud, A.; Hodi, F.S.; Wolchok, J.D.; Kefford, R.; Joshua, A.M.; Patnaik, A.; Hwu, W.J.; Weber, J.S.; et al. Association of pembrolizumab with tumor response and survival among patients with advanced melanoma. *JAMA* **2016**, *315*, 1600–1609. [CrossRef]
11. Daud, A.I.; Wolchok, J.D.; Robert, C.; Hwu, W.J.; Weber, J.S.; Ribas, A.; Hodi, F.S.; Joshua, A.M.; Kefford, R.; Hersey, P.; et al. Programmed death ligand 1 expression and response to the anti-programmed death 1 antibody pembrolizumab in melanoma. *J. Clin. Oncol.* **2016**, *34*, 4102–4109. [CrossRef] [PubMed]
12. Morrison, C.; Pabla, S.; Conroy, J.M.; Nesline, M.K.; Glenn, S.T.; Dressman, D.; Papanicolau-Sengos, A.; Burgher, B.; Andreas, J.; Giamo, V.; et al. Predicting response to checkpoint inhibitors in melanoma beyond PD-L1 and mutational burden. *J. Immunother. Cancer* **2018**, *6*, 32. [PubMed]
13. Gide, T.N.; Quek, C.; Menzies, A.M.; Tasker, A.T.; Shang, P.; Holst, J.; Madore, J.; Lim, S.Y.; Velickovic, R.; Wongchenko, M.; et al. Distinct immune cell populations define response to anti-PD-1 monotherapy and anti-PD-1/anti-CTLA-4 combined therapy. *Cancer Cell* **2019**, *35*, 238–255. [CrossRef]
14. Placke, J.M.; Kimmig, M.; Griewank, K.; Herbst, R.; Terheyden, P.; Utikal, J.; Pföhler, C.; Ulrich, J.; Kreuter, A.; Mohr, P.; et al. Correlation of tumor PD-L1 expression in different tissue types and outcome of PD-1-based immunotherapy in metastatic melanoma—Analysis of the DeCOG prospective multicenter cohort study ADOREG/TRIM. *eBiomedicine* **2023**, *96*, 104774. [CrossRef]

15. Chen, P.L.; Roh, W.; Reuben, A.; Cooper, Z.A.; Spencer, C.N.; Prieto, P.A.; Miller, J.P.; Bassett, R.L.; Gopalakrishnan, V.; Wani, K.; et al. Analysis of immune signatures in longitudinal tumor samples yields insight into biomarkers of response and mechanisms of resistance to immune checkpoint blockade. *Cancer Discov.* **2016**, *6*, 827–837. [CrossRef]
16. Uryvaev, A.; Passhak, M.; Hershkovits, D.; Sabo, E.; Bar-Sela, G. The role of tumor-infiltrating lymphocytes (TILs) as a predictive biomarker of response to anti-PD1 therapy in patients with metastatic non-small cell lung cancer or metastatic melanoma. *Med. Oncol.* **2018**, *35*, 25. [CrossRef]
17. Wong, P.F.; Wei, W.; Smithy, J.W.; Acs, B.; Toki, M.I.; Blenman, K.R.M.; Zelterman, D.; Kluger, H.M.; Rimm, D.L. Multiplex quantitative analysis of tumor-infiltrating lymphocytes and immunotherapy outcome in metastatic melanoma. *Clin. Cancer Res.* **2019**, *25*, 2442–2449. [CrossRef]
18. Adegoke, N.A.; Gide, T.N.; Mao, Y.; Quek, C.; Patrick, E.; Carlino, M.S.; Lo, S.N.; Menzies, A.M.; Pires da Silva, I.; Vargara, I.A.; et al. Classification of the tumor immune microenvironment and associations with outcomes in patients with metastatic melanoma treated with immunotherapies. *J. Immunother. Cancer* **2023**, *11*, e007144. [CrossRef]
19. Daud, A.I.; Loo, K.; Pauli, M.L.; Sanchez-Rodriguez, R.; Munoz Sandoval, P.; Taravati, K.; Tsai, K.; Nosrati, A.; Nardo, L.; Alvarado, M.D.; et al. Tumor immune profiling predicts response to anti-PD1 therapy in human melanoma. *J. Clin. Investig.* **2016**, *126*, 3447–3452. [CrossRef]
20. Cai, L.; Michelakos, T.; Yamada, T.; Fan, S.; Wang, X.; Schwab, J.H.; Ferrone, C.R.; Ferrone, S. Defective HLA class I antigen processing machinery in cancer. *Cancer Immunol. Immunother.* **2018**, *67*, 999–1009. [CrossRef]
21. Zaretsky, J.M.; Garcia-Diaz, A.; Shin, D.S.; Escuin-Ordinas, H.; Hugo, W.; Hu-Lieskovan, S.; Torrejon, D.Y.; Abril-Rodriguez, G.; Sandoval, S.; Barthly, L.; et al. Mutations associated with acquired resistance to PD-1 blockade in melanoma. *N. Engl. J. Med.* **2016**, *375*, 819–829. [CrossRef] [PubMed]
22. Maggs, L.; Sadagopan, A.; Moghaddam, A.S.; Ferrone, S. HLA class I antigen processing machinery defects in antitumor immunity and immunotherapy. *Trends Cancer* **2021**, *7*, 1089–1101. [CrossRef] [PubMed]
23. Johnson, D.B.; Estrada, M.V.; Salgado, R.; Sanchez, V.; Doxie, D.B.; Opalenik, S.R.; Vilgelm, A.E.; Feld, E.; Johnson, A.S.; Greenplate, A.R.; et al. Melanoma-specific MHC-II expression represents a tumour-autonomous phenotype and predicts response to anti-PD-1/PD-L1 therapy. *Nat. Commun.* **2016**, *7*, 10582. [CrossRef] [PubMed]
24. Rodig, S.J.; Gusenleitner, D.; Jackson, D.G.; Gjini, E.; Giobbie-Hurder, A.; Jin, C.; Chang, H.; Lovitch, S.B.; Horak, C.; Weber, J.S.; et al. MHC proteins confer differential sensitivity to CTLA-4 and PD-1 blockade in untreated metastatic melanoma. *Sci. Transl. Med.* **2018**, *10*, eaar3342. [CrossRef] [PubMed]
25. Ladányi, A.; Papp, E.; Mohos, A.; Balatoni, T.; Liszkay, G.; Oláh, J.; Varga, A.; Lengyel, Z.; Emri, G.; Ferrone, S. Role of the anatomic site in the association of HLA class I antigen expression level in metastases with clinical response to ipilimumab therapy in patients with melanoma. *J. Immunother. Cancer* **2020**, *8*, e000209. [CrossRef]
26. Balatoni, T.; Mohos, A.; Papp, E.; Sebestyén, T.; Liszkay, G.; Oláh, J.; Varga, A.; Lengyel, Z.; Emri, G.; Gaudi, I.; et al. Tumor-infiltrating immune cells as potential biomarkers predicting response to treatment and survival in patients with metastatic melanoma receiving ipilimumab therapy. *Cancer Immunol. Immunother.* **2018**, *64*, 141–151. [CrossRef]
27. Wolchok, J.D.; Hoos, A.; O'Day, S.; Weber, J.S.; Hamid, O.; Lebbé, C.; Maio, M.; Binder, M.; Bohnsack, O.; Nichol, G.; et al. Guidelines for the evaluation of immune therapy activity in solid tumors: Immune-related response criteria. *Clin. Cancer Res.* **2009**, *15*, 7412–7420. [CrossRef]
28. Gonzalez-Ericsson, P.I.; Wulfkhule, J.D.; Gallagher, R.I.; Sun, X.; Axelrod, M.L.; Sheng, Q.; Luo, N.; Gomez, H.; Sanchez, V.; Sanders, M.; et al. Tumor-specific major histocompatibility-II expression predicts benefit to anti-PD-1/L1 therapy in patients with HER2-negative primary breast cancer. *Clin. Cancer Res.* **2021**, *27*, 5299–5306. [CrossRef]
29. Bartlett, E.K.; Fetsch, P.A.; Filie, A.C.; Abati, A.; Steinberg, S.M.; Wunderlich, J.R.; White, D.E.; Stephens, D.J.; Marincola, F.M.; Rosenberg, S.A.; et al. Human melanoma metastases demonstrate nonstochastic site-specific antigen heterogeneity that correlates with T-cell infiltration. *Clin. Cancer Res.* **2014**, *20*, 2607–2616. [CrossRef]
30. Gaida, M.M.; Welsch, T.; Herpel, E.; Tschaharganeh, D.F.; Fischer, L.; Schirmacher, P.; Hänsch, G.M.; Bergmann, F. MHC class II expression in pancreatic tumors: A link to intratumoral inflammation. *Virchows Arch.* **2012**, *460*, 47–60. [CrossRef]
31. Kitano, S.; Tsuji, T.; Liu, C.; Hirschhorn-Cymerman, D.; Kyi, C.; Mu, Z.; Allison, J.P.; Gnjatic, S.; Yuan, J.D.; Wolchok, J.D. Enhancement of tumor-reactive cytotoxic $CD4^+$ T cell responses after ipilimumab treatment in four advanced melanoma patients. *Cancer Immunol. Res.* **2013**, *1*, 235–244. [CrossRef] [PubMed]
32. Axelrod, M.L.; Cook, R.S.; Johnson, D.B.; Balko, J.M. Biological consequences of MHC-II expression by tumor cells in cancer. *Clin. Cancer Res.* **2019**, *25*, 2392–2402. [CrossRef] [PubMed]
33. Pyke, R.M.; Thompson, W.K.; Salem, R.M.; Font-Burgada, J.; Zanetti, M.; Carter, H. Evolutionary pressure against MHC class II binding cancer mutations. *Cell* **2018**, *175*, 416–428. [CrossRef] [PubMed]
34. Ott, P.A.; Hu-Lieskovan, S.; Chmielowski, B.; Govindan, R.; Naing, A.; Bhardwaj, N.; Margolin, K.; Awad, M.M.; Hellmann, M.D.; Lin, J.J.; et al. A phase Ib trial of personalized neoantigen therapy plus anti-PD-1 in patients with advanced melanoma, non-small cell lung cancer, or bladder cancer. *Cell* **2020**, *183*, 347–362. [CrossRef]
35. Hall, M.S.; Teer, J.K.; Yu, X.; Branthoover, H.; Snedal, S.; Rodriguez-Valentin, M.; Nagle, L.; Scott, E.; Schachner, B.; Innamarato, P.; et al. Neoantigen-specific $CD4^+$ tumor-infiltrating lymphocytes are potent effectors identified within adoptive cell therapy products for metastatic melanoma patients. *J. Immunother. Cancer* **2023**, *11*, e007288. [CrossRef]

36. Riaz, N.; Havel, J.J.; Makarov, V.; Desrichard, A.; Urba, W.J.; Sims, J.S.; Hodi, F.S.; Martín-Algarra, S.; Mandal, R.; Sharfman, W.H.; et al. Tumor and microenvironment evolution during immunotherapy with nivolumab. *Cell* **2017**, *171*, 934–949. [CrossRef]
37. Edwards, J.; Wilmott, J.S.; Madore, J.; Gide, T.N.; Quek, C.; Tasker, A.; Ferguson, A.; Chen, J.; Hewavisenti, R.; Hersey, P.; et al. $CD103^+$ tumor-resident $CD8^+$ T cells are associated with improved survival in immunotherapy-naïve melanoma patients and expand significantly during anti-PD-1 treatment. *Clin. Cancer Res.* **2018**, *24*, 3036–3045. [CrossRef]
38. Ladányi, A.; Somlai, B.; Gilde, K.; Fejös, Z.; Gaudi, I.; Tímár, J. T-cell activation marker expression on tumor-infiltrating lymphocytes as prognostic factor in cutaneous malignant melanoma. *Clin. Cancer Res.* **2004**, *10*, 521–530. [CrossRef]
39. Petty, J.K.; He, K.; Corless, C.L.; Vetto, J.T.; Weinberg, A.D. Survival in human colorectal cancer correlates with expression of the T-cell costimulatory molecule OX-40 (CD134). *Am. J. Surg.* **2002**, *183*, 512–518. [CrossRef]
40. Massarelli, E.; Lam, V.K.; Parra, E.R.; Rodriguez-Canales, J.; Behrens, C.; Diao, L.; Wang, J.; Blando, J.; Byers, L.A.; Yanamandra, N.; et al. High OX-40 expression in the tumor immune infiltrate is a favorable prognostic factor of overall survival in non-small cell lung cancer. *J. Immunother. Cancer* **2019**, *7*, 351. [CrossRef]
41. Amaria, R.N.; Reddy, S.M.; Tawby, H.A.; Davies, M.A.; Ross, M.I.; Glitza, I.C.; Cormier, J.N.; Lewis, C.; Hwu, W.J.; Hanna, E.; et al. Neoadjuvant immune checkpoint blockade in high-risk resectable melanoma. *Nat. Med.* **2018**, *24*, 1649–1654. [CrossRef] [PubMed]
42. Nelson, B.H. $CD20^+$ B cells: The other tumor-infiltrating lymphocytes. *J. Immunol.* **2010**, *185*, 4977–4982. [CrossRef] [PubMed]
43. Flynn, N.J.; Somasundaram, R.; Arnold, K.M.; Sims-Mourtada, J. The multifaceted roles of B cells in solid tumors: Emerging treatment opportunities. *Targerted Oncol.* **2017**, *12*, 139–152. [CrossRef] [PubMed]
44. Linton, P.J.; Bautista, B.; Biederman, E.; Bradley, E.S.; Harbertson, J.; Kondrack, R.M.; Padrick, R.C.; Bradley, L.M. Costimulation via OX40L expressed by B cells is sufficient to determine the extent of primary CD4 cell expansion and Th2 cytokine secretion in vivo. *J. Exp. Med.* **2003**, *197*, 875–883. [CrossRef] [PubMed]
45. Ladányi, A.; Kiss, J.; Mohos, A.; Somlai, B.; Liszkay, G.; Gilde, K.; Fejős, Z.; Gaudi, I.; Dobos, J.; Tímár, J. Prognostic impact of B-cell density in cutaneous melanoma. *Cancer Immunol. Immunother.* **2011**, *60*, 1729–1738. [CrossRef]
46. Lu, S.; Stein, J.E.; Rimm, D.L.; Wang, D.W.; Bell, J.M.; Johnson, D.B.; Sosman, J.A.; Schalper, K.A.; Anders, R.A.; Wang, H.; et al. Comparison of biomarker modalities for predicting response to PD-1/PD-L1 checkpoint blockade: A systematic review and meta-analysis. *JAMA Oncol.* **2019**, *5*, 1195–1204. [CrossRef]
47. Berry, S.; Giraldo, N.A.; Green, B.F.; Cottrell, T.R.; Stein, J.E.; Engle, E.L.; Xu, H.; Ogurtsova, A.; Roberts, C.; Wang, D.; et al. Analysis of multispectral imaging with the AstroPath platform informs efficacy of PD-1 blockade. *Science* **2021**, *372*, eaba2609. [CrossRef]
48. Attrill, G.H.; Owen, C.N.; Ahmed, T.; Vergara, I.A.; Colebatch, A.J.; Conway, J.W.; Nahar, K.J.; Thompson, J.F.; Pires da Silva, I.; Carlino, M.S.; et al. Higher proportions of CD39+ tumor-resident cytotoxic T cells predict recurrence-free survival in patients with stage III melanoma treated with adjuvant immunotherapy. *J. Immunother. Cancer* **2022**, *10*, e004771. [CrossRef]
49. Toki, M.I.; Merritt, C.R.; Wong, P.F.; Smithy, J.W.; Kluger, H.M.; Syrigos, K.N.; Ong, G.T.; Warren, S.E.; Beechem, J.M.; Rimm, D.L. High-plex predictive marker discovery for melanoma immunotherapy-treated patients using digital spatial profiling. *Clin. Cancer Res.* **2019**, *25*, 5503–5512. [CrossRef]
50. Herbst, R.S.; Soria, J.C.; Kowanetz, M.; Fine, G.D.; Hamid, O.; Gordon, M.S.; Sosman, J.A.; McDermott, D.F.; Powderly, J.D.; Gettinger, S.N.; et al. Predictive correlates of response to the anti-PD-L1 antibody MPDL3280A in cancer patients. *Nature* **2014**, *515*, 563–567. [CrossRef]
51. Tang, H.; Liang, Y.; Anders, R.A.; Taube, J.M.; Qiu, X.; Mulgaonkar, A.; Liu, X.; Harrington, S.M.; Guo, J.; Xin, Y.; et al. PD-L1 on host cells is essential for PD-L1 blockade-mediated tumor regression. *J. Clin. Investig.* **2018**, *128*, 580–588. [CrossRef] [PubMed]
52. Lin, H.; Wei, S.; Hurt, E.M.; Green, M.D.; Zhao, L.; Vatan, L.; Szeliga, W.; Herbst, R.; Harms, P.W.; Fecher, L.A.; et al. Host expression of PD-L1 determines efficacy of PD-L1 pathway blockade-mediated tumor regression. *J. Clin. Investig.* **2018**, *128*, 805–815. [CrossRef] [PubMed]
53. Ladányi, A.; Hegyi, B.; Balatoni, T.; Liszkay, G.; Rohregger, R.; Waldnig, C.; Dudás, J.; Ferrone, S. HLA class I downregulation in progressing metastases of melanoma patients treated with ipilimumab. *Pathol. Oncol. Res.* **2022**, *28*, 1610297. [CrossRef] [PubMed]
54. Hegyi, B.; Csikó, K.G.; Balatoni, T.; Neumark, A.R.; Bőcs, K.; Tóth, E.; Fröhlich, G.; Liszkay, G.; Ferrone, S.; Ladányi, A. Tumor-infiltrating immune cells and HLA expression as potential biomarkers predicting response to PD-1 inhibitor therapy in stage IV melanoma. *Mol. Oncol.* **2024**, *18* (Suppl. S1), 1609. [CrossRef]

biomolecules

MDPI

Article

Non-Coding RNAs: lncRNA, piRNA, and snoRNA as Robust Plasma Biomarkers of Alzheimer's Disease

Ruomin Xin [1,2], Elizabeth Kim [1,2], Wei Tse Li [1,2,3], Jessica Wang-Rodriguez [4,5] and Weg M. Ongkeko [1,2,*]

1 Department of Otolaryngology—Head and Neck Surgery, University of California, La Jolla, San Diego, CA 92093, USA; rxin@ucsd.edu (R.X.); harrison.li@ucsf.edu (W.T.L.)
2 Research Service, VA San Diego Healthcare System, San Diego, CA 92161, USA
3 School of Medicine, University of California, San Francisco, CA 94143, USA
4 Department of Pathology, University of California, La Jolla, San Diego, CA 92093, USA; jwrodriguez@ucsd.edu
5 Pathology Service, VA San Diego Healthcare System, San Diego, CA 92161, USA
* Correspondence: rongkeko@health.ucsd.edu; Tel.: +1-(858)-552-8585 (ext. 7165)

Abstract: Alzheimer's disease (AD) is a leading cause of dementia worldwide. As current diagnostic approaches remain limited in sensitivity and accessibility, there is a critical need for novel, non-invasive biomarkers aiding early detection. Non-coding RNAs (ncRNAs), including long non-coding RNAs (lncRNAs), PIWI-interacting RNAs (piRNAs), and small nucleolar RNAs (snoRNAs), have emerged as promising candidates due to their regulatory roles in gene expression and association with diseases. In this study, we systematically profiled ncRNA expression from RNA sequencing data of 48 AD and 22 control blood tissue samples, aiming to evaluate their utility as biomarkers for AD classification. Differential expression analysis revealed widespread dysregulation of lncRNAs and piRNAs, with over 5000 lncRNAs and nearly 1000 piRNAs significantly upregulated in AD. Weighted gene co-expression network analysis (WGCNA) identified multiple ncRNA modules associated with the AD phenotype. Using supervised machine learning approaches, we evaluated the diagnostic potential of ncRNA expression profiles, including single-gene, multi-gene, and module-level models. Random Forest models trained on individual genes identified 121 ncRNAs with AUROC > 0.8. Feature importance analysis emphasized ncRNAs such as lnc-MYEF2-3, lnc-PRKACB2, and HBII-115 as major contributors to diagnostic accuracy. These findings support the potential of ncRNA signatures as reliable and non-invasive biomarkers for AD diagnosis.

Keywords: Alzheimer's disease; non-coding RNA; disease diagnostics; transcriptomics; gene regulation; liquid biopsy; biomarkers

1. Introduction

Alzheimer's disease (AD) is the seventh leading cause of death, affecting over 6 million individuals in the United States alone [1]. Influenced by multiple risk factors such as age, genetics, lifestyle, cardiovascular health, and history of head trauma, the current understanding of AD's complex pathogenesis is still limited, posing significant challenges to the development of effective AD treatment [2]. The existing U.S. Food and Drug Administration (FDA) approved prescription drugs for AD work towards alleviating the symptoms of AD rather than curing the disease [2]. Popular drugs, including cholinesterase inhibitors, work best only for people in the earlier stages of AD, thus highlighting the critical need for early AD diagnosis to enable effective preventive care [3].

Current methods of diagnosis utilize medical history, physical and mental status examination, neuroimaging, and cerebrospinal fluid (CSF) biomarker tests [3]. The more accurate methods of neuroimaging and CSF tests are often only administered after the initial assessments using behavioral tests. However, behavior observations are prone to misinterpretations and demonstrate underestimation of symptoms when patients become familiar with specific diagnosis tasks, prompting the search for a reliable and non-invasive early diagnostic tool [4]. Recently, Barthélemy et al. constructed a plasma %p-tau217 test that had an area under the curve (AUC) value between 0.95 and 0.98, which was equivalent to the predictive strength of FDA-approved CSF tests (AUC = 0.95–0.97) [5]. While further research is still required to validate this finding, the non-invasive nature of blood-based tests and promising capabilities in AD detection place this method at the center of an investigation to unravel more reliable biomarkers.

Past investigation efforts in AD pathology have focused on amyloid plaques and the development of neurofibrillary tangles. A previously understudied class of biomolecules, non-coding RNAs (ncRNAs), is increasingly being studied for their role in AD progression. Small nucleolar RNAs (snoRNAs), a type of ncRNA, participate in rRNA modification [6]. Studies have linked the dysregulation of snoRNAs to cognitive decline in AD due to their regulatory role in protein synthesis and RNA processing [7,8]. Differential expressions of snoRNAs were observed in plasma extracellular vesicles, providing an accessible type of diagnostic biomarker [7]. Furthermore, in mouse models, dysregulation of snoRNAs was identified before the formation of amyloid plaques, suggesting the promising potential of snoRNA in early diagnosis of AD [9].

PIWI-interacting RNAs (piRNAs), another type of ncRNA, play a role in transposon silencing and may exhibit transgenerational inheritance [10]. Studies reveal more than 103 differentially expressed piRNAs in AD brain tissue and also showed that most of these piRNAs correlated with AD risk SNPs and AD-associated signaling pathways [11,12]. Additionally, genome-wide association studies have revealed a correlation between risk variants and piRNA expression [13]. A combined signature of microRNAs and piRNAs offers a promising strategy for early diagnosis of AD, yet larger and more diverse cohorts are needed to validate the findings [14].

Long non-coding RNAs (lncRNAs), transcripts longer than 200 nucleotides, are another class of ncRNAs that regulate translation, metabolism, and cellular signaling [15]. Studies have shown that dysregulation of specific lncRNAs in AD may serve as diagnostic biomarkers [13,16]. In a study identifying 11 differentially expressed lncRNAs in AD blood samples, the RN7SK LncRNA had a sensitivity and specificity of 50% and 84%, respectively (AUC = 0.658, p = 0.0063) [16]. In AD brain samples, BIN1 has also been shown to transcriptionally modulate mRNA expression and influence the progression of AD [13]. In addition, levels of lncRNA BACE1 in plasma were significantly elevated with AD, showing an AUC of 0.667 [17]. Overall, existing studies underscore the role of snoRNAs, piRNAs, and lncRNAs as novel biomarkers for the early diagnosis of AD.

Hence, this study aims to evaluate the potential of ncRNA expression profiles as predictive biomarkers for AD. RNA sequencing data of AD and control blood samples were downloaded from the NCBI Sequence Read Archive (SRA). The dataset was analyzed to quantify expression levels of lncRNA, piRNA, and snoRNA. Weighted gene co-expression network analysis revealed co-expression modules correlating strongly with AD. Furthermore, ncRNA expression profiles were used to train supervised machine-learning algorithms. A nested cross-validation approach was used to benchmark model performances. We evaluated the diagnostic potential of ncRNA expression profiles, including single-gene, multi-gene, and module-level models. Through this investigation, we provide compelling evidence that ncRNA signatures can serve as reliable biomarkers for predicting AD, potentially offering novel insights into disease progression and therapeutic targets of AD.

2. Materials and Methods

2.1. Data Acquisition

RNA sequencing data were obtained from a publicly available dataset from the NCBI SRA database (https://www.ncbi.nlm.nih.gov/sra) under project accession SRP022043 (https://www.ncbi.nlm.nih.gov/Traces/study/?acc=SRP022043&o=acc_s:a) (accessed on 3 March 2025) and SRP288246 (https://www.ncbi.nlm.nih.gov/bioproject/PRJNA670793) (accessed on 20 May 2025) [18,19]. The main study includes 70 individuals, comprising 48 Alzheimer's disease patients and 22 healthy controls. Plasma samples were sequenced using the Illumina HiSeq 2000 platform (San Diego, CA, USA) in a single-end format. An additional 50 samples, including 25 diseased patients and 22 healthy controls, were processed similarly from the secondary dataset. Postmortem brain samples were sequenced using the Illumina HiSeq 4000 platform in single-end format. FASTQ files were downloaded and extracted using SRA Toolkit v3.2.0 [20].

2.2. Quality Control and Preprocessing

Quality control was performed using FastQC v0.12.1 to assess sequencing read quality [21]. Samples were assessed based on per-base sequence quality, per-sequence quality, sequencing content, overrepresented sequences, and adapter content. Samples SRR837499, SRR837500, and SRR837504 were discarded due to low average per-sequence Phred quality scores. The remainder samples were trimmed using cutadapt v1.18 to remove adapter sequences and to filter reads using a Phred score threshold of 20 and a read length filter of 18 base pairs [22].

2.3. Alignment to Reference Genome and Feature Annotation

Trimmed reads were aligned to the human reference genome (hg38) using STAR v2.7.2b [23]. The resulting BAM files were sorted by genomic coordinates and mapped to the annotated ncRNA feature from the DASHR v2.0 annotation database for lncRNAs, piRNAs, and snoRNAs [24]. Bedtools v2.26.0 was used to calculate coverage per feature, resulting in raw read counts for downstream analysis [25].

2.4. Differential Expression Analysis and WGCNA

Differential expression analysis was performed using DESeq2 v1.46.0 in R [26]. Normalization and dispersion estimation were performed, and differentially expressed genes (DEGs) in AD samples were identified with a threshold of log2 fold change ≥ 0.6 and adjusted p-value ≤ 0.05. Weighted gene co-expression network analysis (WGCNA) was performed using WGCNA 1.73 to identify ncRNA co-expression modules associated with Alzheimer's pathology [27]. Module detection and eigengene computation were conducted using the blockwise modules function with a soft-threshold power of 12, a minimum module size of 30, and hierarchical clustering with dynamic tree cutting. Hypergeometric testing was performed to evaluate the statistical significance of gene overlap between the discovery and validation datasets based on the total number of genes tested and the number of differentially expressed genes in each dataset.

2.5. Machine-Learning Models and Feature Importance

To classify Alzheimer's disease samples based on gene expression profiles, we trained supervised machine-learning Random Forest (RF) models using the Python module scikit-learn 1.6.1 [28]. Models were trained using 70% of the data and evaluated on the remaining 30% using stratified fold cross-validation.

2.6. Pairwise Correlation Analysis and Heatmap Generation

Pairwise Spearman correlation analysis was performed on normalized expression values of top predictive ncRNAs to assess co-expression patterns. Using R (with tidyverse v2.0.0, pheatmap v1.0.12), correlation coefficients and *p*-values were calculated for all gene pairs. Significant correlations ($p \leq 0.05$) were annotated using asterisks [29,30].

3. Results

RNA-seq data from 70 samples (48 AD, 22 controls) were obtained. After quality assessment with FastQC and removal of three low-quality samples (SRR837499, SRR837500, SRR837504), reads were trimmed using cutadapt and aligned to the human genome (hg38) using STAR. Mapped reads were annotated using DASHR v2.0 to quantify ncRNA species (lncRNAs, piRNAs, snoRNAs) with Bedtools. Differential expression analysis was performed with DESeq2, followed by weighted gene co-expression network analysis (WGCNA) to identify modules associated with Alzheimer's disease. Machine-learning models were trained to classify AD vs. control samples and identify key diagnostic features. Model performance was validated using an independent external dataset comprising 50 samples (25 AD, 25 controls) (Figure 1).

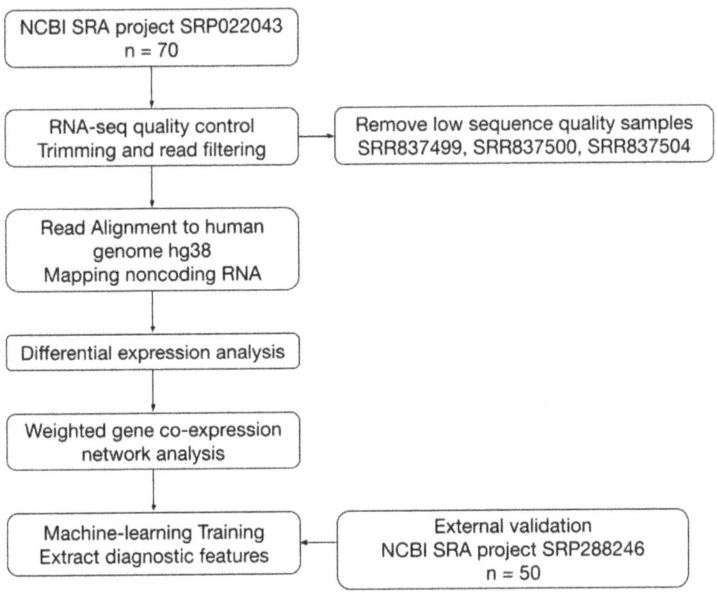

Figure 1. Workflow of the study. Flowchart demonstrates overall methodology employed in this study.

3.1. Differential Expression Analysis of ncRNA

Using DESeq2 (adjusted *p*-value < 0.05), we revealed substantial dysregulation of ncRNA transcripts across 48 AD and 22 control blood samples. A total of 5164 lncRNA transcripts were significantly upregulated, while 290 lncRNA transcripts were significantly downregulated in AD samples (Figure 2A,D). The most strongly upregulated lncRNA—lnc-ARL4C-09—displayed approximately a 250-fold increase ($\log_2$ fold change = 7.98), whereas the most strongly downregulated lncRNA—lnc-CTNNBL1-3—displayed approximately a 260-fold reduction ($\log_2$ fold change = −8.05) in AD. A total of 973 piRNAs were significantly upregulated in AD, while 26 piRNAs were downregulated (Figure 2B,D). Aside from the smaller set of differentially expressed genes, the most significantly dysregulated piRNAs also demonstrate changes in lower magnitudes. The most significantly upregulated piRNA—piR-47795—is enriched approximately 31-fold ($\log_2$ fold change = 4.97) in AD. Similarly, the most significantly downregulated piRNA—piR-48801—was reduced by around 6-fold in AD ($\log_2$FC = −2.48). The fewest number of differentially expressed genes was observed in the class of snoRNA. Analysis revealed six significantly upregulated and one significantly downregulated snoRNA in AD. The most notable changes were observed for HBII-115 ($\log_2$FC = 1.67) and HBII-55 ($\log_2$FC = −1.29), indicating modest yet consistent expression shifts in this class. These findings indicate widespread ncRNA dysregulation in AD, with lncRNAs and piRNAs showing dominant upregulation patterns and a broader range of fold changes compared to snoRNAs.

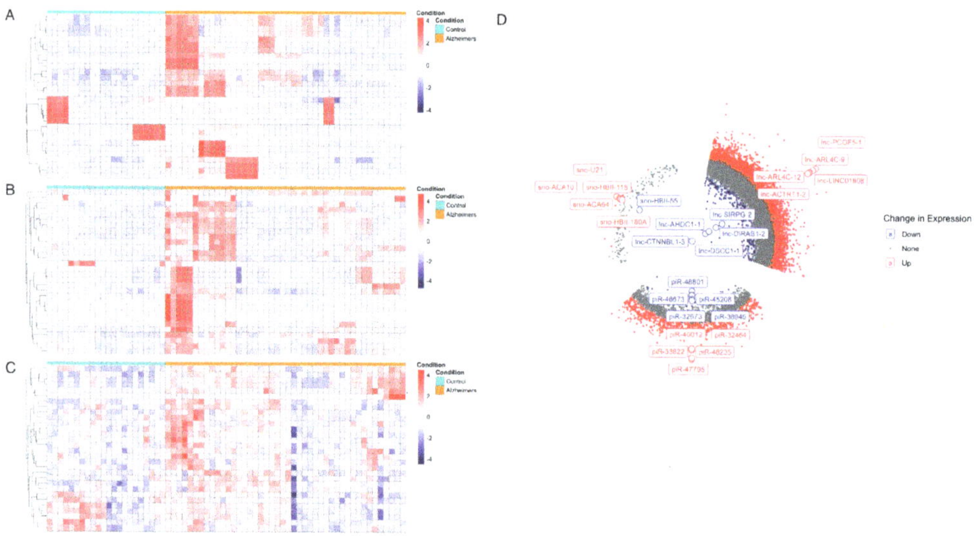

Figure 2. Differential expression of ncRNAs and identification of key features in AD. (**A–C**) Heatmaps depicting the top 100 differentially expressed (**A**) lncRNA, (**B**) piRNA, and (**C**) snoRNA across Alzheimer's (orange) and control (cyan). Normalized expression values (log2-transformed) are expressed on the color scale, where red represents upregulation and blue represents downregulation of expression. (**D**) Circular volcano plot showing the distribution of gene expression across the three classes of ncRNA. Genes with significant upregulation (log2FC $\geq$ 0.6, padj < 0.05) are shown in red, while significantly downregulated genes are shown in blue. Non-significant genes are plotted in grey. The top 5 most differentially upregulated and top 5 most differentially downregulated genes for each ncRNA class are annotated.

3.2. Weighted Gene Co-Expression Network Analysis

To investigate the modular organization of non-coding RNA expression in Alzheimer's disease, we performed weighted gene co-expression network analysis (WGCNA) using variance-stabilized expression values. Samples were first hierarchically clustered using pairwise Euclidean distances to identify intra-dataset heterogeneity (Figure 3A). A static cut height was set visually at 85 to remove sample SRR837506 as an outlier before network construction.

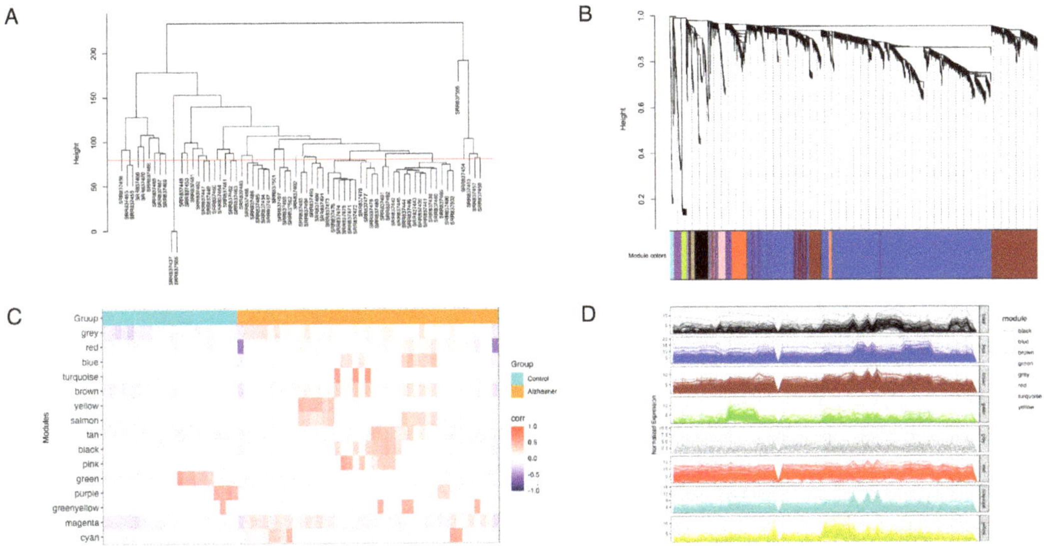

Figure 3. Weighted gene co-expression network analysis (WGCNA) of ncRNAs in AD versus control. (**A**) Sample clustering dendrogram showing hierarchical clustering of samples to detect outliers and assess sample similarity. Samples with similar expression profiles are clustered together based on Euclidean distance. (**B**) Gene clustering dendrogram and module detection illustrate hierarchical clustering of genes based on topological overlap. Module colors are assigned dynamically using the dynamic tree-cut method. Genes with similar expression patterns are grouped into co-expression modules. Modules such as turquoise, blue, and brown contain a high density of correlated genes. (**C**) Module–trait relationship heatmap showing the association between gene modules and sample conditions of Alzheimer's disease vs. control. The color scale represents correlation coefficients, where red indicates positive correlation and blue indicates negative correlation. The top modules (e.g., turquoise, blue, brown, and red) show distinct correlation patterns with Alzheimer's disease. (**D**) Module gene expression profiles line plots showing normalized expression of ncRNA in the selected modules (black, blue, brown, green, grey, red, turquoise, and yellow). Individual lines represent the expression profile of an individual gene across samples.

Following hierarchical clustering and dynamic tree cutting (Figure 3B), we identified 15 co-expression modules represented by distinct colors. The largest modules included the turquoise module (n = 2261 genes) and the blue module (n = 2046 genes), followed by the brown (n = 783), yellow (n = 425), and green (n = 155) modules (Table 1). Following module identification, a module–trait relationship heatmap across selected major modules was performed to assess the association of modules with distinct cohorts (Figure 3C). Expression levels of single genes within each module were plotted across samples to further demonstrate the consistency of expression profiles in AD relative to control. These findings highlight the presence of coordinated ncRNA expression modules that are associated with AD's pathology.

Table 1. Co-expression modules and relative module size by gene count.

Co-Expression Modules (Color)	Module Size (Gene Count)
Turquoise	2261
Blue	2046
Brown	783
Yellow	425
Green	155
Red	149
Grey	139
Black	136
Pink	71
Magenta	70
Purple	67
Green-yellow	48
Tan	40
Salmon	39
Cyan	30

3.3. Supervised Machine-Learning Prediction of Alzheimer's Disease Using ncRNA Expression Profiles

The diagnostic performances of 6459 differentially expressed genes were assessed using a Random Forest (RF) classifier (Figure 4A). Models were trained using single gene expression profiles and evaluated using 5-fold cross-validation to assess the standalone predictive strength of ncRNA. Results revealed substantial variation in the predictive power of individual ncRNA, with an average area under the ROC curve (AUC) of 0.620 $\pm$ 0.099 and 121 differentially expressed genes exceeding an AUC of 0.8. The top five performing candidate genes of each ncRNA class were used to construct multi-gene expression panels for the comprehensive evaluation of the classification performance of different ncRNA subtypes. We performed 10-fold, 100-repeat nested cross-validation using a Random Forest classifier. Across all iterations, the lncRNA panel yielded the highest average AUROC (Mean = 0.974, Median = 1.000), followed closely by the combined ncRNA panel (Mean = 0.972), indicating strong discriminative capability (Figure 4B,D). The piRNA panel also achieved high performance (Mean AUROC = 0.925), while the snoRNA panel showed more variability and overall lower predictive strength (Mean AUROC = 0.802). The combined model notably demonstrated high precision (Mean = 0.960) and sensitivity (Mean = 0.918). Feature importance ranking indicates that snoRNAs have higher importance in the combined ncRNA panel prediction, despite weaker standalone predictive strength and snoRNA panel predictive strength (Figure 4C).

Pairwise correlation analysis was conducted to investigate the relationships among the top predictive ncRNAs. Hierarchical clustering revealed two primary groups of ncRNAs with distinct co-expression patterns (Figure 5). One cluster, consisting of lnc-MYEF2-3, lnc-SI-3, and lnc-TSPYL6-1, exhibited strong positive correlations, suggesting coordinated expression. In contrast, another cluster comprising lnc-PRKACB-2, piR-59350, and piR-32371 showed weaker and more variable correlations, indicative of divergent expression profiles. An intermediate group of ncRNAs also formed a positively correlated cluster, suggesting partial co-expression behavior. These clustering patterns highlight the presence of diverse regulatory dynamics among ncRNAs in Alzheimer's disease.

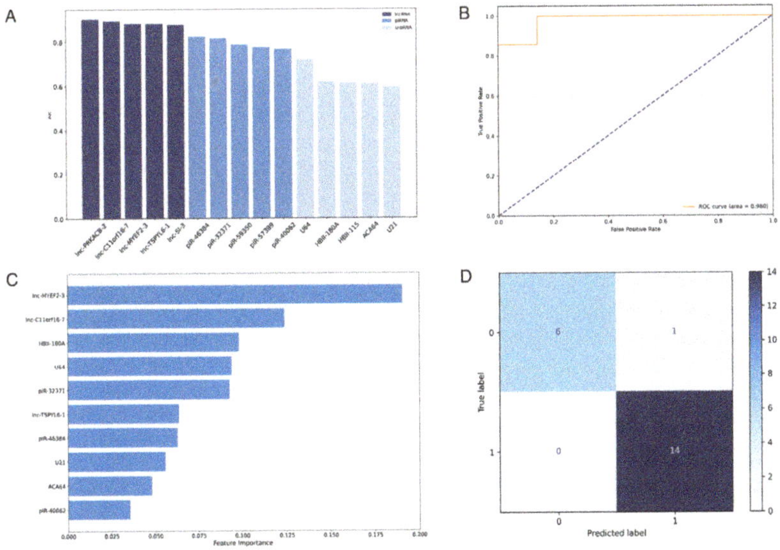

Figure 4. Predictive performance and feature contribution of ncRNA panel for AD diagnosis. (**A**) Area under the curve (AUC) scores of individual non-coding RNA (ncRNA) biomarkers, categorized by RNA type, based on performance in a Random Forest (RF) classification model. (**B**) Receiver-operating characteristic (ROC) curve of the 15-ncRNA panel, demonstrating strong classification performance with an AUC of 0.980. (**C**) Feature importance plot from the trained RF model, highlighting the relative contribution of each ncRNA to predictive accuracy. (**D**) Confusion matrix of model performance on test set, showing high precision (0.933) and perfect recall (1.000) (Label 1 = AD, Label 0 = control).

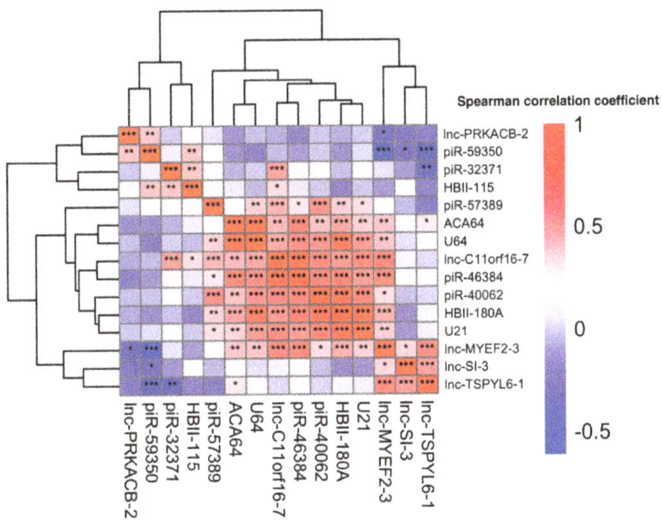

Figure 5. Hierarchical clustering heatmap of pairwise Spearman correlation coefficients among top predictive ncRNAs in Alzheimer's disease. The heatmap displays correlation strength ranging from −1 to +1. Clustering analysis revealed two major co-expression groups: one cluster including lnc-MYEF2-3, lnc-SI-3, and lnc-TSPYL6-1 with strong positive correlations, and another cluster including lnc-PRKACB-2, piR-59350, and piR-32371 with weaker, more variable correlations. Statistical significance of pairwise correlations is annotated using asterisks: $p \leq 0.05$ (*), $p \leq 0.01$ (**), and $p \leq 0.001$ (***).

An alternative Random Forest classifier was trained using the averaged expression profile of 15 co-expression modules constructed through WGCNA. Similarly, this approach was evaluated using 10-fold, 100-repeat nested cross-validation. Feature importance analysis identified a relatively uniform distribution of module importance in establishing the predictive model (Figure 6A). The ROC curve for the final combined model achieved a perfect classification performance with an AUC of 1.000 (Figure 6B), demonstrating the strong discriminative capacity. The confusion matrix portrays a similar high classification accuracy, with 14/14 AD cases and 5/7 controls correctly predicted, corresponding to a precision of 0.875 and a recall of 1.000 for the AD class (Figure 6C). Across iterated cross-validations, module-based prediction demonstrates a narrower distribution of AUC, precision, and recall, indicating greater model consistency.

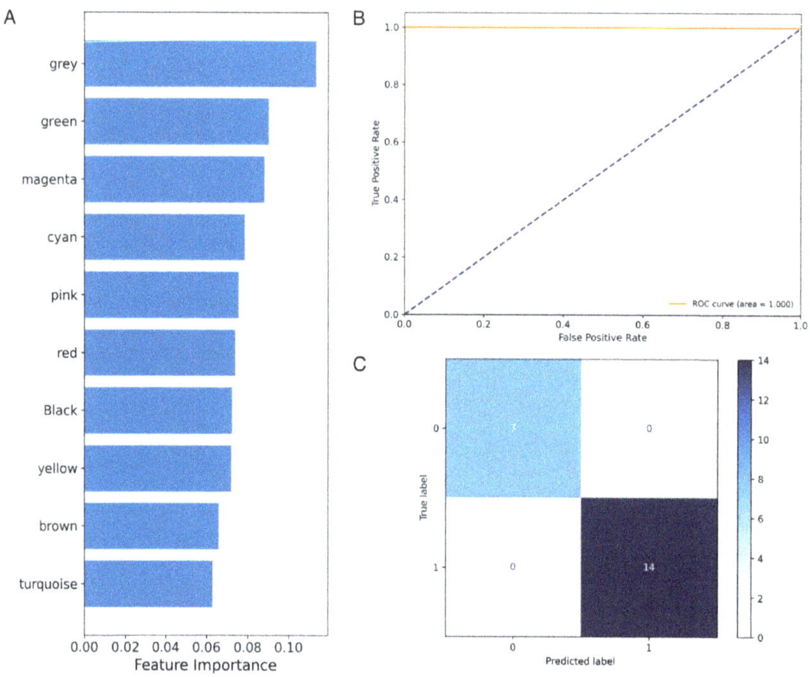

Figure 6. Predictive performance and feature contribution of ncRNA expression modules for Alzheimer's disease diagnosis. (**A**) Feature importance plot of gene co-expression modules (denoted by module color) from the trained Random Forest (RF) model. (**B**) Receiver operating characteristic (ROC) curve showing perfect classification performance with an AUC of 1.000. (**C**) Confusion matrix of predictions on test set, indicating high precision (0.875) and perfect recall (1.000).

To validate the consistency of our differentially expressed ncRNA findings, we analyzed an independent RNA-sequencing dataset (SRP288246) derived from postmortem brain tissue. Differential expression analysis identified 32 ncRNAs that overlapped between our original dataset and the external validation set (Figure 7). A hypergeometric test yielded a p-value of 2.15×10^{-7}, indicating that the observed overlap is highly significant and unlikely to occur by chance. This result supports the robustness and transferability of our findings across independent datasets.

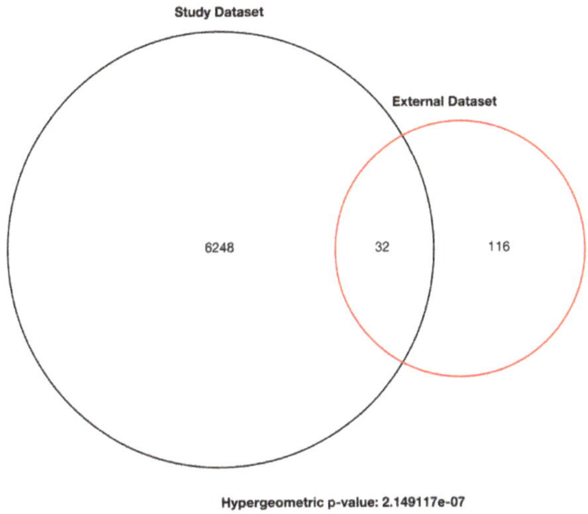

Figure 7. Venn diagram displaying the number of differentially expressed genes common to both the original study dataset and the validation dataset. A total of 32 ncRNAs were found to be commonly expressed in both studies. Statistical testing was performed using the hypergeometric test.

4. Discussion

Our comprehensive analysis revealed significant dysregulation of ncRNAs in the plasma of AD patients compared to controls. Specifically, we identified 5164 upregulated and 290 downregulated lncRNAs, 973 upregulated and 26 downregulated piRNAs, and 6 upregulated and 1 downregulated snoRNAs. The relative abundance of each ncRNA class among the differentially expressed genes mirrors their baseline abundance observed during transcript quantification. This proportionality suggests that the representation of each RNA class is not skewed but rather reflects their natural abundance and expression dynamics in plasma.

ncRNAs have been investigated in multiple prior studies as both biomarkers for AD detection and potential drivers of AD. In a systematic review and meta-analysis of circulating lncRNAs as biomarkers for AD diagnosis, the pooled AUC was 0.86, with a total sample size of over 553 AD patients and 513 controls [31]. Our lncRNA panel and combined ncRNA panel achieved significantly higher AUCs of 0.974 and 0.972, respectively. In a more recent study comprising 192 patients, a six-miRNA signature was only able to achieve an AUC of 0.733 [31]. Additionally, our ncRNA panel demonstrates better performance than previously reported combined miRNA-piRNA signature (AUC = 0.83) to distinguish AD and piRNA signature (AUC = 0.86) to predict the conversion of mild cognitive impairment to AD [14]. To the best of our knowledge, snoRNAs have not been used as a diagnostic signature in past studies. The most accurate plasma AD test, a plasma p-tau217 test, is comparable to established CSF assays and reached AUC values between 0.95 and 0.98 [5]. While protein-based biomarkers offer good accuracy, our RNA-based biomarker approach demonstrated similar predictive accuracy. These results reinforce the potential of ncRNA-based biomarkers as a robust, non-invasive diagnostic tool for AD.

Although the top ncRNAs identified in our analysis did not completely overlap with those reported in prior plasma biomarker studies, several of the highly predictive ncRNAs identified in this study have established links to AD-associated molecular pathways. For example, lnc-PRKACB2 is associated with the PRKACB2 protein, which plays a crucial role in tau phosphorylation—a key pathological hallmark of AD [32]. Similarly, lnc-MYEF2-3 is implicated in AD pathogenesis; the knockdown of its associated gene, MYEF2C, has been shown to enhance cellular apoptosis and increase levels of β-secretase (BACE1), a critical enzyme in amyloid-beta production [33]. Among the top-ranked snoRNAs, HBII-180A and HBII-115 are known to guide site-specific 2′-O-methylation of target RNAs, implicating their potential role in RNA modification processes relevant to AD onset [34]. Although HBII-115 (SNORD23) and HBII-180A (SNORD88A) were among the most dysregulated snoRNAs in our dataset, other C/D box snoRNAs, such as SNORD115 and SNORD116, have been reported in plasma extracellular vesicles from individuals with AD [7]. This suggests a potential role for snoRNA in AD-related pathogenesis. Additionally, a statistically significant overlap between differentially expressed ncRNAs in our dataset and those identified in an external validation dataset derived from postmortem brain samples supports the robustness and transferability of our findings across independent cohorts.

To further investigate coordinated ncRNA expression and perform feature selection for AD prediction in a different approach, we conducted a weighted gene co-expression network analysis (WGCNA), identifying 15 distinct co-expression modules. A Random Forest classifier trained on the averaged expression profiles of these modules achieved perfect classification performance (AUC = 1.000) on the test set. The relatively even distribution of feature importance across modules suggests that multiple ncRNA networks contribute collectively to AD prediction. Compared to single-gene and multi-gene models, the module-based classifier also demonstrated a narrower distribution of AUC values across cross-validation iterations, reinforcing the robustness and stability of this module-based predictive approach.

The dysregulated ncRNA landscape is consistent with previous studies, further supporting the involvement of multiple subtypes of ncRNAs in AD pathogenesis and neurodegeneration [35,36]. The substantial dysregulation observed for piRNAs and snoRNAs, despite being less documented in the context of AD, highlights their previously understudied status and suggests the need for further investigation to fully understand their roles in AD onset and progression. Our findings highlight the biological plausibility of our ncRNA panels with strong diagnostic performance, suggesting the importance of further functional characterization of these ncRNA candidates to elucidate their mechanistic roles in AD and assess their translational potential as plasma-based biomarkers.

5. Conclusions

This study presents compelling evidence for the diagnostic potential of ncRNA expression profiles, including lncRNA, piRNA, and snoRNA, in Alzheimer's disease. Through integrative transcriptomic analysis combining differential expression, co-expression networks, and machine learning, we identified robust ncRNA markers capable of distinguishing AD from control samples. However, limitations such as the small sample size and the absence of parallel mRNA-seq data constrain generalizability and mechanistic interpretation. Future directions include validating these findings in larger, independent cohorts and integrating mRNA expression analysis to determine coding-region interaction counterparts of ncRNA biomarkers, thereby enhancing understanding of their functional roles in AD pathogenesis.

Author Contributions: Conceptualization, W.M.O.; methodology, R.X.; software, R.X. and E.K.; validation, R.X. and E.K.; formal analysis, R.X. and E.K.; investigation, R.X. and E.K.; resources, W.M.O.; data curation, R.X.; writing—original draft preparation, R.X. and E.K.; writing—review and editing, W.T.L., J.W.-R., and W.M.O.; visualization, R.X. and E.K.; supervision, W.M.O.; project administration, W.M.O.; funding acquisition, W.M.O. All authors have read and agreed to the published version of the manuscript.

Funding: This research was funded by the University of California Academic Senate San Diego Division (RG104647).

Institutional Review Board Statement: Not applicable.

Informed Consent Statement: Not applicable.

Data Availability Statement: The data presented in this study are openly available in the NCBI SRA database under project accession SRP022043 (https://www.ncbi.nlm.nih.gov/Traces/study/?acc=SRP022043&o=acc_s:a) (accessed on 3 March 2025) and project accession SRP288246 (https://www.ncbi.nlm.nih.gov/bioproject/PRJNA670793) (accessed on 20 May 2025).

Conflicts of Interest: The authors declare no conflicts of interest.

Abbreviations

The following abbreviations are used in this manuscript:

AD	Alzheimer's disease
ncRNA	Non-coding RNA
lncRNA	Long non-coding RNA
piRNA	PIWI-interacting RNA
snoRNA	Small nucleolar RNA
WGCNA	Weighted gene co-expression network analysis
FDA	Food and Drug Administration
CSF	Cerebrospinal fluid
AUC	Area under the curve
SRA	Sequence Read Archive
RF	Random Forest
ROC	Receiver operating characteristic
DEGs	Differentially expressed genes

References

1. National Institute on Aging. Alzheimer's Disease Fact Sheet. Available online: https://www.nia.nih.gov/health/alzheimers-and-dementia/alzheimers-disease-fact-sheet (accessed on 31 March 2025).
2. Passeri, E.; Elkhoury, K.; Morsink, M.; Broersen, K.; Linder, M.; Tamayol, A.; Malaplate, C.; Yen, F.T.; Arab-Tehrany, E. Alzheimer's Disease: Treatment Strategies and Their Limitations. *Int. J. Mol. Sci.* **2022**, *23*, 13954. [CrossRef] [PubMed]
3. Bomasang-Layno, E.; Bronsther, R. Diagnosis and Treatment of Alzheimer's Disease: An Update. *Dela. J. Public Health* **2021**, *7*, 74–85. [CrossRef] [PubMed]
4. Bellio, M.; Oxtoby, N.P.; Walker, Z.; Henley, S.; Ribbens, A.; Blandford, A.; Alexander, D.C.; Yong, K.X.X. Analyzing Large Alzheimer's Disease Cognitive Datasets: Considerations and Challenges. *Alzheimers Dement. (Amst.)* **2020**, *12*, e12135. [CrossRef]
5. Barthélemy, N.R.; Salvadó, G.; Schindler, S.E.; He, Y.; Janelidze, S.; Collij, L.E.; Saef, B.; Henson, R.L.; Chen, C.D.; Gordon, B.A.; et al. Highly Accurate Blood Test for Alzheimer's Disease Is Similar or Superior to Clinical Cerebrospinal Fluid Tests. *Nat. Med.* **2024**, *30*, 1085–1095. [CrossRef]
6. Webster, S.F.; Ghalei, H. Maturation of Small Nucleolar RNAs: From Production to Function. *RNA Biol.* **2023**, *20*, 715–736. [CrossRef]
7. Fitz, N.F.; Wang, J.; Kamboh, M.I.; Koldamova, R.; Lefterov, I. Small Nucleolar RNAs in Plasma Extracellular Vesicles and Their Discriminatory Power as Diagnostic Biomarkers of Alzheimer's Disease. *Neurobiol. Dis.* **2021**, *159*, 105481. [CrossRef] [PubMed]
8. Chauhan, W.; Sudharshan, S.J.; Kafle, S.; Zennadi, R. SnoRNAs: Exploring Their Implication in Human Diseases. *Int. J. Mol. Sci.* **2024**, *25*, 7202. [CrossRef]
9. Gstir, R.; Schafferer, S.; Scheideler, M.; Misslinger, M.; Griehl, M.; Daschil, N.; Humpel, C.; Obermair, G.J.; Schmuckermair, C.; Striessnig, J.; et al. Generation of a Neuro-Specific Microarray Reveals Novel Differentially Expressed Noncoding RNAs in Mouse Models for Neurodegenerative Diseases. *RNA* **2014**, *20*, 1929–1943. [CrossRef]
10. Iwasaki, Y.W.; Siomi, M.C.; Siomi, H.; Kornberg, R. PIWI-Interacting RNA: Its Biogenesis and Functions. *Annu. Rev. Biochem.* **2015**, *84*, 405–433. [CrossRef]

11. Qiu, W.; Guo, X.; Lin, X.; Yang, Q.; Zhang, W.; Zhang, Y.; Zuo, L.; Zhu, Y.; Li, C.-S.R.; Ma, C.; et al. Transcriptome-Wide piRNA Profiling in Human Brains of Alzheimer's Disease. *Neurobiol. Aging* **2017**, *57*, 170–177. [CrossRef]
12. Roy, J.; Sarkar, A.; Parida, S.; Ghosh, Z.; Mallick, B. Small RNA Sequencing Revealed Dysregulated piRNAs in Alzheimer's Disease and Their Probable Role in Pathogenesis. *Mol. Biosyst.* **2017**, *13*, 565–576. [CrossRef]
13. Guo, X.; Qiu, W.; Garcia-Milian, R.; Lin, X.; Zhang, Y.; Cao, Y.; Tan, Y.; Wang, Z.; Shi, J.; Wang, J.; et al. Genome-Wide Significant, Replicated and Functional Risk Variants for Alzheimer's Disease. *J. Neural Transm.* **2017**, *124*, 1455–1471. [CrossRef] [PubMed]
14. Jain, G.; Stuendl, A.; Rao, P.; Berulava, T.; Pena Centeno, T.; Kaurani, L.; Burkhardt, S.; Delalle, I.; Kornhuber, J.; Hüll, M.; et al. A Combined miRNA–piRNA Signature to Detect Alzheimer's Disease. *Transl. Psychiatry* **2019**, *9*, 250. [CrossRef] [PubMed]
15. Mattick, J.S.; Amaral, P.P.; Carninci, P.; Carpenter, S.; Chang, H.Y.; Chen, L.-L.; Chen, R.; Dean, C.; Dinger, M.E.; Fitzgerald, K.A.; et al. Long Non-Coding RNAs: Definitions, Functions, Challenges and Recommendations. *Nat. Rev. Mol. Cell Biol.* **2023**, *24*, 430–447. [CrossRef] [PubMed]
16. Kazemi, M.; Sadeh, R.N.; Khaniani, M.S.; Rezazadeh, M.; Derakhshan, S.M.; Ghafouri-Fard, S. Identification of RN7SK LncRNA as a Novel Biomarker in Alzheimer's Disease Using Bioinformatics and Expression Analysis. *Sci. Rep.* **2024**, *14*, 31192. [CrossRef]
17. Feng, L.; Liao, Y.-T.; He, J.-C.; Xie, C.-L.; Chen, S.-Y.; Fan, H.-H.; Su, Z.-P.; Wang, Z. Plasma Long Non-Coding RNA BACE1 as a Novel Biomarker for Diagnosis of Alzheimer Disease. *BMC Neurol.* **2018**, *18*, 4. [CrossRef]
18. National Center for Biotechnology Information. SRP022043-Sequence Read Archive. Available online: https://www.ncbi.nlm.nih.gov/sra?term=SRP022043 (accessed on 23 April 2025).
19. National Center for Biotechnology Information. miRNA-Seq Study of Post-Mortem Alzheimer's Disease Brain Samples. Available online: https://trace.ncbi.nlm.nih.gov/Traces/index.html?view=study&acc=SRP288246 (accessed on 19 May 2025).
20. National Center for Biotechnology Information. The NCBI SRA (Sequence Read Archive). Available online: https://github.com/ncbi/sra-tools (accessed on 23 April 2025).
21. Andrews, S. FastQC: A Quality Control tool for High Throughput Sequence Data. Available online: http://www.bioinformatics.babraham.ac.uk/projects/fastqc/ (accessed on 24 April 2025).
22. Martin, M. Cutadapt Removes Adapter Sequences from High-Throughput Sequencing Reads. *EMBnet J.* **2011**, *17*, 10. [CrossRef]
23. Dobin, A.; Davis, C.A.; Schlesinger, F.; Drenkow, J.; Zaleski, C.; Jha, S.; Batut, P.; Chaisson, M.; Gingeras, T.R. STAR: Ultrafast Universal RNA-Seq Aligner. *Bioinformatics* **2013**, *29*, 15–21. [CrossRef]
24. Leung, Y.Y.; Kuksa, P.P.; Amlie-Wolf, A.; Valladares, O.; Ungar, L.H.; Kannan, S.; Gregory, B.D.; Wang, L.-S. DASHR: Database of Small Human Noncoding RNAs. *Nucleic Acids Res.* **2016**, *44*, D216–D222. [CrossRef]
25. Quinlan, A.R.; Hall, I.M. BEDTools: A Flexible Suite of Utilities for Comparing Genomic Features. *Bioinformatics* **2010**, *26*, 841–842. [CrossRef]
26. Love, M.I.; Huber, W.; Anders, S. Moderated Estimation of Fold Change and Dispersion for RNA-Seq Data with DESeq2. *Genome Biol.* **2014**, *15*, 550. [CrossRef] [PubMed]
27. Langfelder, P.; Horvath, S. WGCNA: An R Package for Weighted Correlation Network Analysis. *BMC Bioinform.* **2008**, *9*, 559. [CrossRef] [PubMed]
28. Pedregosa, F.; Varoquaux, G.; Gramfort, A.; Michel, V.; Thirion, B.; Grisel, O.; Blondel, M.; Prettenhofer, P.; Weiss, R.; Dubourg, V.; et al. Scikit-Learn: Machine Learning in Python. *J. Mach. Learn.* **2011**, *12*, 2825–2830.
29. Wickham, H.; Averick, M.; Bryan, J.; Chang, W.; McGowan, L.D.; François, R.; Grolemund, G.; Hayes, A.; Henry, L.; Hester, J.; et al. Welcome to the Tidyverse. *J. Open Source Softw.* **2019**, *4*, 1686. [CrossRef]
30. Kolde, R.; Metsalu, T.; Lizee, A. *pheatmap: Pretty Heatmaps*, R package version 1.0.12; GitHub, Inc.: San Francisco, CA, USA. Available online: https://github.com/raivokolde/pheatmap (accessed on 26 April 2025).
31. Shobeiri, P.; Alilou, S.; Jaberinezhad, M.; Zare, F.; Karimi, N.; Maleki, S.; Teixeira, A.L.; Perry, G.; Rezaei, N. Circulating long non-coding RNAs as novel diagnostic biomarkers for Alzheimer's disease (AD): A systematic review and meta-analysis. *PLoS ONE* **2023**, *18*, e0281784. [CrossRef]
32. Wang, L.; Liu, J.; Wang, Q.; Jiang, H.; Zeng, L.; Li, Z.; Liu, R. MicroRNA-200a-3p Mediates Neuroprotection in Alzheimer- Related Deficits and Attenuates Amyloid-Beta Overproduction and Tau Hyperphosphorylation via Coregulating BACE1 and PRKACB. *Front. Pharmacol.* **2019**, *10*, 806. [CrossRef]
33. Ren, J.; Zhang, S.; Wang, X.; Deng, Y.; Zhao, Y.; Xiao, Y.; Liu, J.; Chu, L.; Qi, X. MEF2C Ameliorates Learning, Memory, and Molecular Pathological Changes in Alzheimer's Disease in Vivo and in Vitro. *Acta Biochim. Biophys. Sin.* **2022**, *54*, 77–90. [CrossRef]
34. Galardi, S.; Fatica, A.; Bachi, A.; Scaloni, A.; Presutti, C.; Bozzoni, I. Purified Box C/D snoRNPs Are Able To Reproduce Site-Specific 2′-O-Methylation of Target RNA In Vitro. *Mol. Cell. Biol.* **2002**, *22*, 6663–6668. [CrossRef]

35. Idda, M.L.; Munk, R.; Abdelmohsen, K.; Gorospe, M. Noncoding RNAs in Alzheimer's Disease. *Wiley Interdiscip. Rev. RNA* **2018**, *9*, e1463. [CrossRef]
36. Salta, E.; De Strooper, B. Noncoding RNAs in Neurodegeneration. *Nat. Rev. Neurosci.* **2017**, *18*, 627–640. [CrossRef]

biomolecules

MDPI

Article

Identification of Two Critical Contact Residues in a Pathogenic Epitope from Tetranectin for Monoclonal Antibody Binding and Preparation of Single-Chain Variable Fragments

Juncheng Wang [1,2,3,†], Meng Liu [2,†], Rukhshan Zahid [2], Wenjie Zhang [2,4], Zecheng Cai [2], Yan Liang [2], Die Li [2], Jiasheng Hao [2] and Yuekang Xu [1,*]

1 Xinjiang Key Laboratory of Biological Resources and Genetic Engineering, College of Life Science and Technology, Xinjiang University, Urumqi 830017, China; 2131010072@ahnu.edu.cn
2 Anhui Provincial Key Laboratory for Conservation and Exploitation of Biological Resources, College of Life Sciences, Anhui Normal University, Wuhu 241000, China; 2121011631@ahnu.edu.cn (M.L.); rukhshanzahid@ahnu.edu.cn (R.Z.); zhangwenjie@yjsyy.com (W.Z.); 2221011740@ahnu.edu.cn (Z.C.); 2321011720@ahnu.edu.cn (Y.L.); 2321021782@ahnu.edu.cn (D.L.); jiashenghao123@ahnu.edu.cn (J.H.)
3 School of Basic Medicine, Nanchang Medical College, Nanchang 330052, China
4 The First Affiliated Hospital of Wannan Medical College, Wuhu 241001, China
* Correspondence: yuekang.xu@hotmail.com
† These authors contributed to this work equally.

Abstract: Sepsis is a fetal disease that requires a clear diagnostic biomarker for timely antibiotic treatment. Recent research has identified a pyroptosis-inducing epitope known as P5-5 in tetranectin (TN), a plasma protein produced by monocytes. Previously, we produced a 12F1 monoclonal antibody against the P5-5 and discovered that it could not only diagnose the presence but also monitor the progress of sepsis in the clinic. In the current study, we further investigated the structure site of the P5-5 and the recognition mechanism between the 12F1 mAb and the P5-5 epitope. To this end, 10 amino acids (NDALYEYLRQ) in the P5-5 were individually mutated to alanine, and their binding to the mAb was tested to confirm the most significant antigenic recognition sites. In the meanwhile, the spatial conformation of 12F1 mAb variable regions was modeled, and the molecular recognition mechanisms in detail of the mAb to the P5-5 epitope were further studied by molecular docking. Following epitope prediction and experimental verification, we demonstrated that the motif "DALYEYL" in the epitope sequence position 2−8 of TN-P5-5 is the major binding region for mAb recognition, in which two residues (4L and 8L) were essential for the interaction between the P5-5 epitope and the 12F1 mAb. Therefore, our study greatly narrowed down the previously reported motif from ten to seven amino acids and identified two Leu as critical contact residues. Finally, a single-chain variable fragment (scFv) from the 12F1 hybridoma was constructed, and it was confirmed that the identified motif and residues are prerequisites for the strong binding between P5-5 and 12F1. Altogether, the data of the present work could serve as a theoretic guide for the clinical design of biosynthetic drugs by artificial intelligence to treat sepsis.

Keywords: sepsis; tetranectin; P5-5; epitope; scFv

1. Introduction

Sepsis, a life-threatening organ dysfunction caused by a dysregulated host response to infection, represents a major worldwide health burden with an estimated annual incidence of 48.9 million cases and 11 million fatalities [1–5]. As pathogens are the predominant cause of sepsis, prompt and accurate diagnosis is crucial due to the rapid disease progression and quick elapse of a critical "golden window" for effective intervention [6]. However, current diagnostic and prognostic approaches face significant limitations. While microbial culture remains the gold standard for sepsis diagnosis, its clinical utility is hampered by a prolonged turnaround time of 2–3 days, delaying timely therapeutic decisions [7]. Rapid alternatives such as C-reactive protein (CRP), procalcitonin (PCT), and interleukin-6 (IL-6) testing provide faster results but suffer from insufficient specificity, often leading to unnecessary broad-spectrum antibiotic use [8,9]. This contributes to antimicrobial resistance (AMR), increased treatment costs, and potential adverse patient outcomes [1].

Furthermore, sepsis prognosis currently relies on clinical scoring systems like the Sequential Organ Failure Assessment (SOFA) and Quick SOFA (qSOFA), which, despite being widely adopted, exhibit notable complexity and misdiagnosis rates of up to 30% [1,10]. The empirical determination of antibiotic therapy duration in sepsis management further exacerbates the risk of antibiotic overuse and AMR. Given these challenges, there is an urgent need for more reliable biomarkers and rapid diagnostic methods to improve sepsis detection, prognostic accuracy, and antimicrobial stewardship, ultimately enhancing patient outcomes in critical care settings.

Recent research revealed that certain plasma and tissue proteins, notably tetranectin (TN), are found in the plasma of patients with sepsis or septic shock [11,12]. TN was discovered in 1986 as an oligomeric fibrinogen-binding protein mostly expressed in the lungs. Further analysis of patients and healthy controls showed that plasma TN levels were reduced by 60–70% in patients with sepsis or septic shock. Consistent with these clinical findings, there was a corresponding reduction in circulating TN levels in experimental sepsis, with a reduction of more than 70% by 24 h after the onset of disease, which is the time point at which some septic animals begin to die. Mechanistically, TN was involved in the pathogen-induced pyroptosis of macrophages for the development of sepsis [13]. Importantly, a crucial epitope in the structure of TN proteins is P5-5, which mediates the interaction of TN with high mobility group box 1 (HMGB1), a protein released by activated leukocytes. Phagocytosis of the TN/HMGB1 complex by macrophages promoted their pyroptosis. leading to significant reduction in macrophage number and depletion of plasma TN levels in septic patients [11]. Therefore, if TN can be utilized as an immunological biomarker for sepsis, the P5-5 epitope in the TN holds promise for quicker clinical diagnosis and symptom assessment of the disease [11,12].

In our previous study, we generated a hybridoma cell line targeting P5-5 and successfully purified the monoclonal antibody 12F1 [14]. Functional analyses revealed that this antibody not only facilitates the assessment of disease severity in patients but also exhibits therapeutic potential by attenuating sepsis progression in murine models [14]. Until now, we only know that the pathogenic epitope P5-5 is composed of 10 amino acids (NDALYEYLRQ) [11]. However, which of the 10 amino acids are the critical residues to mediate the interaction with the therapeutic antibodies is far from clear. This study aims to identify the key contact residues in the pathogenic P5-5 epitope through antigen design and molecular docking. Furthermore, site-directed mutagenesis was employed to elucidate the detailed interactions between P5-5 and the mAb. The effect of the antigen linear epitope on its interaction with the antibody, as well as the mAb's binding mechanism with the pathogenic epitope, was explored and characterized based on the epitope information. Finally, the amplified VH and VL genes, which encode antibodies against P5-5, were employed to create an scFv, and the recombinant antibody was generated through prokaryotic expression to validate our results.

2. Materials and Methods

2.1. Materials

Regal Biology Technology (Shanghai, China) provided the bovine serum albumin (CAS: 9048-46-8) for this study. Sigma Aldrich (St. Lois, MO, USA) supplied ovalbumin (OVA, CAS: A5378), an albumin derived from chicken egg white. The P5-5 epitope (Homo tetranectin peptide NDALYEYLRQ) and mutants (ADALYEYLRQ, NAALYEYLRQ, NDAAYEYLRQ, NDALAEYLRQ, NDALYAYLRQ, NDALYEALRQ, NDALYEYARQ, NDALYEYLAQ, NDALYEYLRA) were synthesized and bought from Sangon (Shanghai, China). HPLC validated the purity of 10 peptides at ≥95.5%. Furthermore, 6 × His, mouse anti-His-Tag Monoclonal antibody (CAS: No. 66005-1-Ig), and rabbit anti-MBP (Maltose binding protein)-Tag Polyclonal antibody (CAS: No. 15089-1-AP) were sourced from Proteintech (Wuhan, China). HRP conjugate and goat anti-mouse IgG (H + L) (CAS: BF03001) and Goat anti-rabbit IgG conjugated with HRP (CAS: BF03008) were acquired from Biodragon (Suzhou, China). Protein G agarose beads for affinity chromatography (CAS: L00209) were obtained from GenScript (Nanjing, China). All other reagents and substances were of analytical grade. They were obtained from regular sources.

2.2. TN-P5-5 Complete Antigen Preparation

The P5-5 and mutants were conjugated to BSA by the EDC conjugating method. Firstly, the P5-5 and other 9 mutant peptide solutions (4 mg/mL in 0.1 M MES buffer, pH 4.7) were separately mixed with BSA solution (10 mg/mL in 0.1 M MES buffer, pH 4.7). Then, the reaction mixtures were all supplemented with 100 μL of EDC solution (10 mg/mL in deionized water) and incubated at 25 °C for 2 h to facilitate conjugation. The haptens were coupled to BSA at a molar ratio of 10:1 [14]. After a two-hour reaction, the resulting mixtures were subjected to dialysis against 0.01 M phosphate-buffered saline (PBS, pH 7.4) for 48 h. The final products were then stored at −20 °C for the following applications. The P5-5 and mutant conjugates were examined using native gel electrophoresis, and the concentration of the complete antigen was determined using the BCA protein assay kit.

2.3. Epitope Prediction of P5-5

The putative epitope location of the P5-5 peptide was further described by B-cell linear epitope prediction analysis using the Immune Epitope Database Analysis Resource service (http://tools.immuneepitope.org/bcell/, accessed on 25 April 2025) [15,16]. Concurrently, the three-dimensional structure of P5-5 was predicted via homology modeling employing the SWISS-MODEL workspace (https://swissmodel.expasy.org/, accessed on 25 April 2025), a web-based framework for modeling protein structural homology [17].

2.4. Sequence Analysis of 12F1 Monoclonal Antibody Variable Regions

Total RNA from mAb 12F1 hybridoma cells was extracted with 1 mL TRIzol Reagent and reverse-transcribed to cDNA using the cDNA Synthesis Kit. Azenta used murine IgG-specific primers and NGS sequencing to acquire the mAb 12F1 genes from the heavy and light chain variable regions (VH and VL, respectively). Antibody sequences were analyzed using the NCBI Ig BLAST (v1.17.0) software to identify the most closely matched germline V-D-J gene sequences within the CDR1, CDR2, and CDR3 regions aligned from the IMGT (International Immunogenetics Information System) database (IMGT, https://www.imgt.org/, accessed on 25 April 2025) [18].

2.5. 12F1 Monoclonal Antibody Homology Model and Docking with P5-5 Epitope

The three-dimensional structure model of scFv based on monoclonal antibody 12F1 was constructed using the SWISS-MODEL homology modeling method. Among the developed models, the one with the highest rank and quality, as measured by the lowest probability density function energy and the highest discrete optimized protein energy, was chosen for the antibody conformation. The validity of the resulting homology model was assessed through Profile-3D and Ramachandran plot analyses. To optimize the complementarity-determining regions (CDRs), the published antibody crystal structures were compared via the IMGT/V-QUEST database (http://www.imgt.org) [19]. To determine the contact forces and critical amino acid positions, the anticipated P5-5 epitope molecules were docked into the cavity produced by the CDR regions in the active conformation utilizing the ZDock device (ZDOCK server, University of Massachusetts Medical School, Boston, MA, USA) [16,20]. The docking outcomes were visualized using the PyMOL program (version 2.2) [21].

2.6. Purification of 12F1 mAb

We previously generated the hybridoma cell line 12F1 in our lab. This cell line stably secreted anti-TN-P5-5 mAb [14]. Antibodies were produced in bulk via in vivo induction in mice to generate ascites. Female BALB/c mice aged 10 weeks bought from Nanjing Junke Bioengineering Ltd. (Nanjing, China) were sensitized by injecting ascites adjuvant into their abdominal cavity. To induce ascites formation, roughly 10^6 positive hybridoma cells were injected into the abdominal cavities of the mice [22–28]. All animal procedures were approved by the Animal Ethics Committee of Anhui Normal University (AHNU-ET2022015), and the 3R principles (Replacement, Reduction, Refinement) were followed. The antibodies were then concentrated and purified from the ascitic fluid using Protein G Resin [16]. Afterwards, we used a BCA protein assay kit to measure the amount of purified antibodies and iELISA to assess the titer [22].

2.7. TN-P5-5 Mutants Design and Synthesis

Based on the research findings regarding antibody binding to specific epitopes, the key residues involved in the interaction between mAb-P5-5 and its epitope were further identified. A total of 10 amino acids within the TN-P5-5 pathogenic epitope were individually mutated to alanine, and these residues were considered critical for the specific antigen–antibody interaction. By sequentially mutating the residues at positions 1, 2, 4, 5, 6, 7, 8, 9, and 10 to alanine, the corresponding mutated protein sequences—ADALYEYLRQ, NAALYEYLRQ, NDAAYEYLRQ, NDALAEYLRQ, NDALYAYLRQ, NDALYEALRQ, NDALYEYARQ, NDALYEYLAQ, and NDALYEYLRA—were successfully generated.

2.8. iELISA and icELISA to Identify the Specific Sites Between 12F1 mAb and TN-P5-5 Interaction

The molecular recognition mechanism of monoclonal antibody 12F1 for P5-5 was determined via two types of ELISA assays [15,16].

Indirect ELISA (iELISA): To identify the key binding epitopes of the prepared mAb to the P5-5 epitope, P5-5-BSA, and P5-5 mutant-BSA, the antigens were coated onto 96-well plates at a concentration of 10 μg/mL [15]. Subsequently, purified 12F1 mAb diluted at 1:2000 in blocking buffer was added to the wells after blocking and washing. Following another washing step, goat anti-mouse IgG-HRP at a 1:5000 dilution in blocking buffer was added and incubated at 37 °C for 1 h. To commence color development, 100 μL of 3,3,5,5-Tetramethylbenzidine (TMB) solution was added to each well, followed by 100 μL of 2 M H_2SO_4. The absorbance at 450 nm was measured using a microplate reader.

Indirect competitive ELISA (ic-ELISA): the inhibition ability of 12F1 mAb against P5-5 and its mutants was assessed through icELISA [16]. A coating antigen P5-5-OVA solution at a concentration of 5 μg/mL was prepared and dispensed (100 μL) into each well of a 96-well ELISA plate. The plates were then left to incubate at 4 °C overnight. Following washing, 5% (*w*/*v*) PBSM was added to each well and incubated at 37 °C for 2 h. The synthesized peptides of P5-5 and its variants were prepared as 1 mg/mL stock solution (dissolved in 0.01M PBS, pH 7.4). These peptides were then serially diluted in PBSM to concentrations of 20 μg/mL, 10 μg/mL, 5 μg/mL, 2.5 μg/mL, 1.25 μg/mL, 625 ng/mL, 312.5 ng/mL, 156.25 ng/mL, 78.13 ng/mL, 39.06 ng/mL, 19.53 ng/mL, 9.77 ng/mL, 4.88 ng/mL, 2.44 ng/mL, 1.22 ng/mL, and 0.61 ng/mL for the competitor in the ic-ELISA assay [29]. Subsequently, 12F1 mAb was mixed thoroughly with the diluted solutions of P5-5 and its mutants at equal volume at 37 °C for 30 min incubation and reaction. Following this, each well was impressed by 100 μL of the mixture, which was then incubated at 37 °C for one hour. The succeeding stages, which included the addition of HRP-goat anti-mouse serum, TMB substrate solution, and stop solution, were carried out as described earlier. The measurement of the OD value was then taken under 450 nm. The IC_{50} values of the 12F1 monoclonal antibody against synthesized P5-5 and nine other mutant peptides were determined using the four-parameter logistic equations in Prism 9.0 GraphPad software (version 9.0 GraphPad Software), following the established methodology.

2.9. Construction of the MBP Fusion scFv Antibody, Soluble Expression, and Purification

Specific primers were designed to amplify the target scFv gene after sequencing the 12F1 hybridoma clone. The pMAL-c2x vector was utilized to create expression vectors containing the scFv protein fused with an MBP tag. The recombinant pMAL-c2X-scFv plasmid was transformed into *E. coli* TB1 through electroporation for scFv expression. To induce the fused scFv expression, 1 mM of isopropyl β-D-thiogalactopyranoside (IPTG) was added to the LB culture overnight at 16 °C until the OD600 reached 0.8 [30]. Subsequently, cell pellets were collected by centrifugation at 4 °C and subjected to ultrasonic fragmentation. The purification of the expressed anti-P5-5 scFv was performed through MBP affinity chromatography, and the resulting proteins were evaluated using SDS-PAGE [31].

2.10. Binding Activity Determination of Anti-P5-5 scFv by iELISA

The evaluation of the binding activity of purified scFv was conducted using the indirect ELISA protocol as described in 2.8. In brief, ELISA plates were coated overnight at 4 °C with serially diluted OVA-P5-5 conjugate antigen. Plates were washed three times with PBST (0.05% Tween-20 in PBS), then blocked with 5% (*w*/*v*) skimmed milk in PBS for two hours at 37 °C. Subsequently, purified scFv fusion proteins at varying concentrations were added to triplicate wells and incubated at 37 °C for two hours. Following subsequent washing processes, a mouse anti-MBP tag primary antibody (1:1000 dilution) was used to identify specific binding, which was then incubated for 1 h at 37 °C with an HRP-conjugated goat anti-mouse IgG secondary antibody (1:5000 dilution). The enzymatic reaction started with a tetramethylbenzidine (TMB) substrate solution and ended after 15 min with a 2 M H_2SO_4 stop solution. Optical density measurements were obtained under 450 nm using a microplate reader. The affinity constant Kaff was calculated based on ELISA absorbance. This procedure involved equal dilutions of the coated antigen and the antibody. Kaff was then determined using the equation below. The ratio of [Ag] to [Agt] was denoted as n, where n = [Ag]/[Agt] and [Ag] and [Ag]t denote the concentrations of the coated antigen. The K_{aff} formula calculated was $K_{aff} = (n - 1)/\{2(n[Abt] - [Ab])\}$ [32,33]. [Ab]t and [Ab] denote the concentration of the scFv at 50% of its OD max [33–35].

2.11. The Sensitivity Determination of Anti-P5-5 scFv

To evaluate the antibody sensitivity to the P5-5 epitope, we conducted ic-ELISA. The competitive antigen (standard P5-5 epitope) was incubated with purified MBP-linker-scFv fusion protein at 37 °C for 2 h [16]. Subsequent steps followed the standard ELISA protocol outlined in Section 2.7. Data analysis was performed using Origin 8.0 software to generate a four-parameter logistic curve. The binding ratio (B/B0) was calculated by analyzing the ratio of antibody binding in the presence versus absence of competing antigen across a concentration gradient of P5-5 [15].

2.12. Statistical Analysis

Experimental datasets were subjected to comprehensive processing and statistical analysis utilizing Origin 8.0 and GraphPad Prism 9.0. The analytical workflow incorporated systematic curve-fitting procedures complemented by analysis of variance (ANOVA) to characterize variable relationships. All experimental determinations were conducted with rigorous replication, maintaining a minimum of three independent experimental replicates for each dataset [36,37].

3. Results

3.1. Epitope Binding Pattern Prediction

To analyze the critical contact residues in the pathogenic P5-5 epitope, a 10-mer peptide with the sequence of NDALYEYLRQ in TN protein, we first utilized the Bepipred-2.0 algorithm for B-cell linear epitope prediction and identified prominent linear epitopes spanning residues 4–8 (motif Leu-Tyr-Glu-Tyr-Leu) within the peptide sequence (Figure 1A). The epitope structure of P5-5 is predicted to consist of an α helix and a binding site covering linear amino acid sequences Leu4 to Leu8 for antibody recognition in three-dimensional conformation (Figure 1B). Collectively, the small P5-5 sequence demonstrated distinct immunogenic properties in comprehensive epitope mapping.

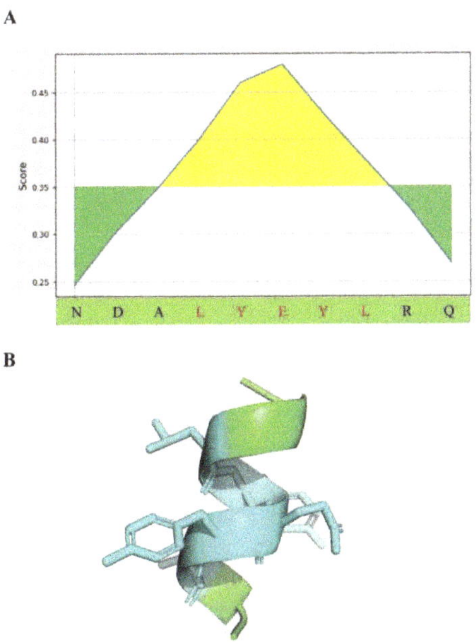

Figure 1. TN-P5-5 epitope prediction and three-dimensional structural analysis. (**A**) The output graph of presented linear B-cell epitopes displays amino acid positions (X-axis) and Bepipred scores (Y-axis). Residues with scores above the default threshold of 0.35 (shown in yellow) are more likely to be part of the epitope. Green implies a decreased likelihood of belonging to the epitope. (**B**) The 3D structure of TN-P5-5 with the predicted epitope from SWISS-MODEL homology modeling. Amino acid residues 4–8 motif (Leu-Tyr-Glu-Tyr-Leu) are shown in light blue.

To be specific, Figure 1B illustrates the three-dimensional structure of P5-5 within the native TN protein, aiming to depict the spatial arrangement of the linear epitope sequence in its original context. The TN-P5-5 amino acid sequence consists of 10 residues, and its corresponding structural conformation is characterized as an α-helix. The antigenic epitope of P5-5 was predicted using the BepiPred B-cell linear epitope prediction tool (http://tools.immuneepitope.org/bcell/, accessed on 25 April 2025). The core amino acid sequence (positions 4–8) of this linear epitope was identified as "LYEYL" (Figure 1A), and its spatial location corresponds to the α-helix region (Figure 1B). In the subsequent experimental phase of this study, the epitope-specific monoclonal antibody was validated through chemical synthesis of the corresponding peptide epitope [15]. The entire 10-amino acid sequence of P5-5, including the LYEYL motif, was synthesized as a linear peptide, which serves as a linear epitope.

3.2. Antibody Variable Region Structure Prediction of 12F1 and Molecular Docking with P5-5 Epitope

With the antigenic information of the targeted peptide in hand, we next turned to its interacting antibody. We began by generating the antibody's cDNA chain via RT-PCR employing total RNA obtained from 12F1 hybridoma cells as the template. Subsequently, the cDNA template was utilized to amplify the antibody gene. The 3D conformation of mAb 12F1 was then precisely predicted by matching its amino acid sequence to reported protein structures in the PDB database. The antibody variable region was subsequently docked with the P5-5 epitope using the ZDOCK online server. As depicted in Figure 2, it was observed that the P5-5 epitope conformation resided within the binding cavity of the murine 12F1 mAb. Moreover, examination of the antigen–antibody interaction complex revealed formation of hydrogen bonds between the residues of the murine antibody and the P5-5 epitope. Specifically, a Leu4 residue in P5-5 and a Tyr184 residue in the mAb were found to establish a hydrogen bond with a bond length of 2.4 Å. Additionally, the Tyr7 residue in P5-5 interacted with two residues (Leu167 and Asn168) in the antibody through hydrogen bonds at distances of 3.40 Å and 3.20 Å, respectively, which are important in the antibody–antigen interaction. Furthermore, a hydrogen bond (2.4 Å) was formed between Leu8 in P5-5 and Arg187 of the antibody. Overall, these four hydrogen bonds were identified as the primarily sensitive and specific binding forces for antibody recognition, thereby enhancing the stability and specificity of the complex.

3.3. Molecular Recognition Mechanism of mAb 12F1 to P5-5

Following theoretic prediction based on the amino acid sequence and their conformational structure, site-directed mutagenesis was performed to determine the critical contact residues in the P5-5 epitope and their mechanism of interaction with the 12F1 mAb. Since alanine is a type of hydrophobic amino acid lacking side groups and has a relatively simple structure, alanine scanning constitutes an extremely effective approach for investigating the key amino acids within the interaction region of antibody and antigen [16,17]. By mutating through alanine scanning and comparing the mutated P5-5 epitope with its reactive mAb in binding affinity, the determination of the amino acid residues and their position to interact with its antibody becomes feasible. Figure 3 depicts the mutant designs, with "A" indicating the locations that were mutated to alanine to scan the critical amino acids involved for binding activity.

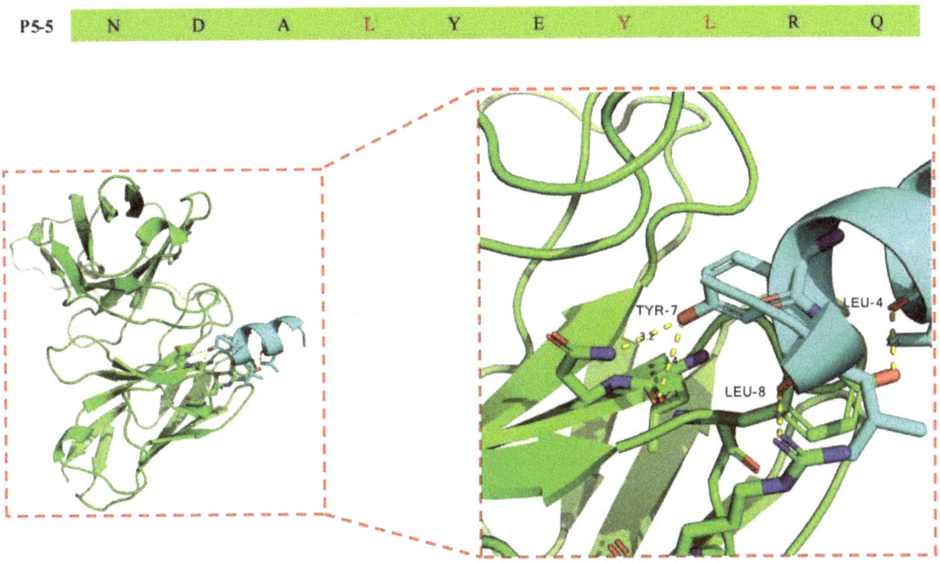

Figure 2. 12F1 mAb homology model and molecular dock. The scFv binding location and interactions between the antibody paratope and the P5-5 epitope were revealed following docking. The conformation prediction of scFv residues (green) to the TN-P5-5 epitope (blue) was depicted as a stick model and labeled. Among them, residues of P5-5 involved in the binding pockets are Leu4, Tyr7, and Leu8.

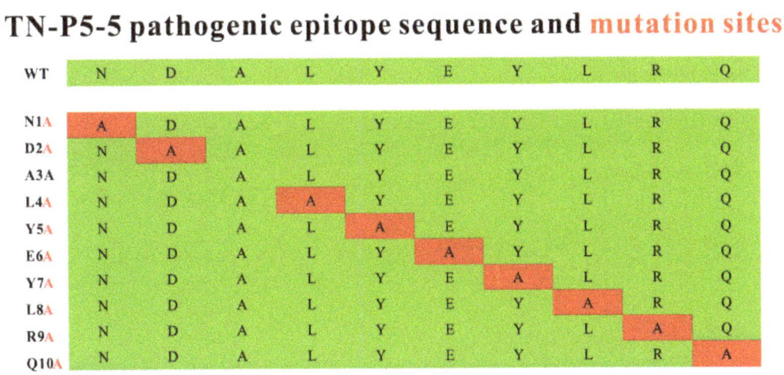

Figure 3. TN-P5-5 pathogenic epitope sequence and mutagenesis map of the P5-5 epitope through alanine (red) scan. To find out the interaction of the 12F1 mAb and different mutants of P5-5 recognition, standard peptides of ADALYEYLRQ, NDALYEYLRQ, NAALYEYLRQ, NDAAYEYLRQ, NDALAEYLRQ, NDALYAYLRQ, NDALYEALRQ, NDALYEYARQ, NDALYEYLAQ, and NDALYEYLRA were chemically synthesized.

Ascites were obtained from BALB/c mice following the injection of cultured hybridoma 12F1 cells. The anti-TN-P5-5 mAb 12F1 (IgG1) was then extracted from the ascites using affinity chromatography and a protein G resin. SDS-PAGE analysis of purified mAb 12F1 (Figure S1) revealed two different protein bands at 50 kDa and 25 kDa, which correspond to the heavy and light chains. This result demonstrates the effective purification of mAb 12F1. To find out the key regions responsible for 12F1 mAb-TN-P5-5 interactions, nine amino acids at the pathogenic epitope of P5-5, with the exception of A3, were mutated to alanine using site-directed mutagenesis. The detecting antigen P5-5-BSA and other mutants-BSA were coated, and the indirect ELISA results in Figure 4 revealed that the binding activities of 12F1 mAb against two mutated amino acids (L4A and L8A) were all significantly reduced to levels similar to the negative control (carrier protein BSA). Other mutants (DA2, Y5A, E6A, and Y7A) also exhibited weak binding signals to 12F1 mAb recognition compared with the WT P5-5 (Figure 4). Meanwhile, ic-ELISA using P5-5 peptides with various alanine substitutions and the 12F1 monoclonal antibody, respectively, was also performed to evaluate antibody inhibitory capacity against the antigens. The results in Figure 5 were consistent with those obtained from indirect ELISA. Following mutation, L4A and L8A entirely lost their ability to bind antibodies. The antigen inhibitory capabilities of the mutated peptides DA2, Y5A, E6A, and Y7A were significantly reduced compared to the WT P5-5, resulting in partial loss of their antibody-binding affinity (Figure 5). Overall, the results showed that the amino acids L4 and L8 in the pathogenic epitope were the most crucial residues for the interaction of 12F1 mAb and P5-5.

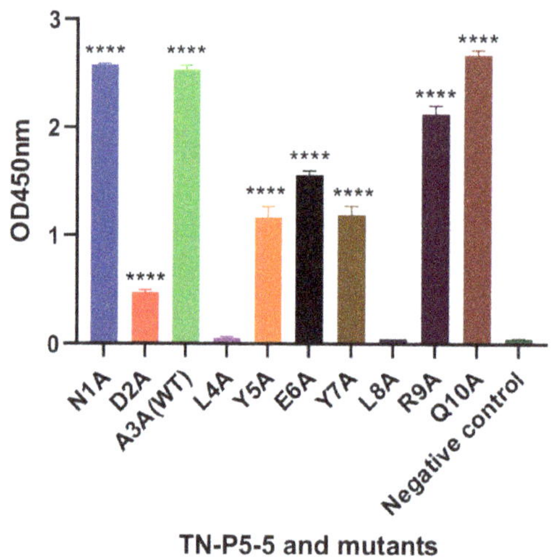

Figure 4. Epitope identification based on P5-5 (WT) and mutants. The indirect ELISA titer for various site mutant peptides. Error bars represent the standard deviation (SD) of six repeated experiments. Data are presented as mean ± SD. **** $p < 0.0001$.

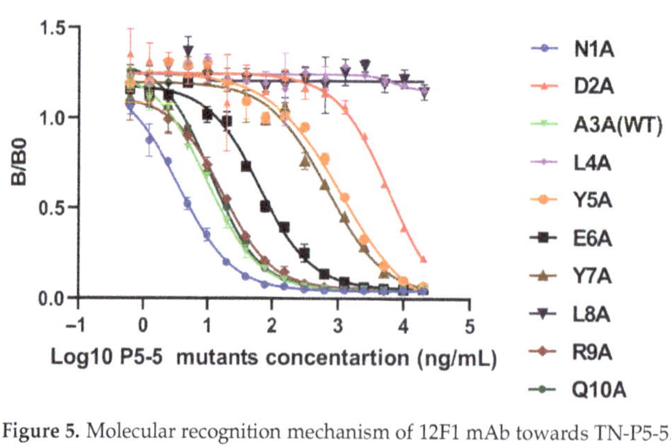

Figure 5. Molecular recognition mechanism of 12F1 mAb towards TN-P5-5. Competitive reactions of P5-5 and mutagenesis. The concentrations of P5-5 (WT) and mutagenesis were diluted to 20 μg/mL, 10 μg/mL, 5 μg/mL, 2.5 μg/mL, 1.25 μg/mL, 625 ng/mL, 312.5 ng/mL, 156.25 ng/mL, 78.13 ng/mL, 39.06 ng/mL, 19.53 ng/mL, 9.77 ng/mL, 4.88 ng/mL, 2.44 ng/mL, 1.22 ng/mL, and 0.61 ng/mL with blocking solution as competition antigen to react with 12F1 mAb, respectively.

3.4. Bioinformatic Analysis of the Anti-P5-5 scFv

Then, we went on to make recombinant anti-P5-5 scFv from the hybridoma firstly to confirm the identified motif and critical contact residues in P5-5 and then preserve the specificity of the valuable theranostic antibody, as hybridoma cell lines may cease antibody production after prolonged cultivation. To achieve this, the VH, VL, and linker genes were assembled into the entire scFv gene via an SOE-PCR (gene splicing by overlap extension PCR) reaction, yielding a 723 bp completed scFv fragment. Figure S2 depicts the scFv's coding sequence, which contains 241 amino acids and includes a (Gly4Ser)3 linker between VH and VL. The three complementarity-determining regions (CDRs) of VH and VL have been determined based on Kabat's approach [38], as illustrated in Figure S2. Meanwhile, Figure S3A displayed the scFv IMGT "collier de perle" graphical two-dimensional (2-D) representations. Sequence alignment of various mouse germline genes using multiple bioinformatics resources (IMGT databases) confirmed that the VH and VL sequences of the scFv correspond to the variable regions of mouse antibody genes, as illustrated in Figure S3B.

3.5. Expression, Purification, and Identification of Anti-P5-5 scFv

Following the successful construction of a fusion expression vector, pMAL-c2x-linker-scFv (sequenced and aligned as Figure S4), with the pMAL-c2x vector serving as a control depicted in Figure 6A, we then introduced it into *E. coli* TB1 to express the scFv. Figure 6B illustrated the expression of the scFv fusion proteins in *E. coli* TB1 cells, their purification using MBP affinity resin, and confirmation through SDS-PAGE analysis. The molecular weight of the produced MBP protein was around 50 kDa, while that of the MBP-linker-scFv fusion proteins increased to around 70 kDa [30,39]. To maintain the structural and functional integrity of the fusion proteins, the recombinant scFvs were purified from the induced bacteria using affinity chromatography under gentle conditions, resulting in protein purity exceeding 90% (Figure 6B).

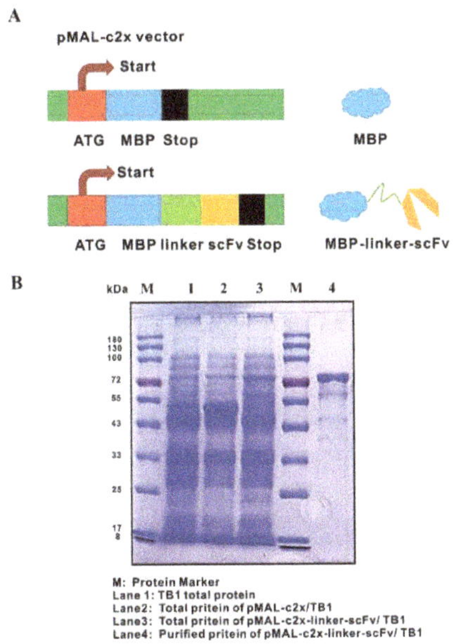

Figure 6. Construction and production of scFv fusion proteins that target TN-P5-5. (**A**) The diagram illustrates the construction of MBP tag and MBP-fused scFv formats against the P5-5 epitope in this

study (MBP, MBP-linker-scFv). ATG, start codon; MBP, maltose binding protein; Linker, $(G_4S)_3$ linker chain; scFv, single-chain variable fragment; Stop, stop codon. (**B**) The production and purification of fused scFv proteins were analyzed based on SDS-PAGE. The molecular weight of the expressed MBP protein was approximately 50 kDa, while the presence of distinct protein bands at 70 kDa confirms the successful purification of the scFv following *E. coli* lysis. M: Protein marker; Lane 1: control *E. coli* TB1 proteins; Lane 2: MBP tag protein (empty vector pMAL-c2x in *E. coli* TB1); Lane 3: Expressed total proteins of MBP-linker-scFv; Lane 4: Purified scFv protein.

Furthermore, an indirect ELISA was performed to demonstrate the scFv's affinity and inhibition against the P5-5 antigen. As depicted in Figure S4, the affinity constant of the scFv was determined to be 1.19×10^7 L/mol (Figure S5), indicating a high-affinity antibody (typically defined as an affinity falling within the range of 10^7 to 10^{12} L/mol). A standard curve for the indirect competitive ELISA (ic-ELISA) targeting the P5-5 epitope was established by utilizing varying concentrations of the P5-5 standard peptide as a competing antigen to interact with the purified MBP-linker-scFv fusion proteins (Figure S6). The relationship between the concentrations of P5-5 and the inhibition value was analyzed using Origin 8 software. The half-maximal inhibitory concentration (IC_{50}) of P5-5 binding to the scFv was calculated to be 359.76 ng/mL (Figure S6A). Moreover, our existing ic-ELISA methods may be used to analyze P5-5 epitope samples in PBS, with a target detection range of 78.125 ng/mL to 1.25 µg/mL (Figure S6B), and the limit of detection (LOD) was 62.86 ng/mL. Collectively, these findings suggest that the expressed and purified MBP-Linker-scFvs may recognize the same motif and exhibit functionalities akin to those of their parental 12F1 monoclonal antibody.

4. Discussion

Studies have demonstrated that C-reactive protein, procalcitonin, tumor necrosis factor α, interleukin-6, and neutrophil surface receptors (CD64) are currently widely utilized as biomarkers for monitoring sepsis [14]. However, these markers lack specificity for sepsis and may also be detected in non-infectious processes or conditions unrelated to sepsis [1]. Chen et al. recently discovered that TN, a plasma and tissue protein found in individuals with sepsis or septic shock, is directly linked to the prevalence of sepsis [11]. Further research has revealed that a pathogenic epitope P5-5 within TN, which contains the peptide sequence NDALYEYLRQ, was critical for TN/HMGB1 complex interaction that led to endocytosis of macrophages and reduction in plasma TN levels [11]. In our previous research, we developed a monoclonal antibody against the P5-5 pathogenic epitope using cell fusion. This antibody can accurately diagnose the presence of sepsis and quantitatively monitor its progression in clinical blood samples [14]. Based on this specific mAb, an ic-ELISA was established as a novel method for sepsis detection. Moreover, the anti-P5-5 mAb was also shown to effectively inhibit sepsis progress in mice by recovering macrophage number and increasing the survival of septic mice.

The formation of an antigen–antibody complex is highly sensitive to even minor alterations in the chemical groups or structures on the antigen surface, which can significantly affect the binding affinity of the antibody. This phenomenon raises a critical scientific question in the P5-5 interaction with mAb: Which specific amino acid within the P5-5 epitope is directly recognized and bound by the therapeutic antibody? To address this question, this study employed alanine scanning mutagenesis, systematically mutating each amino acid of the P5-5 epitope to alanine in sequence. ELISA experiments were subsequently conducted to determine which amino acid mutation abrogated antibody binding, thereby identifying the key amino acids essential for antibody recognition. The results demonstrated that Leu4 and Leu8 are the most critical residues for antibody binding; the loss of either residue completely abolished antibody recognition of the epitope. Additionally, mutations in Asp2,

Tyr5, Glu6, and Tyr7 partially reduced antibody binding affinity, indicating their involvement in modulating the interaction but not as crucial as Leu4 and Leu8. Interestingly, antigen epitope prediction algorithms identified a minimum antigenic determinant of the LYEYL (Leu-Tyr-Glu-Tyr-Leu) motif, corroborated by molecular docking studies, which highlighted Leu4, Tyr7, and Leu8 as the key residues involved in binding capacity. These predictions align closely with the experimental mutation data, validating the accuracy and consistency of the computational models. Furthermore, structural biology analysis provides valuable insights and guidance for understanding antibody–epitope interactions. More importantly, in this study, the ten important residues of the P5-5 epitope were verified by point mutation and reduced to seven amino acids (2–8 motif, DALYEYL), which greatly facilitated future disease treatment and vaccine development.

The P5-5 epitope represents the most critical pathogenic determinant within the TN protein. Our prior research demonstrated that monoclonal antibodies targeting this epitope exhibit therapeutic potential in sepsis-induced murine models. Therefore, elucidating the key amino acid residues involved in the P5-5 epitope is of considerable importance for the design and production of next-generation recombinant antibodies.

The affinity constant of an antibody is a fundamental parameter used to characterize the interaction between an antigen and an antibody. In this study, the Beatty-improved non-competitive ELISA method was utilized to evaluate both antibody purity and the affinity constant, which is expressed by the following formula: $K_{aff} = (n - 1)/\{2(n[Abt] - [Ab])\}$ [32,33]. This assay has been widely adopted and is especially suitable for determining the affinity constant between antibodies and immobilized antigens. The measurable range of the antibody affinity constant (K_{aff}) using this method generally spans from 10^6 to 10^{10} L/mol, which is consistent with the commonly reported affinity range for most monoclonal antibodies (10^7 to 10^9 L/mol). In conclusion, this assay method is simple, practical, and capable of producing accurate and reliable results [15,16,22,23,32,33,40,41]. Furthermore, it is not limited by the molecular weight of the antigen and exhibits broad applicability.

Recombinant antibodies are generated through the expression of specific genes, resulting in high batch-to-batch consistency and overcoming the limitations associated with traditional hybridoma-based antibody production. Unlike conventional methods, recombinant antibodies are not constrained by low repeatability or genetic drift, which ensures their stable binding to pathogenic targets and lays the foundation for highly specific and sensitive detection in disease diagnosis and monitoring. In addition, instead of high manufacturing costs, prolonged production cycles, and limited flexibility for genetic modification as were associated with traditional hybridoma antibody production, the production of recombinant gene-engineered antibodies only requires the preservation of antibody-encoding genes obtained via phage display technology, without the risk of cell line contamination or mutation-induced degradation. Collectively, this approach ensures high-throughput and consistent manufacture of reliable therapeutic bio-agents for the patients who are in urgent need.

5. Conclusions

In this study, the two most critical amino acid residues of the P5-5 pathogenic epitope were identified through alanine scanning for antibody binding. To our best knowledge, this is the first time to report two key residues (4L and 8L) in the P5-5 pathogenic epitope of interaction between therapeutic antibody and antigen. The results demonstrated that minor structural variations in antigens can significantly impact antigen–antibody binding, a process that is vital for both disease research and vaccine development.

Concurrently, this study also prepared a genetically engineered single-chain variable fragment against P5-5, which establishes the basis for future diagnosis and treatment. The sensitivity of our scFv-based ELISA against TN-P5-5 is 62.86 ng/mL. Furthermore, the experiment takes only two hours because the P5-5 antigen has already been precoated and blocked in the ELISA plate. Moreover, our existing ic-ELISA methods may be used to analyze P5-5 epitope samples in PBS, with a target detection range of 78.125 ng/mL to 1.25 μg/mL. The recombinant fusion scFv might be necessary for specific applications. For instance, intricate fusion proteins like antibody–enzyme hybrids or bispecific antibodies necessitated a recombinant approach, as they could not be synthesized with precise stoichiometry through traditional chemical modification methods. Therefore, our work has broad implications for future studies.

Supplementary Materials: The following supporting information can be downloaded at: https://www.mdpi.com/article/10.3390/biom15081100/s1. Figure S1: Purification of anti-TN-P5-5 mAb; Figure S2: Molecular characterization of scFv; Figure S3: Bioinformatics analysis of mAb 12F1 variable regions; Figure S4: The sequencing results alignment of the expression vector pMAL-c2x-linker-12F1-scFv; Figure S5: Affinity determination of 12F1 scFv; Figure S6: 12F1 scFv inhibitory standard curve for TN-P5-5 pathogenic epitope.

Author Contributions: Conceptualization, Y.X., J.W., Z.C. and R.Z.; methodology, J.W., R.Z. and W.Z.; software, J.W.; validation, M.L., Y.L. and D.L.; formal analysis, J.W.; investigation, M.L.; resources, J.H. and Y.X.; data curation, J.W., M.L., R.Z., D.L., and W.Z.; writing—original draft preparation, J.W.; writing—review and editing, Y.X.; visualization, Y.L.; supervision, Y.X.; project administration, Y.X. and J.H.; funding acquisition, Y.X. All authors have read and agreed to the published version of the manuscript.

Funding: This research was funded by the National Natural Science Foundation of China Major Research Plan Project, grant number 91742101; the Anhui International Science and Technology Collaborative Project, China, grant number 1604b0602017; the Natural Science Foundation of Anhui Province, China, grant number 1608085MH160; and the scientific research projects of colleges and universities of Anhui Province, China, grant number 2024AH051938.

Institutional Review Board Statement: The animal study protocol was approved by Anhui Normal University (AHNU-ET2022015) on 8 March 2022 for studies involving animals.

Informed Consent Statement: Not applicable.

Data Availability Statement: The data that support the findings of this study are available from the corresponding author upon reasonable request.

Acknowledgments: We are grateful to Jiasheng Hao, Ruie Nie and Chengyong Su for help and support in providing laboratory.

Conflicts of Interest: The authors declare no conflicts of interest.

References

1. Yang, X.; Pu, X.; Xu, Y.; Zhao, J.; Fang, X.; Cui, J.; Deng, G.; Liu, Y.; Zhu, L.; Shao, M.; et al. A novel prognosis evaluation indicator of patients with sepsis created by integrating six microfluidic-based neutrophil chemotactic migration parameters. *Talanta* **2025**, *281*, 126801. [CrossRef] [PubMed]
2. Singer, M.; Deutschman, C.S.; Seymour, C.W.; Shankar-Hari, M.; Annane, D.; Bauer, M.; Bellomo, R.; Bernard, G.R.; Chiche, J.D.; Coopersmith, C.M.; et al. The Third International Consensus Definitions for Sepsis and Septic Shock (Sepsis-3). *JAMA* **2016**, *315*, 801–810. [CrossRef] [PubMed]
3. Cui, Z.; Wang, L.; Li, H.; Feng, M. Study on immune status alterations in patients with sepsis. *Int. Immunopharmacol.* **2023**, *118*, 110048. [CrossRef]
4. Delano, M.J.; Ward, P.A. Sepsis-induced immune dysfunction: Can immune therapies reduce mortality? *J. Clin. Investig.* **2016**, *126*, 23–31. [CrossRef]

5. Zhu, C.L.; Wang, Y.; Liu, Q.; Li, H.R.; Yu, C.M.; Li, P.; Deng, X.M.; Wang, J.F. Dysregulation of neutrophil death in sepsis. *Front. Immunol.* **2022**, *13*, 963955. [CrossRef]
6. Tseng, Y.T.; Yu, Y.H.; Yeh, Y.Y.; Mai, P.C.; Huang, T.T.; Huang, C.J.; Chau, L.K.; Chen, Y.L. Femtomolar-level detection of procalcitonin using a split aptamer-based fiber optic nanogold-linked sorbent assay for diagnosis of sepsis. *Talanta* **2025**, *293*, 128150. [CrossRef]
7. Chun, K.; Syndergaard, C.; Damas, C.; Trubey, R.; Mukindaraj, A.; Qian, S.; Jin, X.; Breslow, S.; Niemz, A. Sepsis Pathogen Identification. *J. Lab. Autom.* **2015**, *20*, 539–561. [CrossRef]
8. Reddy, B., Jr.; Hassan, U.; Seymour, C.; Angus, D.C.; Isbell, T.S.; White, K.; Weir, W.; Yeh, L.; Vincent, A.; Bashir, R. Point-of-care sensors for the management of sepsis. *Nat. Biomed. Eng.* **2018**, *2*, 640–648. [CrossRef]
9. Bradley, Z.; Bhalla, N. Point-of-care diagnostics for sepsis using clinical biomarkers and microfluidic technology. *Biosens. Bioelectron.* **2023**, *227*, 115181. [CrossRef]
10. Qiu, X.; Lei, Y.P.; Zhou, R.X. SIRS, SOFA, qSOFA, and NEWS in the diagnosis of sepsis and prediction of adverse outcomes: A systematic review and meta-analysis. *Expert Rev. Anti Infect. Ther.* **2023**, *21*, 891–900. [CrossRef] [PubMed]
11. Chen, W.; Qiang, X.; Wang, Y.; Zhu, S.; Li, J.; Babaev, A.; Yang, H.; Gong, J.; Becker, L.; Wang, P.; et al. Identification of tetranectin-targeting monoclonal antibodies to treat potentially lethal sepsis. *Sci. Transl. Med.* **2020**, *12*, eaaz3833. [CrossRef] [PubMed]
12. Li, J.; Bao, G.; Wang, H. Time to Develop Therapeutic Antibodies Against Harmless Proteins Colluding with Sepsis Mediators? *Immunotargets Ther.* **2020**, *9*, 157–166. [CrossRef]
13. Paterson, C.W.; Ford, M.L.; Coopersmith, C.M. Breaking the bond between tetranectin and HMGB1 in sepsis. *Sci. Transl. Med.* **2020**, *12*, eabb2575. [CrossRef]
14. Wang, J.; Liu, M.; Cai, Z.; Zahid, R.; Zhang, W.; Ma, D.; Li, D.; Liang, Y.; Zha, L.; Zhou, Y.; et al. Pathogenic epitope-specific monoclonal antibody-based immunoassay for accurate diagnosis and monitoring of tetranectin in sepsis. *Int. Immunopharmacol.* **2024**, *143*, 113473. [CrossRef]
15. Tang, H.; Liu, H.; Gao, Y.; Chen, R.; Dong, M.; Ling, S.; Wang, R.; Wang, S. Detection of alphaB-Conotoxin VxXXIVA (alphaB-CTX) by ic-ELISA Based on an Epitope-Specific Monoclonal Antibody. *Toxins* **2022**, *14*, 166. [CrossRef]
16. Tang, H.; Wang, Q.; Yang, M.; Jia, R.; Yuan, J.; Wang, R. Development of sensitive immunoassay for identification and detection of mu-KIIIA-CTX: Insights into antibody discovery, molecular recognition, and immunoassay. *Int. J. Biol. Macromol.* **2025**, *310*, 143346. [CrossRef] [PubMed]
17. Wang, R.; Gu, X.; Zhuang, Z.; Zhong, Y.; Yang, H.; Wang, S. Screening and Molecular Evolution of a Single Chain Variable Fragment Antibody (scFv) against Citreoviridin Toxin. *J. Agric. Food Chem.* **2016**, *64*, 7640–7648. [CrossRef]
18. Wang, R.; Fang, S.; Wu, D.; Lian, J.; Fan, J.; Zhang, Y.; Wang, S.; Lin, W. Screening for a single-chain variable-fragment antibody that can effectively neutralize the cytotoxicity of the Vibrio parahaemolyticus thermolabile hemolysin. *Appl. Environ. Microbiol.* **2012**, *78*, 4967–4975. [CrossRef]
19. Martviset, P.; Thanongsaksrikul, J.; Geadkaew-Krenc, A.; Chaimon, S.; Glab-Ampai, K.; Chaibangyang, W.; Sornchuer, P.; Srimanote, P.; Ruangtong, J.; Prathaphan, P.; et al. Production and immunological characterization of the novel single-chain variable fragment (scFv) antibodies against the epitopes on Opisthorchis viverrini cathepsin F (OvCatF). *Acta Trop.* **2024**, *254*, 107199. [CrossRef]
20. Jalilzadeh-Razin, S.; Mantegi, M.; Tohidkia, M.R.; Pazhang, Y.; Pourseif, M.M.; Barar, J.; Omidi, Y. Phage antibody library screening for the selection of novel high-affinity human single-chain variable fragment against gastrin receptor: An in silico and in vitro study. *Daru* **2019**, *27*, 21–34. [CrossRef] [PubMed]
21. Ren, Z.; Zhang, H.; Yang, L.; Wang, Z.; Xiong, J.; Zheng, P.; Wang, J.; Jiang, H. Targeted preparation and recognition mechanism of broad-spectrum antibody specific to Aconitum alkaloids based on molecular modeling and its application in immunoassay. *Anal. Chim. Acta* **2022**, *1222*, 340011. [CrossRef]
22. Tang, H.; Liu, H.; Chen, R.; Gao, Y.; Dong, M.; Ling, S.; Wang, R.; Wang, S. Development of Immunochromatographic Strip for Detection of alphaB-VxXXIVA-Conotoxin Based on 5E4 Monoclonal Antibody. *Toxins* **2022**, *14*, 191. [CrossRef]
23. Wang, R.; Wang, J.; Liu, H.; Gao, Y.; Zhao, Q.; Ling, S.; Wang, S. Sensitive immunoassays based on specific monoclonal IgG for determination of bovine lactoferrin in cow milk samples. *Food Chem.* **2021**, *338*, 127820. [CrossRef]
24. Chang, W.; Fang, J.; Han, S.; Sun, H.; Zhai, T.; Wang, L.; Qi, X.; Xue, Q.; Wang, J. Screening and identification of LSDV-specific monoclonal antibodies to establish a double-antibody sandwich ELISA for distinguishing LSDV from SPPV and GTPV. *Int. J. Biol. Macromol.* **2025**, *307*, 142242. [CrossRef] [PubMed]
25. Huang, J.; Lin, A.; Gu, Y.; Pan, X.; Ma, X.; Qing, Y.; Li, J. Fluorescence-activated cell sorting-based efficient screening of monensin monoclonal antibodies and applications in lateral flow immunoassay. *Talanta* **2025**, *293*, 128128. [CrossRef] [PubMed]
26. Liu, Y.; Jin, Z.; Sun, D.; Xu, B.; Lan, T.; Zhao, Q.; He, Y.; Li, J.; Cui, Y.; Zhang, Y. Preparation of hapten and monoclonal antibody of hesperetin and establishment of enzyme-linked immunosorbent assay. *Talanta* **2025**, *281*, 126912. [CrossRef] [PubMed]

27. Wang, J.; Sun, C.; Hu, Z.; Wang, F.; Chang, J.; Gao, M.; Ye, D.; Jia, Q.; Zou, H.; Willems, L.; et al. Development of a novel monoclonal antibody-based competitive ELISA for antibody detection against bovine leukemia virus. *Int. J. Biol. Macromol.* **2024**, *267*, 131446. [CrossRef]
28. Peng, Y.; Jin, Y.; Sun, D.; Jin, Z.; Zhao, Q.; He, Y.; Jiao, B.; Cui, Y.; Zhang, Y. Monoclonal antibody-based icELISA for sensitive monitoring fenpyroximate residue by hydrolysis conversion. *Talanta* **2024**, *268*, 125288. [CrossRef]
29. Wang, R.; Zhong, Y.; Wang, J.; Yang, H.; Yuan, J.; Wang, S. Development of an ic-ELISA and immunochromatographic strip based on IgG antibody for detection of omega-conotoxin MVIIA. *J. Hazard. Mater.* **2019**, *378*, 120510. [CrossRef]
30. Yang, H.; Zhong, Y.; Wang, J.; Zhang, Q.; Li, X.; Ling, S.; Wang, S.; Wang, R. Screening of a ScFv Antibody with High Affinity for Application in Human IFN-gamma Immunoassay. *Front. Microbiol.* **2018**, *9*, 261. [CrossRef]
31. Chen, Y.; Feng, C.; Huang, C.; Shi, Y.; Omar, S.M.; Zhang, B.; Cai, G.; Liu, P.; Guo, X.; Gao, X. Preparation of polyclonal antibodies to chicken P62 protein and its application in nephropathogenic infectious bronchitis virus-infected chickens. *Int. J. Biol. Macromol.* **2024**, *271*, 132515. [CrossRef] [PubMed]
32. Beatty, J.D.; Beatty, B.G.; Vlahos, W.G. Measurement of monoclonal antibody affinity by non-competitive enzyme immunoassay. *J. Immunol. Methods* **1987**, *100*, 173–179. [CrossRef] [PubMed]
33. Wang, R.; Zeng, L.; Yang, H.; Zhong, Y.; Wang, J.; Ling, S.; Saeed, A.F.; Yuan, J.; Wang, S. Detection of okadaic acid (OA) using ELISA and colloidal gold immunoassay based on monoclonal antibody. *J. Hazard. Mater.* **2017**, *339*, 154–160. [CrossRef]
34. Yang, M.; Jia, R.; Liu, Y.; Tang, H.; Wu, H.; Yuan, J.; Wang, R. High-sensitivity and rapid immunoassays for furosine detection based on monoclonal antibody 1C3. *Int. J. Biol. Macromol.* **2025**, *309*, 143126. [CrossRef] [PubMed]
35. Xu, J.; Sun, J.; Lu, X.; Wang, Y.; Zhang, Y.; Sun, X. A highly sensitive fluorescence immunochromatography strip for thiacloprid in fruits and vegetables using recombinant antibodies. *Talanta* **2023**, *256*, 124258. [CrossRef]
36. Wang, F.; Liu, M.; Ma, D.; Cai, Z.; Liu, L.; Wang, J.; Zhang, W.; Zhao, L.; Zhai, C.; Xu, Y. Dendritic cell-expressed IDO alleviates atherosclerosis by expanding $CD4^{+}CD25^{+}Foxp3^{+}$Tregs through IDO-Kyn-AHR axis. *Int. Immunopharmacol.* **2023**, *116*, 109758. [CrossRef]
37. Wang, F.; Liu, L.; Wang, J.; Liu, M.; Zhang, W.; Zhao, L.; Zhai, C.; Xu, Y. Gain-of-function of IDO in DCs inhibits T cell immunity by metabolically regulating surface molecules and cytokines. *Exp. Ther. Med.* **2023**, *25*, 234. [CrossRef]
38. Dondelinger, M.; Filee, P.; Sauvage, E.; Quinting, B.; Muyldermans, S.; Galleni, M.; Vandevenne, M.S. Understanding the Significance and Implications of Antibody Numbering and Antigen-Binding Surface/Residue Definition. *Front. Immunol.* **2018**, *9*, 2278. [CrossRef]
39. Liang, Y.; Wang, Y.; Wang, F.; Li, J.; Wang, C.; Dong, J.; Ueda, H.; Xiao, Z.; Shen, Y.; Xu, Z.; et al. An enhanced open sandwich immunoassay by molecular evolution for noncompetitive detection of Alternaria mycotoxin tenuazonic acid. *Food Chem.* **2021**, *361*, 130103. [CrossRef]
40. Cai, P.; Wang, R.; Ling, S.; Wang, S. A high sensitive platinum-modified colloidal gold immunoassay for tenuazonic acid detection based on monoclonal IgG. *Food Chem.* **2021**, *360*, 130021. [CrossRef]
41. Wang, R.; Huang, A.; Liu, L.; Xiang, S.; Li, X.; Ling, S.; Wang, L.; Lu, T.; Wang, S. Construction of a single chain variable fragment antibody (scFv) against tetrodotoxin (TTX) and its interaction with TTX. *Toxicon* **2014**, *83*, 22–34. [CrossRef] [PubMed]

biomolecules

MDPI

Article

Glycan Signatures on Neutrophils in an Equine Model for Autoimmune Uveitis

Carolin J. Sprenzel, Barbara Amann, Cornelia A. Deeg and Roxane L. Degroote *

Chair of Physiology, Department of Veterinary Sciences, Ludwig-Maximilians-University Munich, D-82152 Martinsried, Germany
* Correspondence: r.degroote@lmu.de

Abstract: Glycosylation of surface proteins is a crucial post-translational modification that reflects the activation status of neutrophils, the predominant leukocyte subset in humans and horses. Neutrophils have emerged as active contributors to diseases mediated by the adaptive immune system, such as equine recurrent uveitis (ERU), a sight-threatening disease in horses and a unique model for studying the pathogenesis of autoimmune uveitis in humans. Since changes in surface glycosylation can impact neutrophil function, we were interested in the surface glycosylation landscape on neutrophils from healthy horses and the potential changes in surface glyco-signatures in ERU. Using 35 different plant lectins, we outlined a profile of surface-exposed glycan moieties on equine neutrophils and detected significantly increased O-glycosylation in a diseased state through Jacalin (JAC) binding via flow cytometry. Subsequent molecular weight comparison of JAC pull-down assay data and neutrophil proteomics indicated the surface proteins Integrin beta-2 and CUB domain-containing protein 1 as potential anchors for increased O-glycan levels in ERU. These findings give novel insights into neutrophil surface glycosylation in health and disease and propose O-glycosylation as a possible biomarker for autoimmune uveitis.

Keywords: O-glycosylation; surface glycome; Jacalin; neutrophils; equine recurrent uveitis; ERU; autoimmune uveitis

1. Introduction

The cell surface is characterized by membrane proteins, which fulfill various functions like transport, signaling, cell–cell-communication, enzymatic activity or identification of cell subsets via glycoproteins. For example, human blood types can be determined by glycoproteins on the cell surface of erythrocytes [1]. After transcription, translation, and folding, membrane proteins usually undergo post-translational modifications (PTM). One of the most important PTMs of membrane proteins is glycosylation [2].

Glycosylation describes the attachment of glycans to proteins. The two major types of glycosylation are N- and O-glycosylation [3,4]. While N-glycosylation is characterized by glycan-attachment to the nitrogen atom of asparagine residues, O-glycosylation describes the attachment of glycans to the oxygen atom of serine or threonine residues [5]. The role of cell surface glycosylation in autoimmune diseases is an emerging area of interest. In patients suffering from systemic lupus erythematosus, alterations in numerous glycans were discovered in different immune cell subsets. These alterations were described to correlate with the severity of systemic lupus erythematosus, indicating a connection between surface glycome composition and disease pathogenesis in humans [6]. Aberrant surface glycosylation may even serve as a biomarker for autoimmune diseases like multiple

https://doi.org/10.3390/biom15101444

sclerosis in humans [7]. Furthermore, PTMs might be involved in autoimmune uveitis (AU) pathogenesis. For example, Yin Yang 1 lactylation, a PTM describing the addition of a lactate molecule to the protein Yin Yang 1, was described to enhance microglial functions in the retina of a murine experimental autoimmune uveitis model by upregulation of inflammatory genes [8].

AU is characterized by chronic recurrent inflammation of the eye, caused by invading immune cells from the peripheral bloodstream, which overcome the blood–retinal barrier and cause severe destruction of inner eye structures [9]. Treatment options are limited to glucocorticoids and immunosuppressants, which can cause detrimental side effects in patients [10]. Equine recurrent uveitis (ERU) resembles AU in many aspects of pathophysiology and serves as a valuable spontaneous model [11–13]. Unlike AU, treatment of horses affected by ERU is performed only during active uveitis [13–15]. Therefore, the stages of quiescence allow investigation of untreated immune cells in ERU. Although ERU and AU are described as mainly T-cell driven diseases [16,17], the presences of neutrophils in eyes from induced mouse and rat models for AU [18–20] and in ocular tissue from ERU horses [21] indicates a possible role of these cells in disease pathogenesis, especially since these cells show a pre-activated phenotype even in the quiescent stage of disease [22].

Neutrophils are the most abundant leukocyte subset in human and equine blood. As the first line of defense against pathogens, these cells play a crucial role in inflammation [23] by performing their well-described repertoire of defense mechanisms: phagocytosis, degranulation, production of reactive oxygen species or neutrophil extracellular traps [24,25]. Excessive neutrophil effector functions can lead to hyperinflammation, resulting in tissue damage. This explains why an adequate regulation of neutrophil immune response is essential for health [26].

Since PTMs such as glycosylation can modulate immune cell function, we aimed to unravel the surface glycome of equine neutrophils and investigate potential changes in ERU, which may provide insights into novel activation mechanisms of neutrophils in autoimmune uveitis. Aberrant glycan moieties on neutrophils from horses with ERU may be suitable to serve as biomarkers for disease prediction and monitoring or even as targets for therapeutic intervention.

2. Materials and Methods

2.1. Animals and Cell Preparation

Blood was drawn from 13 control horses and 22 equine recurrent uveitis (ERU) cases for all experiments within this study. Some of the horses in both groups were used in multiple assays. ERU was diagnosed by experienced clinicians from the Equine Clinic at LMU Munich according to typical clinical signs of recurrent uveitis in addition to a documentation of at least two recent episodes of inflammation in the affected eye [27]. Horses with end-stage recurrent uveitis (blindness, enucleation) were excluded from the study. During blood withdrawal, all ERU cases were in a quiescent stage of disease to undergo a therapeutic procedure (pars plana vitrectomy). Horses were classified as ophthalmologically healthy when no history of ocular inflammation was reported by the owner, and a standard clinical examination performed by a veterinarian revealed no evidence of pathophysiological alterations within the eye. Collection of blood was permitted by the local authority, Regierung von Oberbayern (Permit number: ROB-55.2-2532.Vet_03-22-37).

All equine whole blood samples were taken from the vena jugularis and stored in tubes containing heparin sodium (50 I.U. per ml blood; Ratiopharm, Ulm, Germany). Following initial sedimentation of erythrocytes at room temperature, the leukocyte-enriched plasma was carefully layered onto Pancoll separation solution (PanBiotech, Aidenbach,

Germany), and leukocyte subsets were separated using density gradient centrifugation (room temperature (RT), 350× g, 25 min, brake off). The neutrophils were extracted from the sediment at the bottom of the density gradient. After removing the remaining layers of plasma and pancoll, the sediment was carefully resuspended in phosphate-buffered saline (PBS; NaCl 136.9 mM (Sigma-Aldrich, Taufkirchen, Germany), $Na_2HPO_4 \times 2H_2O$ 8.1 mM (AppliChem, Darmstadt, Germany), KH_2PO_4 1.4 mM (Sigma-Aldrich, Taufkirchen, Germany), KCl 2.6 mM (Sigma-Aldrich, Taufkirchen, Germany); pH 7.4). The remaining erythrocytes were lysed twice with 20 mL 0.2% sodium chloride (NaCl, Sigma-Aldrich, Taufkirchen, Germany) for 30 s, followed by 20 mL 1.6% NaCl to recover isotonicity and a washing step (RT, 400× g, 10 min). The resulting neutrophils were resuspended in PBS. Cell counts for neutrophils were determined in a Neubauer improved counting chamber using trypan blue solution (VWR Life Science, Darmstadt, Germany) for exclusion of dead cells.

2.2. Analysis of Surface Glycosylation by Flow Cytometry

Neutrophils from seven controls and eleven ERU cases were seeded into a 96-well U-bottom plate with a cell count of 5×10^5 per well after excluding dead cells. All lectins were incubated for 15 min at RT and subsequently washed with PBS (RT, 400× g, 2 min). To exclude dead cells, Viobility 400/452 Fixable Dye (Miltenyi Biotec, Bergisch Gladbach, Germany) was used. Cells were stained with 35 biotinylated plant lectins (1 µL/mL) to assess glycan-binding profiles (lectins used are from Biozol, Eching, Germany, or Linaris, Dossenheim, Germany): Aleuria aurantia lectin (AAL), Agaricus bisporus lectin (ABL), Amaranthus caudatus lectin (ACL), Banana lectin (BanLec), Bauhinia purpurea lectin (BPL), Concanavalin A (ConA), Dolichos biflorus agglutinin (DBA), Datura stramonium lectin (DSL), Erythrina cristagalli lectin (ECL), Galanthus nivalis lectin (GNL), Griffonia simplicifolia lectin I (GSL-I), Griffonia simplicifolia lectin I Isolectin B4 (GSL-I-B4), Griffonia simplicifolia lectin II (GSL-II), Hippeastrum hybrid lectin (HHL), Jacalin (JAC), Lens culinaris agglutinin (LCA), Lycopersicon esculentum (tomato) lectin (LEL), Lotus tetragonolobus lectin (LTL), Maackia amurensis lectin II (MAL-II), Narcissus pseudonarcissus lectin (NPL), Phaseolus vulgaris erythroagglutinin (PHA-E), Phaseolus vulgaris leucoagglutinin (PHA-L), Peanut agglutinin (PNA), Pisum sativum agglutinin (PSA), Psophocarpus tetragonolobus lectin (PTL), Ricinus communis agglutinin I (RCA-I), Soybean agglutinin (SBA), Sophora japonica agglutinin (SJA), Sambuccus nigra agglutinin (SNA), Solanum tuberosum (potato) lectin (STL), succinylated Wheat germ agglutinin (sWGA), Ulex europaeus agglutinin I (UEA-I), Vicia villosa lectin (VVA), Wisteria floribunda lectin (WFL), Wheat germ agglutinin (WGA). More detailed information about the plant lectins is displayed in Table 1. Lectin binding was visualized using Streptavidin iFluor 488 and Streptavidin iFluor 555 (Biomol, Hamburg, Germany, 1:200). Cells were subsequently fixed with 1% paraformaldehyde (PFA; Sigma-Aldrich, Darmstadt, Germany).

To assess the glycosylation pattern of the neutrophil cell surface, measurements were performed using a NovoCyte Quanteon 4016 flow cytometer (Agilent Technologies, Waldbronn, Germany). Data were analyzed with Flowlogic Software V7 (Inivai Technologies, Mentone Victoria, Australia) and NovoExpress Software version 1.5.0 (Agilent Technologies). Lectin binding expression intensities were quantified using mean fluorescence intensity (MFI) values and normalized to respective unstained cells per animal.

2.3. Immunocytochemistry

Neutrophils (5×10^5 cells per well) from three controls and three ERU cases were seeded into a 96-well U-bottom plate and incubated with the biotinylated plant lectins DBA, BanLec and JAC (Biozol, Eching, Germany, 1 µg/mL each) for 15 min at RT. Neutrophils were then washed (RT, 400× g, 2 min). To visualize plant lectin binding on the cell surface,

Streptavidin iFluor488 (Biomol, Hamburg, Germany, 1:200) was added and incubated for 15 min at RT, followed by a final washing step (RT, 400× *g*, 2 min). Unspecific binding of Streptavidin iFluor488 was assessed via single staining without lectins. Cell nuclei were stained with 4′, 6-Diamidin-2-phenylindol (DAPI; Invitrogen, Karlsruhe, Germany; 1:1000) and fixed in PBS containing 1% PFA. Half of the cell suspension (2.5 × 10^5 cells) was transferred to microscope slides (RT, 300× *g*, 10 min) and embedded in fluoromount medium (Serva, Heidelberg; Germany), adding coverslips. Visualization was achieved with a Leica DMi8 microscope (Leica Microsystems, Wetzlar, Germany). Background removal was performed via THUNDER Computational Clearing (Leica Microsystems, Wetzlar, Germany) (Supplementary Materials: RawData ICC). Fluorescent images were quantified with ImageJ (Version 1.54g) [28]. For ImageJ analysis, signals from cell nucleus staining were excluded and fluorescent images of the lectin signals were converted into a greyscale format. After background subtraction, the pixel intensity per lectin signal on each cell was measured as mean gray value. A total of 10 cells per lectin were used for quantification.

For evaluation of O-glycan prediction from proteome data, neutrophils from one control and two ERU cases were used. Staining and visualization was performed as described (see above) with the following antibodies: anti-CD18 (Integrin beta-2; Clone CLB-LFA-1/1; Thermo Fisher Scientific, Ulm, Germany; 1:50), goat anti-mouse IgG Alexa568 (Invitrogen, Karlsruhe, Germany; 1:500), biotinylated plant lectin JAC (Biozol, Eching, Germany; 1 µg/mL), Streptavidin Alexa488 (Invitrogen, Karlsruhe, Germany; 1:500) (Supplementary Materials: RawData ICC).

2.4. JAC Pull-Down

Neutrophils from three controls and three ERU cases were lysed in CHAPS-lysis buffer (2% CHAPS (Carl Roth, Karlsruhe, Germany) in Tris-buffered saline with 1× Roche complete protease inhibitor (Sigma-Aldrich, Taufkirchen, Germany)) using 500 µL per 5 × 10^5 cells. CHAPS-lysis buffer was incubated for 30 min on ice and mixed repeatedly. Lysed cells were loaded twice onto a QIAshredder homogenizer (Quiagen, Stockach, Germany) and centrifuged (23 °C, 16,100 × *g*, 2 min). Protein concentration was determined by Bicinchonin Acid-Assay (Thermo Fisher Scientific, Ulm, Germany). Agarose bound JAC (Biozol, Eching, Germany) was transferred to low-binding cups (Sarstedt, Nümbrecht, Germany) and washed (RT, 1800× *g*, 30 s) five times with 250 µL TRIS-buffer (175 mM TRIS (Carl Roth, Karlsruhe, Germany), pH 7.5). Agarose beads were mixed with 1 mg neutrophil lysate, and the mixture was rotated overnight at 4 °C. The next day, samples were centrifuged (RT, 1000× *g*, 2 min) and the supernatants were transferred to new low-binding cups. The precipitates were washed three times with PBS (RT, 1000× *g*, 1 min) before adding Laemmli buffer (without dithiothreitol and glycerin: 50 mM Tris-HCl, 1% SDS (both AppliChem, Darmstadt, Germany), 10% glycerol (SERVA, Heidelberg, Germany), 4 mM 2-mercaptoethanol, bromophenol blue (both Sigma-Aldrich, Taufkirchen, Germany)) and heating the samples to 95 °C for 10 min while shaking (400× *g*). Samples were centrifuged (RT, 1000× *g*, 2 min) and the eluates were collected.

Proteins were separated by 1D SDS-PAGE on 10% gels with either 20 µL of eluate, 2 µL of supernatant, or 5 µg of lysate per slot. Serva Triple Color Protein Standard III (Serva Electrophoresis, Heidelberg, Germany) was used as a molecular weight marker. After electrophoresis, gels were briefly washed in Aqua bidest and proteins were detected overnight by a modified Coomassie Brilliant Blue G250 (Fluka, Buchs, Switzerland)-based staining method [29]. Visualization of stained SDS gels was performed with Amersham Imager680, Analysis2.0 (GE Healthcare, Freiburg, Germany) (Supplementary Materials: RawData WB). The exact molecular weight of protein bands of interest was determined by generating a graph based on the migration distance of the protein marker bands (distance protein marker band migration divided by total migration distance on gel) and subsequently allocating the migration distance of protein bands of interest on the graph (Figure S1).

Quantification was performed with ImageJ (Version 1.54g) [28]. Prior to ImageJ analysis, the scanned gel images were converted into a grayscale format. Pixel intensity per band was measured after background subtraction.

2.5. JAC Inhibition

Neutrophils from seven horses were seeded into a 96-well U-bottom plate with a cell count of 5×10^5 per well. To inhibit JAC binding to neutrophil surface glycans, FITC-conjugated JAC (Biozol, Eching, Germany) was incubated with the saccharides galactose, mannose, and sucrose (Merck KGaA, Darmstadt, Germany), diluted to a concentration of 20% in PBS for 30 min. Neutrophils were then stained with pre-treated JAC (1 µL/mL) for 15 min at RT and subsequently washed with PBS (RT, $400\times g$, 2 min).

Measurements were performed using a NovoCyte Quanteon 4016 flow cytometer (Agilent Technologies, Waldbronn, Germany). Data were analyzed with Flowlogic Software V7 (Inivai Technologies, Mentone Victoria, Australia) and NovoExpress Software version 1.5.0 (Agilent Technologies). JAC binding intensities (MFI) of saccharide-pre-treated JAC were normalized to binding intensities of untreated JAC.

2.6. Prediction of O-GalNAc Glycosylation and Evaluation of Neutrophil Proteome Data

Mass spectrometry analysis, protein identification and label-free quantification were conducted using a dataset previously generated and published by our group [22]. For the identification of candidate proteins, we reanalyzed this published dataset (raw data have been deposited to the ProteomeXchange Consortium via the PRIDE repository with the dataset identifier: PXD013648 [30]). This approach ensured consistency and reproducibility based on our established workflows. Prediction of O-GalNAc glycosylation was performed using NetOGlyc 4.0 [31]. Proteins were classified as O-glycosylated if they contained multiple predicted glycosylation sites with a probability score ≥0.5.

2.7. Statistical Analysis

To assess Gaussian distribution of the data, the Kolmogorov–Smirnov (KS) test was applied. For data showing normal distribution (KS $p > 0.05$), statistical differences between two groups were evaluated using Student's t-test. Equality of variances was tested using the F-test; in cases of significantly different variances or unequal group sizes, Welch's correction was applied. For data not following a normal distribution, the non-parametric Mann–Whitney U test was used. Results were considered statistically significant at $p \leq 0.05$. Statistical significance was indicated as follows: * $p \leq 0.05$, ** $p \leq 0.01$ and *** $p \leq 0.001$, **** $p \leq 0.0001$. All statistical analyses and graphical representations were performed using GraphPad Prism software (version 5.04; GraphPad Software, San Diego, CA, USA). Outliers were identified using Grubbs' test and excluded from further analysis if flagged as significant.

3. Results

3.1. Defining the Surface Glycome of Equine Neutrophils

To assess the surface glycosylation of neutrophils from control horses, a panel of 35 plant lectins was employed. These lectins specifically bind to well-defined glycan structures, enabling a detailed profiling of surface-exposed carbohydrate moieties. The screening revealed a heterogeneous glycosylation pattern: while certain glycan structures exhibited high surface abundance, others were barely detectable or entirely absent. Notably, very low binding was observed for lectins recognizing terminal poly- or multiantennary type 2 LacNAc or LacdiNAc (SJA, VVA), carbohydrate structures containing (α-1,3) linked Gal and (α-1,2) linked fucose (PTL), terminal GalNAc-α1,3-GalNAc (DBA), oligosaccharides with terminal α- or β-linked GalNAc (SBA, VVA), α- or β-linked GlcNAc (GSL II), (α-1,2) linked fucose to (β-1,4) linked galactose (UEA-1), (α-1,3) linked fucose containing oligosaccharides (LTL), and terminal α-mannose structures (HHL) (Figure 1, Tables 1 and S1). This absence of binding may indicate either a true lack of these glycan structures on the neutrophil surface or a spatial conformation that masks lectin accessibility.

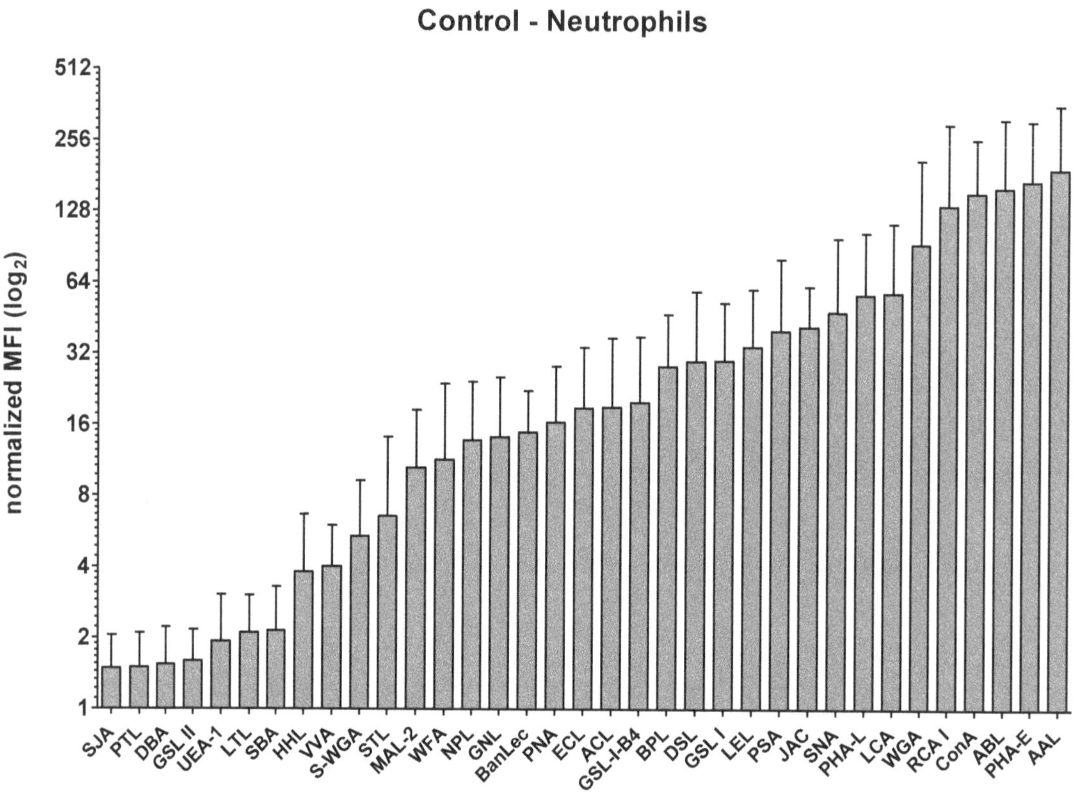

Figure 1. Surface glycosylation landscape of neutrophils from healthy control horses ($n = 7$; except AAL, ABL, ACL, GSL II, STL, WGA: $n = 6$ and DSL, JAC, LCA, LEL, PSA, RCA, SNA: $n = 5$; data was obtained in six independent experiments). Binding intensity of all 35 biotinylated plant lectins used in flow cytometry assays is shown with bar graphs. Mean fluorescence intensity with standard deviation (MFI ± SD) of lectin binding was normalized to respective unstained cells per animal. The y-axis is displayed on a $\log_2$ scale to account for differences in binding intensity across several orders of magnitude. Lectins are arranged from low binding to high binding intensity along the x-axis.

In contrast to the absent glycan structures, several carbohydrate motifs were accessible on the neutrophil surface and yielded strong lectin-binding signals. These included β-galactose, terminal α-linked mannose or biantennary N-glycans with multiple extensions, core 2 o-glycans, bisecting GlcNAc preferred with type 2 LacNAc, as well as α-linked fucose in specific linkages (Figure 1, Table 1 and Table S1). The observed lectin binding (e.g., by RCA I, ConA, ABL, PHA-E, and AAL; see Table 1) suggests the presence or accessibility of corresponding glycan motifs. However, due to the limited specificity of plant lectins and the influence of glycan context and conformation, these findings reflect the probable occurrence of terminal glycan features rather than complete structural elucidation. All other glycan epitopes showed intermediate signal intensities, further emphasizing the complexity and heterogeneity of surface glycosylation in equine neutrophils.

For a better overview of surface glycans, lectins were grouped according to their glycan binding motifs, resulting in seven distinct groups [32] (Figure 2): lectins with O-glycan binding, N-glycan binding, mannose binding, fucose binding, terminal galactose and LacNAc binding, GalNAc binding and GlcNAc, and sulfate and galactose binding motifs.

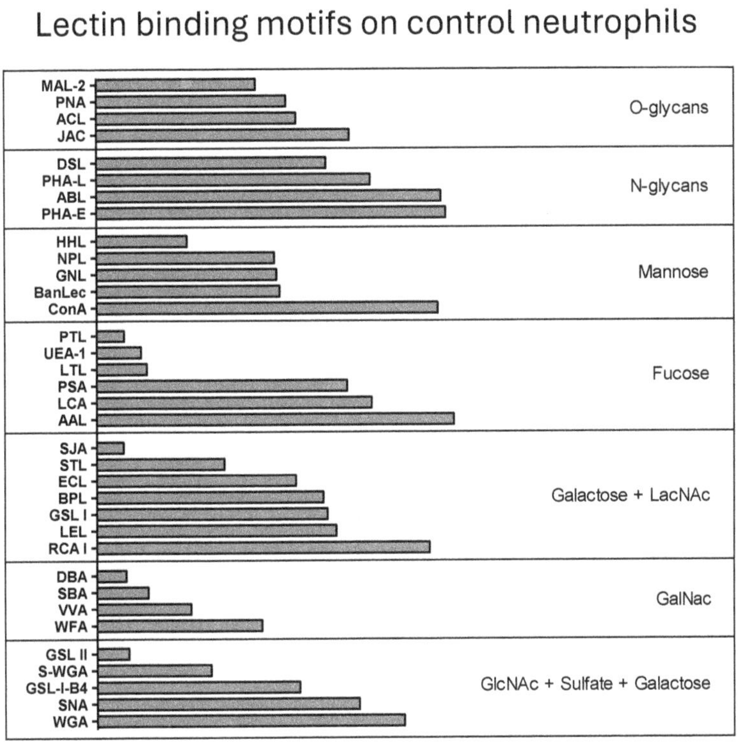

Figure 2. Lectin binding motifs on neutrophils from healthy control horses ($n = 7$; except AAL, ABL, ACL, GSL II, STL, WGA: $n = 6$ and DSL, JAC, LCA, LEL, PSA, RCA, SNA: $n = 5$; data was obtained in six independent experiments). Binding intensity of all 35 biotinylated plant lectins used in flow cytometry assays is shown with bar graphs analogous to Figure 1. Lectins are grouped according to their binding motif and arranged from low binding to high binding intensity within each group.

3.2. *Differentially Expressed Glycans on ERU Neutrophils*

To investigate whether surface glycosylation of neutrophils is altered in equine recurrent uveitis (ERU), lectin binding profiles of neutrophils from ERU cases were compared to those from healthy controls. Thirty-three of the thirty-five lectins used for the identification of equine surface glycosylation showed no significantly changed binding (Figure 3B,D–G), whereas two lectins showed statistically significant differences in binding intensity between controls and ERU cases. These lectins belonged to the O-glycan-binding (Figure 3A) and mannose-binding groups (Figure 3C).

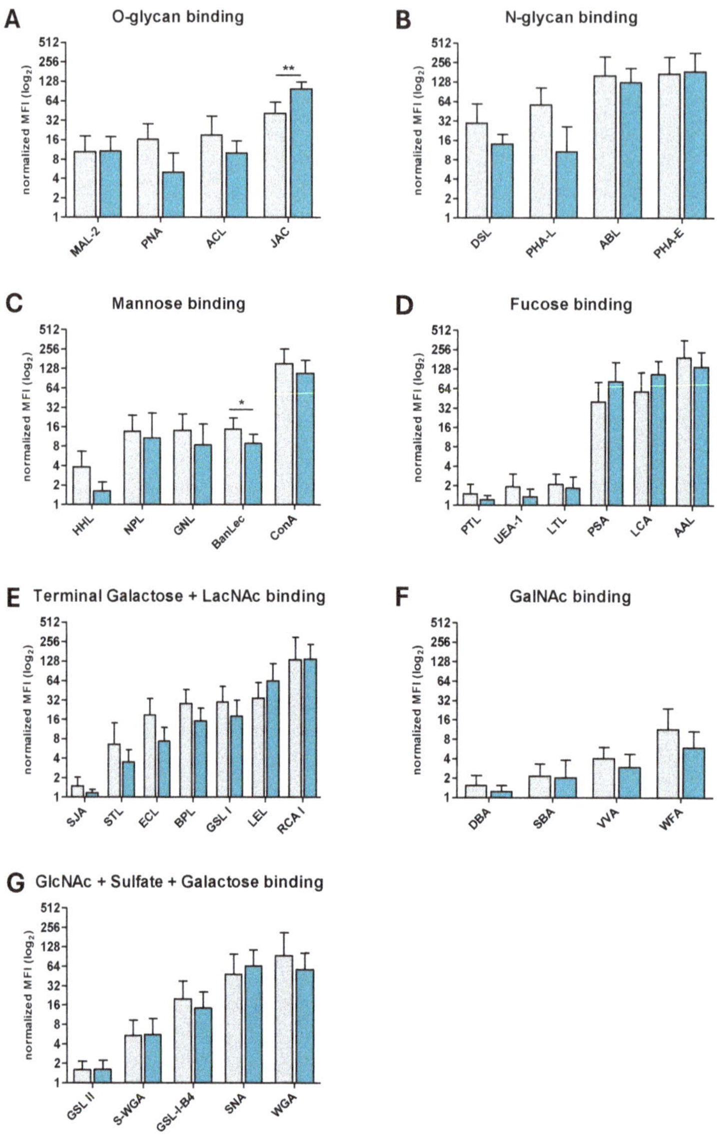

Figure 3. Surface glycosylation pattern of neutrophils from ERU cases ($n = 11$; except DBA, GSL II, HHL, LTL, MAL-2, NPL, PNA, UEA-1: $n = 10$ and ABL, DSL, JAC, LCA, LEL, PSA, PTL, RCA-I, SBA, SJA, SNA: $n = 9$) compared to controls ($n = 7$; except AAL, ABL, ACL, GSL II, STL, WGA: $n = 6$ and DSL, JAC, LCA, LEL, PSA, RCA, SNA: $n = 5$; data was obtained in six independent experiments). Binding intensities of the 35 biotinylated plant lectins used in flow cytometry assays are shown with bar graphs (gray: controls, blue: ERU) and are arranged according to the grouped binding motifs in Figure 2: (**A**) O-glycans with significant 2.4-fold increase in JAC reactive O-glycans on neutrophils from ERU cases (Student's *t*-test with Welch's correction, ** $p = 0.0012$), (**B**) N-glycans, (**C**) mannose with significant 1.7-fold decrease in BanLec reactive glycans on neutrophils from ERU cases (Mann–Whitney test, * $p = 0.0163$), (**D**) fucose, (**E**) galactose and LacNAc, (**F**) GalNac, and (**G**) GlcNAc and sulfate and galactose. Mean fluorescence intensity (MFI ± SD) of lectin binding was normalized to respective unstained cells per animal. Outliers were excluded via Grubb's test. The *y*-axis is displayed on a $\log_2$ scale to account for differences in binding intensity across several orders of magnitude.

Five of the plant lectins used in the surface glycome screening were grouped as mannose binding: BanLec, ConA, GNL, HHL, and NPL (Figures 2 and 3C). The lectin BanLec has binding sites for internal (α-1,3) glucosyl and mannosyl residues as well as reducing terminal (α 1,3) and α (1,6) glucosyl residues. HHL primarily detects terminally bound α-mannoses, which are also bound by ConA. However, ConA additionally detects biantennary N-glycans with multiple extensions. The plant lectins GNL and NPL preferably bind to α1-6-linked mannose, but an affinity for α1-3-linked mannose and for N-glycans with terminated short LacNAc has also been reported. The terminally bound α-mannoses detected by HHL were only detectable to a very small extent on the neutrophils in both comparison groups (Figure 1). ConA-bound structures were highly abundant on equine neutrophils (Figure 1). The glycan structures identified by BanLec, GNL, and NPL showed moderate surface abundance (Figure 1). Among the five mannose-binding lectins in this group, only the glycan structures containing internal α-(1,3)-linked glucose and mannose residues and reducing terminal (α 1,3) and α (1,6) glucosyl residues detected by BanLec showed significantly lower abundance on the cell surface of ERU neutrophils (1.7-fold, * $p = 0.0163$) (Figure 3C, Table 1).

Four of the plant lectins used for surface glycan screening could be used to detect O-glycans, namely ACL, JAC, MAL-2, and PNA (Figures 2 and 3A). ACL recognizes core 1 and core 2 O-glycans and poly Lewis structures. JAC, a lectin with a high affinity for mucin-type O-glycans, also binds to Galβ1-3GalNAc structures but prefers core 1 and core 3 glycans, and even tolerates mono- and disialylation. In contrast, PNA only detects terminally accessible core 1 and core 2 O-glycans. The lectin MAL-2 interacts primarily with α 2-3-sialylated O-glycans. The O-glycans bound by ACL, JAC, MAL-2, and PNA could be detected on the cell surface with moderate to high abundance (Figure 1) with no significant differences between controls and ERU (Figure 3A, Table 1). In contrast, JAC-bound O-glycans showed a significantly increased abundance on neutrophils from ERU cases (2.4-fold, ** $p = 0.0012$) (Figure 3A, Table 1).

While JAC is commonly used as a tool to detect O-glycoproteins due to its affinity for galactosyl (β-1,3) N-acetylgalactosamine, its specificity has limitations. JAC does not bind α-linked N-acetylgalactosamine if the sugar is modified at the C6 position or capped by sialic acid [33]. Thus, while the observed increase reflects altered O-glycosylation, it may underestimate total mucin-type glycan complexity on ERU neutrophils. To evaluate JAC specificity, we performed lectin inhibition via its monosaccharide agonist galactose as well as sucrose and mannose (Figure S2). Pre-incubation of JAC with galactose caused strongly decreased binding to equine neutrophils, which did not occur after JAC inhibition with sucrose and mannose (Figure S2).

Table 1. Alternations in surface glycosylation of equine neutrophils sorted by *p*-value. Normalized mean fluorescence intensities (MFI) show differences in lectin binding to specific glycan structures on neutrophils from controls ($n = 5$–7) and ERU cases ($n = 9$–11). (a) Lectin acronym, (b) binding specificity according to Bojar et al. [32], Lescar et al. [34], Singh et al. [35], and Monsigny et al. [36], (c) normalized mean fluorescence intensity (MFI) measured for binding of lectins on granulocytes from controls and ERU cases, with standard deviation (SD), (d) *p*-value of differential lectin binding intensity in controls compared to ERU cases.

Acronym [a]	Specificity [b]	Norm. MFI ± SD [c]		*p*-Value Ctr vs. ERU [d]
		Controls	ERU	
PHA-E	bisecting GlcNac preferred with type 2 LacNAc	172.6 ± 136.9	185.4 ± 176.2	1.000

Table 1. *Cont.*

Acronym [a]	Specificity [b]	Norm. MFI ± SD [c]		*p*-Value Ctr vs. ERU [d]
		Controls	ERU	
RCA I	terminal ß-galactose residues, with a preference for type 2 LacNAc	135.6 ± 164.8	139.4 ± 95.0	0.9638
S-WGA	Oligosaccarides containing terminal GlcNAc	5.4 ± 3.9	5.5 ± 4.2	0.9396
AAL	α-linked fucose; preferring (α-1,2)-linked fucose terminated structures on type 2 LacNAc	135.6 ± 71.8	139.8 ± 95.2	0.8802
STL	internal type 2 LacNAc as linear glycans; type 2 LacdiNAc; Chitin oligomers	3.9 ± 3.1	3.5 ± 1.9	0.8802
ABL	biantennary N-glycans terminating with ß-GlcNAc or LacNAc; core 2 O-glycans	108.7 ± 73.0	126.0 ± 82.3	0.6764
MAL-2	(α-2,3) sialylated O-glycans	10.5 ± 7.9	8.7 ± 3.5	0.6100
GSL II	α- or β-linked GlcNAc, preferring GlcNac capped LacNAc in multiantennary N-glycans	1.6 ± 0.6	1.5 ± 0.3	0.5944
SNA	(α-2,6) sialylated LacNAc or LacdiNAc	48.0 ± 50.6	64.5 ± 49.9	0.5746
SBA	Oligosaccharide structures with terminal α- or β-linked GalNAc	2.1 ± 1.2	2.2 ± 1.9	0.5220
WGA	multiple terminal N-acetylstructures; oligosaccharides containing terminal α- or β-linked GlcNac	52.8 ± 54.1	57.1 ± 45.2	0.5136
LTL	(α-1,3) linked fucose containing oligosaccharides	2.1 ± 0.9	1.6 ± 0.4	0.4927
GSL-I-B4	terminal α-galactose residues	19.9 ± 17.8	14.3 ± 11.1	0.4777

Table 1. *Cont.*

Acronym [a]	Specificity [b]	Norm. MFI ± SD [c]		*p*-Value Ctr vs. ERU [d]
		Controls	ERU	
PTL	Carbohydrate structures containing (α-1,3) linked galactose and (α-1,2) linked fucose	1.5 ± 0.6	1.2 ± 0.2	0.4482
DSL	(β-1,4) and (β-1,6) linked GlcNAc oligomers, preferring chitobiose or chitotriose; branching structures with several type 2 LacNac repeats	18.4 ± 10.2	14.1 ± 5.8	0.4271
ConA	terminal α-linked mannose, biantennary N-glycans with multiple extensions	153.1 ± 105.8	108.6 ± 64.1	0.3450
DBA	terminal GalNAc-α1,3-GalNAc	1.5 ± 0.7	1.2 ± 0.2	0.3166
SJA	terminal poly- or multiantennary type 2 LacNAc or LacdiNAc; bloodgroup B	1.5 ± 0.6	1.2 ± 0.2	0.2804
WFA	Carbohydrate structures terminating in α- or β-linked GalNAc, terminal LacNAc	11.4 ± 12.5	5.9 ± 4.7	0.2766
VVA	terminal α- or β-linked GalNac, terminal ß-LacdiNAc	4.0 ± 2.0	2.9 ± 1.8	0.2716
ACL	core 1 and core 2 O-glycans; poly Lewis structures	12.6 ± 7.2	10.0 ± 5.4	0.2272
PSA	core fucose structures	40.0 ± 40.5	82.7 ± 80.0	0.2123
LEL	type 2 polyLacNac; type 2 LacdiNAc; Chitin oligomers	34.1 ± 25.8	63.6 ± 54.5	0.1978
LCA	Core fucose; terminal (α-1,2) mannose	58.0 ± 55.8	106.9 ± 64.3	0.1714
HHL	terminal α-mannose structures	3.8 ± 2.9	2.0 ± 0.6	0.1711
GSL I	α-GalNac residues and α-galactose residues	29.9 ± 22.7	18.0 ± 14.0	0.1600

Table 1. *Cont.*

Acronym [a]	Specificity [b]	Norm. MFI ± SD [c]		p-Value Ctr vs. ERU [d]
		Controls	ERU	
UEA-1	(α-1,2) linked fucose to (β-1,4) linked galactose	1.9 ± 1.1	1.3 ± 0.4	0.1521
BPL	oligosaccharides with terminal β-linked galactose or β-linked GalNac	28.2 ± 18.8	15.1 ± 8.9	0.1282
ECL	terminal type 2 LacNAc or LacdiNAc	18.8 ± 15.1	7.3 ± 4.6	0.0993
NPL	terminal (α-1,6) and (α-1,3) mannose residues; N-glycans with terminated short LacNAc	13.8 ± 10.5	6.3 ± 4.6	0.0878
PHA-L	(ß-1,6)-branched N-glycans	56.9 ± 47.0	27.2 ± 19.6	0.0853
GNL	terminal (α-1,6) and (α-1,3) mannose residues; N-glycans with terminated short LacNAc	14.2 ± 11.2	8.2 ± 8.9	0.0700
PNA	core 1 and core 2 O-glycans	15.4 ± 11.1	5.0 ± 5.0	0.0636
BanLec	Internal (α-1,3) glucosyl- and mannosyl residues; reducing terminal (α-1,3) and (α-1,6) glucosyl residues	21.1 ± 17.8	8.8 ± 3.6	0.0163
JAC	core 1 and core 3 O-glycans; 3-substituents α-GalNAc	41.5 ± 20.3	98.9 ± 27.2	0.0012

3.3. BanLec-Bound Glycans Show Significantly Decreased Abundance on the Surface of ERU Neutrophils, Whereas JAC-Bound Glycans Are Significanty More Abundant

To explore these differences in more detail, we selected the significantly different surface glycans detected via Banana Lectin (BanLec; Mannose-binding) and Jacalin (JAC; O-glycan binding) for further validation and quantification with immunocytochemistry. Dolichos Biflorus Agglutinin (DBA) was used as a negative control since DBA-reactive glycans were absent in controls and ERU in the flow cytometry screening with all 35 plant lectins (Figures 1 and 3). In the immunocytochemistry experiments, DBA-reactive glycans were also not detected, confirming the consistent lack of detectable terminal GalNAc-α1,3-GalNAc residues (Figure 4A).

BanLec-reactive structures, which are associated with internal α-1,3-glucosyl and mannosyl residues, visibly decreased in the immunocytochemistry experiments (Figure 4B). Quantification revealed a significant reduction in BanLec-reactive structures on the surface of ERU neutrophils compared to controls (**** $p < 0.0001$; Figure 4B). In contrast, JAC, which binds to galactosyl (β-1,3) N-acetylgalactosamine structures, showed significantly increased binding intensity to ERU neutrophils, which is clearly evident in the immunocytochemistry images (**** $p < 0.0001$; Figure 4C). These findings highlight surface glycan abundance differences on the cellular level, supporting the disease-associated redistribution of specific O-glycan and mannose motifs detected in our initial screening assays.

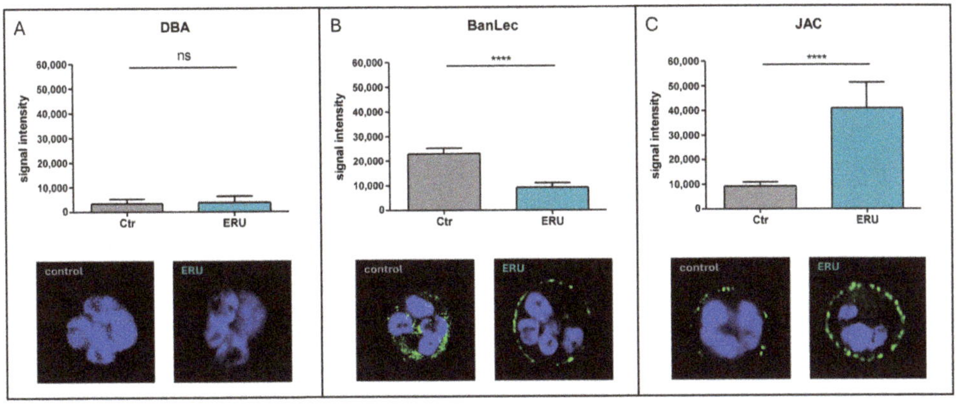

Figure 4. ERU-associated changes in neutrophil surface glycosylation (ERU: $n = 3$, controls: $n = 3$; data was obtained in five independent experiments). Immunocytochemistry analyses visualize binding intensity of each represented lectin to cell surface glycan structures (green). Cell nuclei were counterstained with DAPI (blue). Bar graphs display quantification of lectin signal intensity measured via immunocytochemisty (gray: controls, blue: ERU). (**A**) DBA-reactive glycans were not detected in either control or ERU cases since no lectin binding can be observed (no green visible) (Student's *t*-test, ns). (**B**) BanLec-reactive glycans were significantly decreased on neutrophils from ERU cases reflected by decreased lectin binding (green) and corresponding quantification data shown in bar graphs (Student's *t*-test, **** $p < 0.0001$). (**C**) JAC-reactive glycans were significantly increased on the surface of neutrophils from ERU cases as shown by increased lectin binding (green) and corresponding quantification data (Student's *t*-test with Welch's correction, **** $p < 0.0001$).

Together, the immunocytochemistry results confirm the selected changes in surface glycosylation identified by flow cytometry in the initial surface glycosylation screening assay, suggesting that ERU alters the surface glycan landscape of circulating neutrophils in a structure-specific manner.

3.4. *JAC Lectin Pull-Down Reveals ERU-Associated O-Glycosylated Proteins*

To further characterize the increased JAC-reactivity observed in ERU neutrophils, we performed a JAC lectin pull-down assay using agarose bound JAC to enrich for surface-accessible O-glycoproteins. SDS-PAGE analysis of lysates, supernatants, and eluates revealed multiple protein bands, among which a prominent band at 94.5 kDa was consistently observed in both ERU cases and controls (Figure 5A). This band appeared more intense in ERU, and densitometric analysis confirmed a 1.4-fold higher signal in ERU neutrophils (Figure 5B), although the difference did not reach statistical significance. To clarify whether this band might correspond to proteins with increased overall abundance in ERU neutrophils, we searched previously published proteome data for candidate proteins that (i) fall within the 90–100 kDa molecular weight range, (ii) are located on the cell surface, (iii) possess predicted O-glycosylation sites, and (iv) were detected in all six mass spectrometry samples. We then filtered for proteins with an abundance ratio $\geq$1.1 in ERU cases (Table 2). It should be noted that this approach specifically identifies proteins with increased abundance in ERU neutrophils but does not assess changes in glycosylation per protein molecule. Therefore, the identified candidates represent surface proteins that are not only potentially O-glycosylated but also more abundant in ERU and could thus account for the observed increase in JAC binding. Among the five proteins fulfilling all criteria, CUB domain-containing protein 1 and Integrin beta-2 were approximately 2-fold more abundant in ERU cases, as displayed in flow cytometry analysis, and fell within the relevant molecular weight range. These proteins are plausible contributors to the increased JAC-reactive band at 94.5 kDa. Subsequent immunocytochemistry analysis revealed a noticeable increase in Integrin beta-2 on ERU neutrophils compared to controls (Figure 6C,G), as well as an increased presence of JAC-bound O-glycans (Figure 6B,F). Further, we detected colocalization of JAC-bound surface glycosylation and Integrin beta-2 (Figure 6D,H), underlining Integrin beta-2 as a candidate O-glycosylated protein on neutrophils with increased abundance and/or O-glycosylation in ERU.

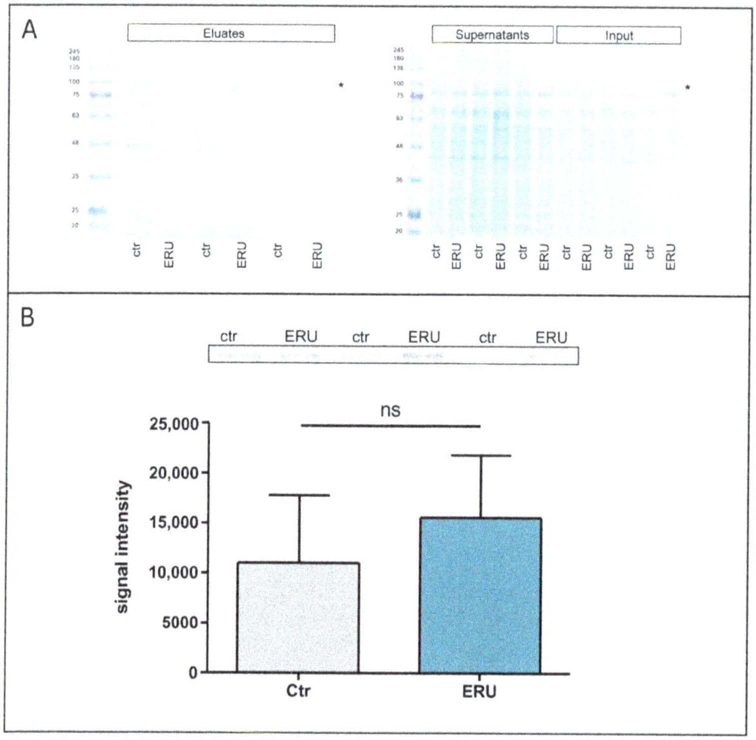

Figure 5. Proteins that were enriched via JAC lectin pull-down assay (controls: $n = 3$, ERU: $n = 3$; shown data were obtained in one experiment). (**A**) 1D-SDS-gels (10%) run with eluates from the JAC lectin pull-down assay showing bands at 94.5 kDa in all lanes (asterisk), indicating the molecular weight of O-glycosylated proteins bound by JAC in equine neutrophils (left), and 1D-SDS-gels (10%) run with supernatants from the JAC lectin pull-down assay and lysates that were used as input (right) show no or weak bands at 94.5 kDa (asterisk). (**B**) Statistical analysis of the 94.5 kDa band intensity in controls (gray) and ERU cases (blue). Data are represented as mean ± SD. Band intensities in ERU cases were 1.4-fold higher compared to controls (Mann–Whitney test, ns $p > 0.05$). The band of interest is shown more detailed in greyscale above the bar chart.

Table 2. Proteins of control and ERU neutrophils identified in all samples (6 of 6) of previously published proteome data, filtered after for molecular weight between 90 and 100 kDa, predicted O-glycosylation sites, and higher abundance (≥1.1) in ERU cases compared to controls. All listed proteins are located on the cell surface. Potential O-glycosylated JAC targets are highlighted in bold. (a) Protein name. (b) Name of the corresponding gene. (c) Protein accession number according to the Ensembl horse database (Version 75, www.ensembl.org (accessed on 2 July 2025)). (d) Ratio of protein abundance differences in ERU cases compared to controls. (e) Molecular weight of listed proteins according to uniprot database (https://www.uniprot.org (accessed on 2 July 2025)).

Protein [a]	Gene Name [b]	Accession Number [c]	Ratio ERU/Controls [d]	Molecular Weight [e]
CUB Domain Containing Protein 1	*CDCP1*	ENSECAP00000022716	2.2	93.0
Integrin, Beta 2	*ITGB2*	ENSECAP00000021220	1.9	94.8
Adhesion G Protein-Coupled Receptor E5	*ADRE5*	ENSECAP00000008739	1.4	90.7
Oxysterol Binding Protein-Like 8	*OSBPL8*	ENSECAP00000013659	1.1	95.6
Glucosidase, alpha	*GANAB*	ENSECAP00000012762	1.1	96.3

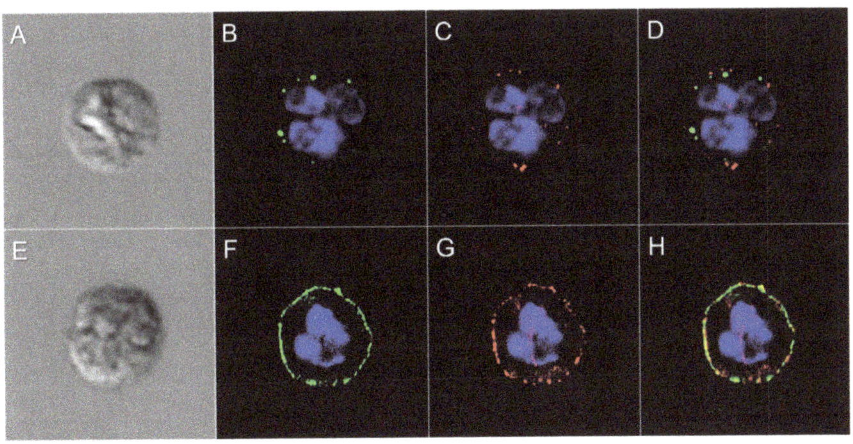

Figure 6. Representative images of JAC and anti-Integrin beta-2 binding on the surface of neutrophils from controls (**A–D**) and ERU horses (**E–H**) (controls: $n = 1$, ERU: $n = 2$; data were obtained in one experiment). (**A**,**E**) DIC image of equine neutrophils, (**B**,**F**) JAC staining on the cell surface indicating O-glycosylation (green), (**C**,**G**) anti-Integrin beta-2 staining indicating the presence of Integrin beta-2 (red), (**D**,**H**) overlay of JAC staining (green) and anti-Integrin beta-2 staining (red) shows colocalization of JAC-bound O-glycans with Integrin beta-2 (yellow). JAC and anti-Integrin beta-2 show increased binding to neutrophils from ERU cases (**D–F**) compared to controls (**A–C**). Cell nuclei were stained with DAPI (blue).

4. Discussion

Neutrophils are highly abundant cells of the innate immune system with a wide variety of effector functions and implications in diseases. Neutrophil activation status is mirrored in the cell surface protein landscape and can be presented through protein abundance changes as well as post-translational modifications (PTM) of surface proteins [2]. One of the crucial PTMs during cell activation is glycosylation, which can shape immune response through impact on cell–cell interactions such as antigen presentation and modification of signal transduction [37,38].

Surface glycosylation of neutrophils has previously been described for specific proteins, for single glycans or in the context of certain diseases. For example, it is well-described on human neutrophils that toll-like receptor 4, a member of one of the most important cell receptor families of the innate immune system, shows several N-glycosylation sites [39,40]. Further, N-glycome profiling of isolated neutrophil granules and vesicular populations has been described for human neutrophils [41]. Neutrophil granules and vesicles contain a large amount of highly glycosylated glycoproteins. Once neutrophils are activated, as is the case in equine recurrent uveitis (ERU) [22,42], they can express these intracellularly stored glycoproteins on the cell surface to modulate their immune response [43]. Of note, protein biosynthesis is low in neutrophils [44], meaning that the majority of neutrophil glycoproteins is already present within the cell and can be re-located to the cell surface in the course of activation. Knowledge of the neutrophil surface glycosylation landscape, independent of specific proteins, glycan moieties, or disease, is scarce to date, especially in large animal models. Therefore, we investigated the cell surface of equine neutrophils for the presence of a variety of different glycans and are the first to describe a comprehensive surface glycome for equine neutrophils. To achieve this, we used 35 different plant lectins, each with a specific glycan structure as binding target. The resulting lectin binding pattern underlines the heterogeneity of surface glycosylation on equine neutrophils. To date, the full spectrum of neutrophil surface glycosylation remains largely unknown. Given that mammals express an immense diversity of surface glycans, our study provides a valuable initial overview of the equine neutrophil surface glycome, based on the binding specificities of 35 plant lectins. These important aspects of the neutrophil surface glycan profile in its "healthy" state have not been described before and are useful as a basis for further investigations on the composition of surface-exposed glycan moieties on neutrophils. Since protein glycosylation impacts neutrophil effector functions [43], we were especially interested in possible changes in surface glyco-signatures of neutrophils from horses with ERU. ERU is a painful, sight-threatening disease in horses, with a remitting–relapsing character and increasingly severe inflammatory attacks over time [45]. While its exact etiology remains unclear, ongoing research continues to explore potential triggers, biomarkers for pathophysiology, and causative treatment options. Described as primarily driven by autoreactive CD4+ T cells, this cell type dominates among the immune cell populations found in ocular infiltrates in ERU [46,47]. However, pure neutrophil infiltrates can be observed in spontaneous ERU and interphotoreceptor retinoid-binding protein-induced cases, even in early stages of uveitis [21,48]. In addition, circulating neutrophils show an activated phenotype in ERU, even between uveitic attacks [22,49] and more readily release neutrophil extracellular traps, which have been shown to associate with disease severity [25,50], pointing to active involvement in ERU pathogenesis. Besides its importance for veterinary medicine, ERU is a promising model for autoimmune-mediated recurrent uveitis in humans, given notable parallels in clinical and pathophysiological characteristics as well as immune system composition and function [11,14,51,52]. The predominance of neutrophils among the blood-derived leukocyte population in both humans and horses makes the horse especially interesting for investigations on early disease-initiating events which may be neutrophil-mediated. This allows for better extrapolation of findings to human disease pathomechanisms as opposed to rodent models for autoimmune uveitis, which have a very different immune system organization [53]. Although autoimmune uveitis is described as T-cell driven, the active involvement neutrophils in disease pathogenesis is becoming increasingly evident as neutrophils were recently suggested to have a regulatory effect on T cells via the formation of neutrophil extracellular traps [54]. These observations in humans, rodent models, and horses suggest neutrophils to be active participants in shaping adaptive immune responses in AU and ERU, which emphasizes the role of the innate immune system in this sight-threatening disease.

A few of the plant lectins used in our study, including HHL, LTL, PSA, PTL, UEA-1, SJA, DBA, SBA, VVA, and GSL II, showed weak to no binding to equine neutrophil surface glycans (Figure 1, Table 1). This indicates that said surface glycans may not be part of the equine neutrophil surface glycome. To determine whether this originates in absence or in poor lectin-accessibility of glycans [55], further analyses can be performed under different experimental conditions that determine how the glycan-binding sites are presented [56]. Most of the glycan moieties that we detected on the surface of neutrophils from healthy horses were similarly abundant in neutrophils of ERU cases. For example, core fucosylation—a type of N-glycosylation, catalyzed by fucosyltransferase 8, where fucose is attached to the N-glycan through an α-1,6-linkage [57]—which we detected via the plant lectins AAL, LCA, and PSA, showed equally strong presence on the surface of neutrophils from controls and ERU cases (Figures 1 and 3, Table 1). Interestingly, surface fucosylation seems to specifically change on different leukocyte subtypes in other autoimmune diseases. For instance, increased core fucosylation can be observed on the surface of activated T cells from patients with systemic lupus erythematosus [6], indicating that defined glycan alterations specifically depend on cell type and disease. In humans, core fucosylation can be found in every subtype of neutrophil granules, with a particularly high abundance in azurophil granules [58], but to date there is no implication of core fucosylation on neutrophils in autoimmune disease pathogenesis.

We previously described an altered glycosylation profile on other cells in the context of ERU: In retinal Müller glia cells, the main macroglia cells of the retina with a pivotal role in maintaining retinal function and integrity, we could demonstrate a loss of α2-3- and α2-6-linked terminal sialic acids in the uveitic state via lectin-binding (MAL-2, SNA, WGA) in immunohistochemistry analyses of equine retinae [59]. These terminal sialic acids were also identified on neutrophils in the present study, but with equal abundance in health and disease (Figures 1 and 3, Table 1). This suggests that terminal sialic acids may belong to the basic glycan signature on neutrophils, which remains unaltered on activated neutrophils in ERU. Interestingly, we identified two other glycan structures that were significantly altered on the neutrophil surface in ERU: glycans that are bound by the plant lectin BanLec and glycans that are bound by JAC (Figure 3, Table 1). Since neutrophils continue to emerge as a heterogeneous leukocyte population with a variety of finely tuned effector functions [60–62], which may also be influenced by the nature of their surface glycosylation, particularly in different inflammatory environments and with regard to their activation status, these significant changes in the surface glycome signature are especially interesting in the context or ERU.

Notably, we detected a significant decrease in mannose or glucose linked N-glycans through binding to the plant lectin Banana Lectin (BanLec) on the surface of neutrophils from ERU cases in flow cytometry and immunochemistry (Table 1, Figures 3C and 4B), which is a new finding on equine neutrophils. Interestingly, we detected that binding to ConA, GNL, und NPL was also decreased in ERU cases, but not significantly. This may be explained by their similar binding specificity, since these lectins can be grouped as mannose-binding, whereas BanLec additionally binds internal (α-1,3) glucosyl residues and reducing terminal (α 1,3) and α (1,6) glucosyl residues [32,35]. Mannose linked N-glycans are described in healthy human neutrophil granules [58] but changed surface abundance has not been described for neutrophils in autoimmune diseases so far. In contrast, the plant lectin JAC was the only lectin with significantly elevated binding to neutrophils from ERU cases (Table 1, Figures 3A and 4C). JAC has a highly specific binding affinity for core 1 and core 3 O-glycans, and it is therefore widely used as a reliable marker for

O-glycosylation [32,33,63,64]. Interestingly, the observed binding alteration in ERU could not be detected for the other lectins with O-glycan binding specificity, such as PNA, ACL, and MAL-2. JAC predominantly binds to core 1 and core 3 O-glycans, whereas ACL and PNA are also able to bind core 1 and core 2 O-glycans, but not to the core 3 O-glycans, and MAL-2 only recognizes α2,3-sialylated O-glycans [32]. This emphasizes the unique selectivity of JAC to bind certain types of O-glycans on neutrophils of ERU cases and makes it a promising biomarker candidate. Although our study is an important first step towards elucidating disease-associated changes in neutrophil surface O-glycosylation, further experimental approaches—such as enzymatic trimming of O-glycans, glycomic mass spectrometry, or site-specific analysis of glycoproteins—are needed to provide more detailed structural insight. Divergent O-glycosylation on neutrophils in autoimmune uveitis is a novel discovery that has not been investigated so far. The increased presence of O-glycans may reflect the presence of a generally increased abundance of JAC-specific O-glycosylation on the neutrophil membrane or may occurdue to increased abundance of specific O-glycosylated proteins. Distinguishing between these two possibilities is critical, as they imply fundamentally different modes of regulation—one at the level of glycan processing and the other at the level of protein expression and turnover—that may influence neutrophil activation states and effector functions in different ways. Regardless of the origin, the functional meaning of the increased O-glycosylation on neutrophils in ERU remains unclear and needs further in-depth investigations.

To further characterize the specific O-glycosylation which we identified via Jacalin (JAC) binding, we aimed at finding potentially O-glycosylated proteins targeted by JAC. To achieve this, we enriched O-glycoproteins via JAC pull-down and detected several different protein bands. One particular band was present in all JAC pull-down eluates, at a molecular weight of 95.4 kDa. This specific protein band was more intense in ERU (Figure 5). After reinvestigating neutrophil proteomics data, filtering for the defined criteria molecular weight 90–100 kDa, cell surface proteins, possibility of O-glycosylation, identification in all measured samples, as well as higher abundance in ERU, we found two possible membrane protein candidates as anchors for O-glycosylation: CUB domain-containing protein 1 and Integrin beta-2 (Table 2). Although these two candidate proteins may be a promising starting point for glycoprotein biomarker discovery, we need to keep in mind that the identification of these proteins is based on an indirect approach, conducted by molecular weight comparison of pull-down data and proteomics analyses. Moreover, other O-glycosylated proteins may also be present, even if not (yet) identified, and more in-depth investigations are needed to determine a more comprehensive O-glycan surface landscape on equine neutrophils with potential alterations in disease.

CUB domain-containing protein 1 is an O-glycosylated protein with a molecular weight of 93 kDa. It is located at the cell membrane, characterized as a single-pass membrane protein, and is 2.2-fold more abundant in proteomics data from ERU neutrophils. CUB domain-containing protein 1 is a ligand for CD6, which is a type I transmembrane glycoprotein expressed by T-lymphocytes and has been described to be involved in the pathophysiology of autoimmune-mediated diseases as multiple sclerosis and rheumatoid arthritis [65]. Interestingly, CUB domain-containing protein 1is highly expressed by retinal pigment epithelial cells (RPE), and CUB domain-containing protein 1-CD6 interaction has been shown to facilitate T cell transmigration through the RPE monolayer in mice, resulting in uveitis [66]. The role of CUB domain-containing protein 1 in neutrophils has not previously been investigated. However, regarding its implication in induced uveitis in the mouse via the CUB domain-containing protein 1-CD6 axis, the presence and increased abundance of CUB domain-containing protein 1 in neutrophils is in alignment with the generally more activated equine neutrophils in ERU [22].

Integrin beta-2 is a protein with predicted O-gylcosylation, which was calculated via NetOGlyc 4.0 [31]. It has a molecular weight of 94.8 kDa, which is a good fit to the protein bald precipitated with JAC at 95.4 kDa; it is located at the cell membrane and was detected in the neutrophil proteome with 1.9-fold higher abundance in ERU. Integrin beta-2 is a member of the transmembrane receptor family composed of non-covalently associated heterodimers formed by one α and one β subunit. In neutrophils, integrin beta-2 mediates firm adhesion to the vascular endothelium, a prerequisite for transendothelial migration into inflamed tissues. Additionally, it facilitates key effector functions including chemotaxis, phagocytosis, and the formation of immunological synapses, thereby playing a central role in innate immune responses [67]. Due to the strong binding of the plant lectin JAC to neutrophils of ERU horses, we showed increased O-glycosylation in the diseased state, which might be associated with increased Integrin beta-2 abundance. Although we could show a higher abundance of Integrin beta-2 on neutrophils in ERU cases and colocalization of Integrin beta-2 and JAC binding on equine neutrophils via immunocytochemistry, more investigations are needed to confirm O-glycosylation of this candidate protein. Deviant glycan moieties on the Integrin beta-2 receptor molecule, such as terminal fucose and GlcNAc glycans as well as sialylated glycans, play a role in regulating neutrophil function such as transepithelial migration, degranulation, apoptosis, and superoxide generation in inflammatory mucosal diseases [68–70]. This implies that deviant O-glycosylation on the Integrin beta-2 receptor of neutrophils can have an impact in the regulation of inflammation. Since Integrin beta-2 bound glycans have been described as potential targets for selective manipulation of neutrophil function [69], the possibility of O-glycosylated Integrin beta-2 to serve as a biomarker for ERU merits further assessment.

5. Conclusions

In this study, we provide the first characterization of the equine neutrophil surface glycome, revealing a largely conserved glycosylation profile across healthy horses and horses suffering from equine recurrent uveitis (ERU), a spontaneous equine model for autoimmune uveitis. Despite these similarities, we detected aberrant glyco-signatures in ERU, especially through increased binding of the plant lectin Jacalin, which points to increased O-glycosylation in the diseased state. Alignment with previously published neutrophil proteomic data suggests that the surface proteins Integrin beta-2 and CUB domain-containing protein 1 may serve as possible O-glycosylated protein biomarker candidates. While additional studies are required to fully disentangle the relative contributions of glycan modifications or glycoprotein expression, our data demonstrate disease-associated alterations in the neutrophil surface glycome in equine autoimmune uveitis.

Supplementary Materials: The following supporting information can be downloaded at: https://www.mdpi.com/article/10.3390/biom15101444/s1, Table S1: lectin binding intensity raw data from flow cytometry-based screening analyses with 35 plant lectins; Figure S1: Graph for molecular weight calculation of precipitated proteins; Figure S2: JAC inhibition via monosaccharides. Original images: RawData ICC and RawData WB.

Author Contributions: Conceptualization, C.A.D.; methodology, C.A.D., C.J.S., and R.L.D.; formal analysis, C.J.S., R.L.D., and C.A.D.; investigation, C.J.S.; B.A. and R.L.D.; resources, C.A.D.; data curation, C.J.S. and B.A.; writing—original draft preparation, C.J.S., R.L.D., and C.A.D.; writing—review and editing, C.J.S. and R.L.D.; visualization, C.J.S. and B.A.; supervision, C.A.D. and R.L.D.; project administration, C.A.D.; funding acquisition, C.A.D. All authors have read and agreed to the published version of the manuscript.

Funding: This research was funded by Deutsche Forschungsgemeinschaft, DFG DE 719/4-4 (to C.A.D.).

Institutional Review Board Statement: No experimental animals were used in this study. Horses were treated according to the ethical principles and guidelines in the ARVO statement for the use of animals in ophthalmic and vision research. Collection of blood samples was permitted by the local authority (Regierung von Oberbayern, permit number: ROB-55.2-2532.Vet_03-22-37, date of approval: 16 January 2023, permit effective from 20 January 2023).

Informed Consent Statement: Not applicable.

Data Availability Statement: The raw data supporting the conclusions of this article will be made available by the authors on request. The raw proteome data have been deposited to the ProteomeXchange Consortium via the PRIDE repository with the dataset identifier: PXD013648 [30].

Acknowledgments: The authors want to thank the staff from LMU Munich equine clinic for supplying equine blood samples, Tanja Weisser, Ursula Cramer, Lydia Hoffmann, and Jonas Liedl for excellent technical assistance, and Adrian Schmalen, Ulrich Wilk, Salome Fischer, and Jana Schreff for critical discussions.

Conflicts of Interest: The authors declare no conflicts of interest.

Abbreviations

The following abbreviations are used in this manuscript:

PTM	Post-translational Modifications
AU	Autoimmune Uveitis
ERU	Equine Recurrent Uveitis
RT	Room Temperature
PBS	Phosphate-buffered Saline
NaCl	Sodium Chloride
AAL	Aleuria Aurantia Lectin
ABL	Agaricus Bisporus Lectin
ACL	Amaranthus Caudatus Lectin
BanLec	Banana Lectin
BPL	Bauhinia Purpurea Lectin
ConA	Concanavalin A
DBA	Dolichos Biflorus Agglutinin
DSL	Datura Stramonium Lectin
ECL	Erythrina Cristagalli Lectin
GNL	Galanthus Nivalis Lectin
GSL I	Griffonia Simplicifolia Lectin I
GSL-I-B4	Griffonia Simplicifolia Lectin I Isolectin B4
GSL II	Griffonia Simplicifolia Lectin II
HHL	Hippeastrum Hybrid Lectin
JAC	Jacalin
LCA	Lens Culinaris Agglutinin
LEL	Lycopersicon Esculentum Lectin
LTL	Lotus Tetragonolobus Lectin
MAL-2	Maackia Amurensis Lectin II
NPL	Narcissus Pseudonarcissus Lectin
PHA-E	Phaseolus Vulgaris Erythroagglutin
PHA-L	Phaseolus Vulgaris Leucoagglutinin
PNA	Peanut Agglutinin
PSA	Pisum Sativum Agglutinin
PTL	Psophocarpus Tetragonolobus Lectin
RCA I	Ricinus Communis Agglutitin I
SBA	Soybean Agglutinin

SJA	Sophora Japonica Agglutinin
SNA	Sambucus Nigra Lectin
STL	Solanum Tuberosum Lectin
S-WGA	Succinylated Wheat Germ Agglutinin
UEA-1	Ulex Europaeus Agglutinin I
VVA	Vicia Villosa Lectin
WFA	Wisteria Floribunda Lectin
WGA	Wheat Germ Agglutinin
PFA	Paraform aldehyde
MFI	Mean Fluorescence Intensity
DAPI	4′, 6-Diamidin-2-phenylindol
KS	Kolmogorov–Smirnov Test
CDCP1	CUB Domain-containing Protein 1
ITGB2	Integrin β2
RA	Rheumatoid Arthritis
RPE	Retinal Pigment Epithelial Cells

References

1. de Haan, N.; Pucic-Bakovic, M.; Novokmet, M.; Falck, D.; Lageveen-Kammeijer, G.; Razdorov, G.; Vuckovic, F.; Trbojevic-Akmacic, I.; Gornik, O.; Hanic, M.; et al. Developments and perspectives in high-throughput protein glycomics: Enabling the analysis of thousands of samples. *Glycobiology* **2022**, *32*, 651–663. [CrossRef]
2. He, M.; Zhou, X.; Wang, X. Glycosylation: Mechanisms, biological functions and clinical implications. *Signal Transduct. Target. Ther.* **2024**, *9*, 194. [CrossRef]
3. Marrero Roche, D.E.; Chandler, K.B. Clinical glycoprotein mass spectrometry: The future of disease detection and monitoring. *J. Mass Spectrom.* **2024**, *59*, e5083. [CrossRef]
4. Schjoldager, K.T.; Narimatsu, Y.; Joshi, H.J.; Clausen, H. Global view of human protein glycosylation pathways and functions. *Nat. Rev. Mol. Cell Biol.* **2020**, *21*, 729–749. [CrossRef]
5. Kissel, T.; Toes, R.E.M.; Huizinga, T.W.J.; Wuhrer, M. Glycobiology of rheumatic diseases. *Nat. Rev. Rheumatol.* **2023**, *19*, 28–43. [CrossRef]
6. Szabo, E.; Farago, A.; Bodor, G.; Gemes, N.; Puskas, L.G.; Kovacs, L.; Szebeni, G.J. Identification of immune subsets with distinct lectin binding signatures using multi-parameter flow cytometry: Correlations with disease activity in systemic lupus erythematosus. *Front. Immunol.* **2024**, *15*, 1380481. [CrossRef] [PubMed]
7. Sahin, F.; Kaya, Z.Z.; Serteser, M.; Ozturk, H.U.; Baykal, A.T. Glycan profiling of multiple sclerosis oligoclonal bands with MALDI-TOF. *Anal. Methods* **2025**, *17*, 850–858. [CrossRef] [PubMed]
8. Huang, J.; Wang, X.; Li, N.; Fan, W.; Li, X.; Zhou, Q.; Liu, J.; Li, W.; Zhang, Z.; Liu, X.; et al. YY1 Lactylation Aggravates Autoimmune Uveitis by Enhancing Microglial Functions via Inflammatory Genes. *Adv. Sci.* **2024**, *11*, e2308031. [CrossRef] [PubMed]
9. Wiedemann, C.; Amann, B.; Degroote, R.L.; Witte, T.; Deeg, C.A. Aberrant Migratory Behavior of Immune Cells in Recurrent Autoimmune Uveitis in Horses. *Front. Cell Dev. Biol.* **2020**, *8*, 101. [CrossRef]
10. Peng, X.; Li, H.; Zhu, L.; Zhao, S.; Li, Z.; Li, S.; Wu, D.; Chen, J.; Zheng, S.; Su, W. Single-cell sequencing of the retina shows that LDHA regulates pathogenesis of autoimmune uveitis. *J. Autoimmun.* **2024**, *143*, 103160. [CrossRef]
11. Deeg, C.A.; Hauck, S.M.; Amann, B.; Pompetzki, D.; Altmann, F.; Raith, A.; Schmalzl, T.; Stangassinger, M.; Ueffing, M. Equine recurrent uveitis--a spontaneous horse model of uveitis. *Ophthalmic Res.* **2008**, *40*, 151–153. [CrossRef]
12. Gilger, B.C. Immune Relevant Models for Ocular Inflammatory Diseases. *ILAR J.* **2018**, *59*, 352–362. [CrossRef]
13. Soth, R.; Hoffmann, A.L.C.; Deeg, C.A. Enhanced ROS Production and Mitochondrial Metabolic Shifts in CD4(+) T Cells of an Autoimmune Uveitis Model. *Int. J. Mol. Sci.* **2024**, *25*, 11513. [CrossRef]
14. Malalana, F.; Stylianides, A.; McGowan, C. Equine recurrent uveitis: Human and equine perspectives. *Vet. J.* **2015**, *206*, 22–29. [CrossRef] [PubMed]
15. Gerding, J.C.; Gilger, B.C. Prognosis and impact of equine recurrent uveitis. *Equine Vet. J.* **2016**, *48*, 290–298. [CrossRef] [PubMed]
16. Li, Z.; Liu, X.; Li, Z.; Xiao, Z.; Chen, G.; Li, Y.; Huang, J.; Hu, Y.; Huang, H.; Zhu, W.; et al. *STING* Deficiency Promotes Th17-Like Tfh to Aggravate the Experimental Autoimmune Uveitis. *Investig. Ophthalmol. Vis. Sci.* **2025**, *66*, 8. [CrossRef]
17. Barfusser, C.; Wiedemann, C.; Hoffmann, A.L.C.; Hirmer, S.; Deeg, C.A. Altered Metabolic Phenotype of Immune Cells in a Spontaneous Autoimmune Uveitis Model. *Front. Immunol.* **2021**, *12*, 601619. [CrossRef] [PubMed]

18. Caspi, R.R.; Chan, C.C.; Fujino, Y.; Najafian, F.; Grover, S.; Hansen, C.T.; Wilder, R.L. Recruitment of antigen-nonspecific cells plays a pivotal role in the pathogenesis of a T cell-mediated organ-specific autoimmune disease, experimental autoimmune uveoretinitis. *J. Neuroimmunol.* **1993**, *47*, 177–188. [CrossRef]
19. Kerr, E.C.; Copland, D.A.; Dick, A.D.; Nicholson, L.B. The dynamics of leukocyte infiltration in experimental autoimmune uveoretinitis. *Prog. Retin. Eye Res.* **2008**, *27*, 527–535. [CrossRef]
20. Kerr, E.C.; Raveney, B.J.; Copland, D.A.; Dick, A.D.; Nicholson, L.B. Analysis of retinal cellular infiltrate in experimental autoimmune uveoretinitis reveals multiple regulatory cell populations. *J. Autoimmun.* **2008**, *31*, 354–361. [CrossRef]
21. Deeg, C.A.; Kaspers, B.; Gerhards, H.; Thurau, S.R.; Wollanke, B.; Wildner, G. Immune responses to retinal autoantigens and peptides in equine recurrent uveitis. *Investig. Ophthalmol. Vis. Sci.* **2001**, *42*, 393–398.
22. Weigand, M.; Hauck, S.M.; Deeg, C.A.; Degroote, R.L. Deviant proteome profile of equine granulocytes associates to latent activation status in organ specific autoimmune disease. *J. Proteom.* **2021**, *230*, 103989. [CrossRef]
23. Reno, F.; Pagano, C.A.; Bignotto, M.; Sabbatini, M. Neutrophil Heterogeneity in Wound Healing. *Biomedicines* **2025**, *13*, 694. [CrossRef]
24. Sheahan, B.J.; Schubert, A.G.; Schubert, W.; Sheats, M.K.; Schnabel, L.V.; Gilbertie, J.M. Equine neutrophils selectively release neutrophil extracellular traps in response to chemical and bacterial agonists. *Front. Vet. Sci.* **2025**, *12*, 1512343. [CrossRef] [PubMed]
25. Fingerhut, L.; Ohnesorge, B.; von Borstel, M.; Schumski, A.; Strutzberg-Minder, K.; Morgelin, M.; Deeg, C.A.; Haagsman, H.P.; Beineke, A.; von Kockritz-Blickwede, M.; et al. Neutrophil Extracellular Traps in the Pathogenesis of Equine Recurrent Uveitis (ERU). *Cells* **2019**, *8*, 1528. [CrossRef] [PubMed]
26. Maier-Begandt, D.; Alonso-Gonzalez, N.; Klotz, L.; Erpenbeck, L.; Jablonska, J.; Immler, R.; Hasenberg, A.; Mueller, T.T.; Herrero-Cervera, A.; Aranda-Pardos, I.; et al. Neutrophils-biology and diversity. *Nephrol. Dial. Transplant.* **2024**, *39*, 1551–1564. [CrossRef] [PubMed]
27. Gilger, B.C.; Michau, T.M. Equine recurrent uveitis: New methods of management. *Vet. Clin. N. Am. Equine Pract.* **2004**, *20*, 417–427. [CrossRef]
28. Schneider, C.A.; Rasband, W.S.; Eliceiri, K.W. NIH Image to ImageJ: 25 years of image analysis. *Nat. Methods* **2012**, *9*, 671–675. [CrossRef]
29. Kang, D.H.; Gho, Y.S.; Suh, M.K.; Kang, C.H. Highly sensitive and fast protein detection with coomassie brilliant blue in sodium dodecyl sulfate-polyacrylamide gel electrophoresis. *B Korean Chem. Soc.* **2002**, *23*, 1511–1512.
30. Perez-Riverol, Y.; Csordas, A.; Bai, J.; Bernal-Llinares, M.; Hewapathirana, S.; Kundu, D.J.; Inuganti, A.; Griss, J.; Mayer, G.; Eisenacher, M.; et al. The PRIDE database and related tools and resources in 2019: Improving support for quantification data. *Nucleic Acids Res.* **2019**, *47*, D442–D450. [CrossRef]
31. Steentoft, C.; Vakhrushev, S.Y.; Joshi, H.J.; Kong, Y.; Vester-Christensen, M.B.; Schjoldager, K.T.; Lavrsen, K.; Dabelsteen, S.; Pedersen, N.B.; Marcos-Silva, L.; et al. Precision mapping of the human O-GalNAc glycoproteome through SimpleCell technology. *EMBO J.* **2013**, *32*, 1478–1488. [CrossRef] [PubMed]
32. Bojar, D.; Meche, L.; Meng, G.; Eng, W.; Smith, D.F.; Cummings, R.D.; Mahal, L.K. A Useful Guide to Lectin Binding: Machine-Learning Directed Annotation of 57 Unique Lectin Specificities. *ACS Chem. Biol.* **2022**, *17*, 2993–3012. [CrossRef] [PubMed]
33. Tachibana, K.; Nakamura, S.; Wang, H.; Iwasaki, H.; Tachibana, K.; Maebara, K.; Cheng, L.; Hirabayashi, J.; Narimatsu, H. Elucidation of binding specificity of Jacalin toward O-glycosylated peptides: Quantitative analysis by frontal affinity chromatography. *Glycobiology* **2006**, *16*, 46–53. [CrossRef]
34. Lescar, J.; Loris, R.; Mitchell, E.; Gautier, C.; Chazalet, V.; Cox, V.; Wyns, L.; Pérez, S.; Breton, C.; Imberty, A. Isolectins I-A and I-B of *Griffonia(Bandeiraea)* simplicifolia: Crystal structure of metal-free GS I-B_4 and molecular basis for metal binding and monosaccharide specificity. *J. Biol. Chem.* **2002**, *277*, 6608–6614. [CrossRef] [PubMed]
35. Singh, S.S.; Devi, S.K.; Ng, T.B. Banana lectin: A brief review. *Molecules* **2014**, *19*, 18817–18827. [CrossRef]
36. Monsigny, M.; Sene, C.; Obrenovitch, A.; Roche, A.C.; Delmotte, F.; Boschetti, E. Properties of succinylated wheat-germ agglutinin. *Eur. J. Biochem.* **1979**, *98*, 39–45. [CrossRef]
37. van Kooyk, Y.; Rabinovich, G.A. Protein-glycan interactions in the control of innate and adaptive immune responses. *Nat. Immunol.* **2008**, *9*, 593–601. [CrossRef]
38. Wolfert, M.A.; Boons, G.J. Adaptive immune activation: Glycosylation does matter. *Nat. Chem. Biol.* **2013**, *9*, 776–784. [CrossRef]
39. Ohnishi, T.; Muroi, M.; Tanamoto, K. MD-2 is necessary for the toll-like receptor 4 protein to undergo glycosylation essential for its translocation to the cell surface. *Clin. Diagn. Lab. Immunol.* **2003**, *10*, 405–410. [CrossRef]
40. da Silva Correia, J.; Ulevitch, R.J. MD-2 and TLR4 N-linked glycosylations are important for a functional lipopolysaccharide receptor. *J. Biol. Chem.* **2002**, *277*, 1845–1854. [CrossRef]
41. Kawahara, R.; Ugonotti, J.; Chatterjee, S.; Tjondro, H.C.; Loke, I.; Parker, B.L.; Venkatakrishnan, V.; Dieckmann, R.; Sumer-Bayraktar, Z.; Karlsson-Bengtsson, A.; et al. Glycoproteome remodeling and organelle-specific N-glycosylation accompany neutrophil granulopoiesis. *Proc. Natl. Acad. Sci. USA* **2023**, *120*, e2303867120. [CrossRef]

42. Hoffmann, A.L.C.; Hauck, S.M.; Deeg, C.A.; Degroote, R.L. Pre-Activated Granulocytes from an Autoimmune Uveitis Model Show Divergent Pathway Activation Profiles upon IL8 Stimulation In Vitro. *Int. J. Mol. Sci.* **2022**, *23*, 9555. [CrossRef] [PubMed]
43. Ugonotti, J.; Chatterjee, S.; Thaysen-Andersen, M. Structural and functional diversity of neutrophil glycosylation in innate immunity and related disorders. *Mol. Asp. Med.* **2021**, *79*, 100882. [CrossRef] [PubMed]
44. Hughes, V.; Humphreys, J.M.; Edwards, S.W. Protein synthesis is activated in primed neutrophils: A possible role in inflammation. *Biosci. Rep.* **1987**, *7*, 881–890. [CrossRef]
45. McMullen, R.J., Jr.; Fischer, B.M. Medical and Surgical Management of Equine Recurrent Uveitis. *Vet. Clin. N. Am. Equine Pract.* **2017**, *33*, 465–481. [CrossRef]
46. Gilger, B.C.; Malok, E.; Cutter, K.V.; Stewart, T.; Horohov, D.W.; Allen, J.B. Characterization of T-lymphocytes in the anterior uvea of eyes with chronic equine recurrent uveitis. *Vet. Immunol. Immunopathol.* **1999**, *71*, 17–28. [CrossRef] [PubMed]
47. Deeg, C.A.; Ehrenhofer, M.; Thurau, S.R.; Reese, S.; Wildner, G.; Kaspers, B. Immunopathology of recurrent uveitis in spontaneously diseased horses. *Exp. Eye Res.* **2002**, *75*, 127–133. [CrossRef]
48. Deeg, C.A.; Thurau, S.R.; Gerhards, H.; Ehrenhofer, M.; Wildner, G.; Kaspers, B. Uveitis in horses induced by interphotoreceptor retinoid-binding protein is similar to the spontaneous disease. *Eur. J. Immunol.* **2002**, *32*, 2598–2606. [CrossRef]
49. Degroote, R.L.; Schmalen, A.; Hauck, S.M.; Deeg, C.A. Unveiling Differential Responses of Granulocytes to Distinct Immunostimulants with Implications in Autoimmune Uveitis. *Biomedicines* **2023**, *12*, 19. [CrossRef]
50. Fingerhut, L.; Yucel, L.; Strutzberg-Minder, K.; von Kockritz-Blickwede, M.; Ohnesorge, B.; de Buhr, N. Ex Vivo and In Vitro Analysis Identify a Detrimental Impact of Neutrophil Extracellular Traps on Eye Structures in Equine Recurrent Uveitis. *Front. Immunol.* **2022**, *13*, 830871. [CrossRef]
51. Horohov, D.W. The equine immune responses to infectious and allergic disease: A model for humans? *Mol. Immunol.* **2015**, *66*, 89–96. [CrossRef] [PubMed]
52. Deeg, C.A.; Raith, A.J.; Amann, B.; Crabb, J.W.; Thurau, S.R.; Hauck, S.M.; Ueffing, M.; Wildner, G.; Stangassinger, M. *CRALBP* is a highly prevalent autoantigen for human autoimmune uveitis. *Clin. Dev. Immunol.* **2007**, *2007*, 39245. [CrossRef]
53. Zschaler, J.; Schlorke, D.; Arnhold, J. Differences in innate immune response between man and mouse. *Crit. Rev. Immunol.* **2014**, *34*, 433–454. [CrossRef]
54. Li, Z.; Li, Z.; Hu, Y.; Xie, Y.; Shi, Y.; Chen, G.; Huang, J.; Xiao, Z.; Zhu, W.; Huang, H.; et al. Neutrophil extracellular traps potentiate effector T cells via endothelial senescence in uveitis. *JCI Insight* **2025**, *10*, e180248. [CrossRef]
55. Varki, A. Biological roles of glycans. *Glycobiology* **2017**, *27*, 3–49. [CrossRef]
56. Andre, S.; Kaltner, H.; Kayser, K.; Murphy, P.V.; Gabius, H.J. Merging carbohydrate chemistry with lectin histochemistry to study inhibition of lectin binding by glycoclusters in the natural tissue context. *Histochem. Cell Biol.* **2016**, *145*, 185–199. [CrossRef]
57. Sun, Y.; Li, X.; Wang, T.; Li, W. Core Fucosylation Regulates the Function of Pre-BCR, BCR and IgG in Humoral Immunity. *Front. Immunol.* **2022**, *13*, 844427. [CrossRef] [PubMed]
58. Venkatakrishnan, V.; Dieckmann, R.; Loke, I.; Tjondro, H.C.; Chatterjee, S.; Bylund, J.; Thaysen-Andersen, M.; Karlsson, N.G.; Karlsson-Bengtsson, A. Glycan analysis of human neutrophil granules implicates a maturation-dependent glycosylation machinery. *J. Biol. Chem.* **2020**, *295*, 12648–12660. [CrossRef] [PubMed]
59. Lorenz, L.; Amann, B.; Hirmer, S.; Degroote, R.L.; Hauck, S.M.; Deeg, C.A. NEU1 is more abundant in uveitic retina with concomitant desialylation of retinal cells. *Glycobiology* **2021**, *31*, 873–883. [CrossRef]
60. Silvestre-Roig, C.; Fridlender, Z.G.; Glogauer, M.; Scapini, P. Neutrophil Diversity in Health and Disease. *Trends Immunol.* **2019**, *40*, 565–583. [CrossRef]
61. Ng, L.G.; Ostuni, R.; Hidalgo, A. Heterogeneity of neutrophils. *Nat. Rev. Immunol.* **2019**, *19*, 255–265. [CrossRef]
62. Wigerblad, G.; Kaplan, M.J. Neutrophil extracellular traps in systemic autoimmune and autoinflammatory diseases. *Nat. Rev. Immunol.* **2023**, *23*, 274–288. [CrossRef]
63. Jeyaprakash, A.A.; Katiyar, S.; Swaminathan, C.P.; Sekar, K.; Surolia, A.; Vijayan, M. Structural basis of the carbohydrate specificities of jacalin: An X-ray and modeling study. *J. Mol. Biol.* **2003**, *332*, 217–228. [CrossRef]
64. Rubin, J.R.; Taylor, S.K.; Rudchenko, S.; Stojanovic, M.N.; Hess, H. Single Molecule Kinetic Fingerprinting of Glycans on IgA1 Antibodies. *Anal. Chem.* **2025**, *97*, 14388–14396. [CrossRef] [PubMed]
65. Gurrea-Rubio, M.; Fox, D.A.; Castresana, J.S. CD6 in Human Disease. *Cells* **2025**, *14*, 272. [CrossRef]
66. Zhang, L.; Borjini, N.; Lun, Y.; Parab, S.; Asonye, G.; Singh, R.; Bell, B.A.; Bonilha, V.L.; Ivanov, A.; Fox, D.A.; et al. *CDCP1* regulates retinal pigmented epithelial barrier integrity for the development of experimental autoimmune uveitis. *JCI Insight* **2022**, *7*, e157038. [CrossRef]
67. Wen, L.; Moser, M.; Ley, K. Molecular mechanisms of leukocyte β2 integrin activation. *Blood* **2022**, *139*, 3480–3492. [CrossRef]
68. Brazil, J.C.; Sumagin, R.; Cummings, R.D.; Louis, N.A.; Parkos, C.A. Targeting of Neutrophil Lewis X Blocks Transepithelial Migration and Increases Phagocytosis and Degranulation. *Am. J. Pathol.* **2016**, *186*, 297–311. [CrossRef] [PubMed]

69. Brazil, J.C.; Kelm, M.; Lehoux, S.; Azcutia, V.; Cummings, R.D.; Nusrat, A.; Parkos, C.A. Regulation of Neutrophil Function by Selective Targeting of Glycan Epitopes Expressed on the Integrin CD11b/CD18. *FASEB J.* **2020**, *34*, 1. [CrossRef]
70. Azcutia, V.; Kelm, M.; Fink, D.; Cummings, R.D.; Nusrat, A.; Parkos, C.A.; Brazil, J.C. Sialylation regulates neutrophil transepithelial migration, CD11b/CD18 activation, and intestinal mucosal inflammatory function. *JCI Insight* **2023**, *8*, e167151. [CrossRef]

biomolecules

MDPI

Review

The Role of Circulating Biomarkers in Patients with Coronary Microvascular Disease

Rossella Quarta [1,2], Giovanni Martino [3], Letizia Rosa Romano [2,3], Giovanni Lopes [1], Francesco Fabio Greco [4], Carmen Anna Maria Spaccarotella [5], Ciro Indolfi [1], Antonio Curcio [1,2,*,†] and Alberto Polimeni [1,4,*,†]

1 Department of Pharmacy, Health and Nutritional Sciences, University of Calabria, 87036 Rende, Italy
2 Division of Cardiology, Annunziata Hospital, 87100 Cosenza, Italy
3 Department of Medical and Surgical Sciences, Magna Graecia University, 88100 Catanzaro, Italy
4 Division of Interventional Cardiology, Annunziata Hospital, 87100 Cosenza, Italy
5 Division of Cardiology, Department of Advanced Biomedical Sciences, Federico II University, 80134 Naples, Italy
* Correspondence: antonio.curcio.cardio@unical.it (A.C.); alberto.polimeni@unical.it (A.P.)
† These authors contributed equally to this work.

Abstract: Coronary microvascular disease (CMD) comprises a spectrum of conditions characterized by the functional and structural abnormalities of coronary microcirculation, affecting vessels typically smaller than 500 μm. Despite its clinical significance as a contributor to myocardial ischemia, CMD frequently remains underdiagnosed due to the limitations of current diagnostic approaches. Invasive testing, including coronary reactivity assessment, is considered the gold standard, but it is resource-intensive and not always accessible. Non-invasive methods, such as positron emission tomography (PET) and transthoracic Doppler echocardiography (TTDE), offer alternatives but are limited by varying accuracy and accessibility. Amid these diagnostic challenges, there is increasing interest in circulating biomarkers as adjuncts in CMD evaluation. Biomarkers associated with endothelial dysfunction, inflammation, and oxidative stress, detectable through routine blood tests, may assist in CMD diagnosis, risk stratification, and therapeutic monitoring. These biomarkers can offer insights into CMD pathogenesis and enable early, non-invasive screening to identify patients who may benefit from more invasive investigations. This narrative review examines studies assessing biomarkers in CMD patients with diagnoses confirmed through invasive techniques. Our objective is to focus on circulating biomarkers linked to the invasive evaluation of coronary microcirculation, aiming to advance the understanding of the underlying mechanisms of this prevalent condition and enhance diagnostic accuracy and the clinical management of affected patients.

Keywords: circulating biomarkers; coronary microvascular dysfunction (CMD); coronary flow reserve (CFR); index of microvascular resistance (IMR)

1. Introduction

Coronary microvascular dysfunction (CMD), which affects vessels usually smaller than 500 μm in diameter, such as pre-arterioles, arterioles, and capillaries, is becoming more widely acknowledged as a serious pathological condition marked by structural and functional abnormalities in coronary microcirculation [1,2]. These tiny vessels are crucial regulators of coronary blood flow, accounting for about 70% of coronary resistance. Inward arteriolar remodeling, capillary rarefaction, intravascular plugging, perivascular fibrosis or infiltration, and extramural compression (e.g., myocardial hypertrophy, increased left ventricular end-diastolic pressure [LVEDP]) are structural changes in the coronary

https://doi.org/10.3390/biom15020177

microcirculation [3]. Meanwhile, functional changes involve impaired vasodilator response and increased vasoconstrictor activity, such as microvascular spasms, which primarily result from endothelial and smooth muscle cell dysfunction [3].

CMD is a key contributor to myocardial ischemia and presents clinically as various conditions, including angina with non-obstructive coronary artery disease (ANOCA), ischemia with non-obstructive coronary artery disease (INOCA) [4], and myocardial infarction with non-obstructive coronary artery disease (MINOCA) [5–7]. In addition, CMD plays an important role in other cardiac and systemic diseases, such as hypertrophic cardiomyopathy, systemic autoimmune diseases, diabetes mellitus, and heart failure with preserved ejection fraction (HFpEF) [2,8–10].

To enhance the understanding of CMD, a clinical–pathogenetic classification is proposed, dividing CMD into four main categories: (1) CMD occurring in the absence of myocardial diseases or obstructive coronary artery disease (CAD), (2) CMD associated with myocardial diseases, (3) CMD related to obstructive epicardial CAD, and (4) iatrogenic CMD [11]. This classification underscores the diverse etiologies of CMD and highlights the need for tailored diagnostic and therapeutic approaches [12].

The pathogenesis of CMD is complex and multifactorial, involving mechanisms such as endothelial dysfunction, inflammation [13], oxidative stress, and altered vasoreactivity [14]. These processes collectively contribute to impaired coronary flow reserve, microvascular spasms, and increased coronary microvascular resistance, leading to the clinical manifestations of CMD [7]. Given the heterogeneity of its mechanisms, CMD is prevalent in up to 50% of patients with chronic coronary syndromes and over 20% of those presenting with acute coronary syndrome (ACS) [9].

Despite its considerable impact on patient outcomes, CMD remains underdiagnosed primarily due to challenges associated with its detection. Invasive coronary reactivity testing, including the assessment of coronary physiology, remains the gold standard for diagnosing CMD [15,16]. Due to its limited availability and the potential risks associated with invasive testing, alternative noninvasive techniques such as positron emission tomography (PET), transthoracic Doppler echocardiography (TTDE), stress echocardiography [17], and cardiac magnetic resonance imaging (CMR) have been employed, although their effectiveness can vary [18] (Figure 1).

Given these diagnostic challenges, there is a growing interest in identifying circulating biomarkers that could enhance the diagnosis, risk assessment, and management of patients with CMD. Biomarkers related to key processes like endothelial dysfunction, inflammation, and oxidative stress offer promising avenues for improving CMD detection and treatment strategies [19].

This narrative review aims to provide a comprehensive analysis of the role of circulating biomarkers in patients with CMD. Specifically, we focus on key biomarkers involved in fundamental pathological processes such as endothelial dysfunction, inflammation, and oxidative stress. A targeted search was performed using major databases, including PubMed and Scopus, to identify studies evaluating circulating biomarkers in patients diagnosed with CMD through invasive methods, such as coronary flow reserve (CFR) and the index of microcirculatory resistance (IMR). The novelty of this review lies in its emphasis on studies employing these advanced techniques, ensuring high diagnostic accuracy and providing deeper insights into the underlying pathophysiological mechanisms of CMD.

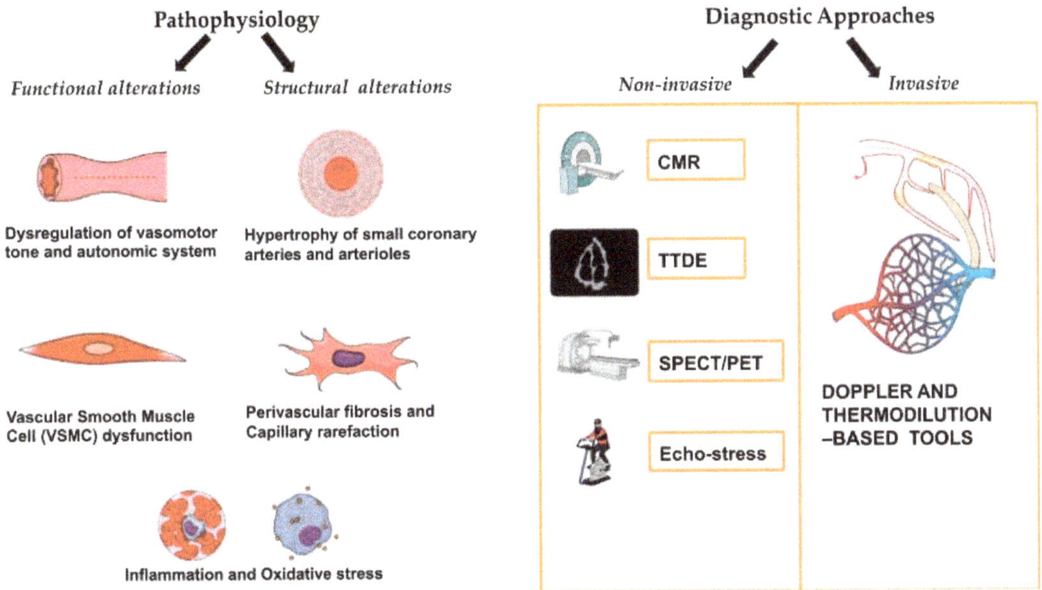

Figure 1. Pathophysiology and diagnostic approaches to CMD.

Therefore, for each pathophysiological mechanism involved in CMD, we prioritize studies in which CMD was evaluated invasively (Figure 2). This approach ensures an analysis based on the current gold standard, allowing for differentiation between the endothelium-dependent and endothelium-independent forms of CMD.

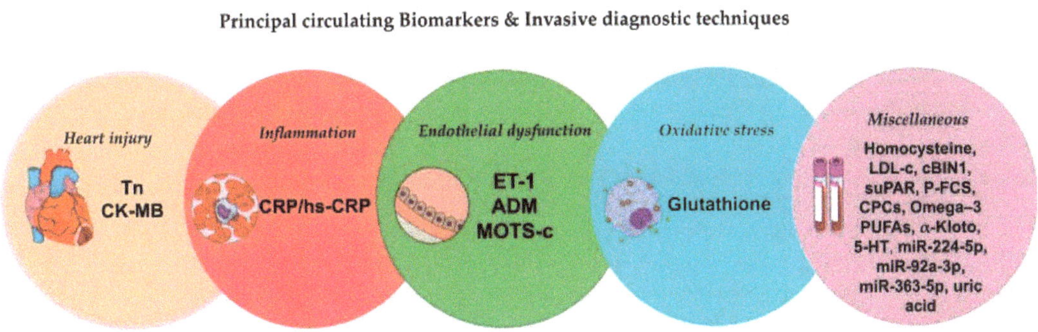

Figure 2. Principal circulating biomarkers evaluated with invasive diagnostic techniques.

2. Pathophysiology of Coronary Microvascular Disease

The proximal coronary compartment consists of epicardial coronary arteries, which are conductance vessels with a cross-sectional diameter greater than 500 µm. These vessels, visible on coronary angiography, are the primary sites for obstructive atherosclerosis. Under normal conditions, the epicardial arteries account for less than 10% of coronary vascular resistance and only become hemodynamically significant when more than 70% of the arterial lumen is narrowed. The distal compartment includes coronary microcirculation, encompassing all vessels with a cross-sectional diameter smaller than 500 µm. This includes pre-arteriolar vessels (100–500 µm in diameter), intramural arterioles (diameter less than 100 µm), and capillaries [1,2].

CMD is characterized by a multifaceted interplay of functional and structural disruptions within the coronary microcirculation, resulting in impaired coronary blood flow (CBF) and myocardial ischemia, even in the absence of significant epicardial coronary artery stenosis. The pathophysiology of CMD can generally be divided into functional and structural abnormalities, each of which plays a crucial role in reducing coronary flow reserve and contributing to the clinical manifestations of the disease [14,15] (Figure 1).

2.1. Functional Alterations

The functional disturbances inherent to CMD primarily involve the dysregulation of vasomotor tone, presenting as impaired vasodilation and the enhanced vasoconstriction of coronary microvasculature [14,20]. Endothelial dysfunction is central to these functional impairments. By producing vasodilatory mediators such as prostacyclin, nitric oxide (NO), and endothelium-derived hyperpolarizing factors (EDHFs), the endothelium controls vascular tone under normal physiological conditions [1,2]. However, in CMD, there is a marked reduction in the bioavailability of these mediators, particularly NO, largely due to increased oxidative stress. Reactive oxygen species (ROS) play a crucial role in this process by degrading NO, thereby diminishing its vasodilatory effect and simultaneously enhancing the vasoconstrictive actions of endothelin-1 (ET-1). This imbalance contributes to impaired vasodilation and increased microvascular resistance.

Endothelium-independent mechanisms, though less well understood, are thought to involve the impaired relaxation of vascular smooth muscle cells (VSMCs), the increased sensitivity of VSMCs to typical vasoconstrictor stimuli, and abnormal autonomic regulation. Additionally, ET-1 may play a role in these processes [11,21].

Inflammatory processes involved in CMD further exacerbate functional dysregulation. Inflammatory mediators such as interleukin-6 (IL-6), tumor necrosis factor-alpha (TNF-α), and C-reactive protein (CRP) promote endothelial activation and reduce NO bioavailability, aggravating the impaired vasomotor response. These inflammatory pathways, combined with oxidative stress-induced NO degradation, result in a diminished ability of the microvasculature to respond adequately to increased myocardial oxygen demand, leading to ischemic symptoms even in the absence of epicardial coronary artery stenosis [21,22].

In this context, diabetes mellitus (DM) significantly contributes to CMD primarily through oxidative stress, and it is characterized by reduced NO bioavailability, increased reactive oxygen species (ROS), elevated endothelin, and systemic inflammation. CMD often precedes hyperglycemia in type 2 diabetes, highlighting the role of insulin resistance and inflammation in its progression. Diabetes-associated CMD manifests systemically, affecting multiple organs, including the heart, retina, and kidneys [9].

Furthermore, coronary microvascular tone, regulated by myogenic mechanisms, metabolic control, endothelial function, circulating factors, and autonomic innervation, is significantly influenced by adrenergic innervation during exercise or the presence of endothelial dysfunction while contributing minimally at rest in healthy coronary circulations. Adrenergic receptors (α and β) modulate coronary circulation, with β2 receptors primarily mediating vasodilation, whereas α-receptor stimulation induces vasoconstriction, except for α2 receptors on endothelial cells, which promote vasodilation [9].

Autonomic dysregulation significantly contributes to CMD pathophysiology. An imbalance between sympathetic and parasympathetic activity exacerbates coronary vasoconstriction, particularly in clinical scenarios such as diabetes mellitus, hypertension, and post-myocardial infarction, where intrinsic vasodilatory mechanisms are already compromised. This autonomic imbalance, coupled with the increased vasoconstrictive actions of ET-1 and reduced vasodilatory capacity, further impairs coronary microvascular function and exacerbates myocardial ischemia. Overall, the interplay between oxidative stress, inflammation, and autonomic dysfunction creates a vicious cycle of impaired coronary vasoreactivity, contributing to the functional alterations seen in CMD [1].

2.2. Structural Alterations

The structural remodeling of coronary microvasculature is another critical aspect of CMD. This remodeling includes hypertrophic changes in small coronary arteries and arterioles, characterized by luminal narrowing and increased medial wall thickening due to smooth muscle hypertrophy and enhanced collagen deposition. Perivascular fibrosis and capillary rarefaction, manifested as a reduction in capillary density, are also prominent features. These structural alterations are particularly severe in conditions associated with increased left ventricular mass, such as hypertrophic cardiomyopathy and hypertensive heart disease, where they significantly impair coronary microvascular function [2]. In both primary and secondary left ventricular hypertrophy (LVH), CMD is marked by capillary rarefaction and the adverse remodeling of intramural arterioles, leading to increased wall-to-lumen ratios and progressive ischemic damage. This contributes to fibrosis, ventricular remodeling, and heightened risks of systolic dysfunction and heart failure. In HFpEF, CMD exacerbates oxygen demand mismatches and myocardial injury, even without obstructive coronary disease, driven by endothelial dysfunction, systemic inflammation, and oxidative stress, especially in patients with cardiometabolic risk factors [23].

The molecular mechanisms underlying these structural changes are complex, involving a cascade of pathological processes driven by oxidative stress and chronic inflammation. Key signaling pathways, such as the RhoA/Rho-kinase pathway, are implicated in promoting vascular smooth muscle contraction, pathological vascular remodeling, and perpetuating endothelial dysfunction [24]. Furthermore, epigenetic mechanisms, such as DNA methylation and histone modifications, further regulate vascular inflammation and aging, offering the potential for more precise diagnostics and management [25].

Given the complex and multifactorial nature of CMD, there is a growing interest in the potential role of circulating biomarkers as tools for the early detection, risk stratification, and management of this condition. Circulating biomarkers could provide insights into the underlying pathophysiological processes, such as endothelial dysfunction, oxidative stress, and inflammation, which are central to CMD [14]. By reflecting the biochemical milieu associated with CMD, these biomarkers could serve as non-invasive indicators of disease activity, helping identify patients at risk, monitor disease progression, and potentially guide therapeutic interventions. Indeed, the identification of specific biomarkers linked to key molecular pathways, such as NO bioavailability or ROS production, could facilitate the development of targeted therapies aimed at mitigating the functional and structural impairments seen in CMD. In this context, circulating biomarkers hold promise as valuable adjuncts in the comprehensive assessment and management of patients with CMD.

3. Diagnostic Approaches for Coronary Microvascular Disease

The Coronary Vasomotion Disorders International Study Group (COVADIS) established standardized international criteria for diagnosing CMD. These criteria include the presence of clinical symptoms, the absence of obstructive epicardial coronary artery disease, and evidence of myocardial ischemia, ascertained through both non-invasive tests and invasive assessments of coronary microvascular impairment [26–31].

In this context, both invasive and non-invasive diagnostic techniques have been developed to assess the functional integrity of the coronary microvasculature and to rule out obstructive coronary artery disease (CAD) as the underlying cause of symptoms. Despite significant advances in diagnostic methods, diagnosing CMD remains challenging.

Invasive methods, such as coronary reactivity testing, are currently considered the most accurate and reliable approach for diagnosing CMD and are regarded as the gold standard due to their capacity to provide detailed insights into microvascular function.

3.1. Invasive Diagnostic Techniques

Invasive testing is considered the gold standard for diagnosing CMD, especially in patients with a high clinical suspicion of microvascular dysfunction after obstructive CAD has been excluded. Different tools that are Doppler- and thermodilution-based are available for assessing microvascular function. A comprehensive evaluation of coronary microvascular function encompasses both endothelium-independent and endothelium-dependent assessments [30,32].

Endothelium-independent microvascular function: This is primarily assessed using CFR, the index of microvascular resistance (IMR), and hyperaemic microvascular resistance (hMR). CFR is calculated as the ratio of maximal coronary blood flow during pharmacologically induced hyperemia (commonly with adenosine) to resting flow, reflecting the combined vasodilatory capacity of both epicardial and microvascular coronary arteries. IMR, on the other hand, specifically evaluates microvascular resistance. Using thermodilution techniques, it is calculated as the product of distal coronary pressure and the mean transit time during hyperemia, providing a precise measure independent of epicardial artery function. The pathological values of CFR below 2.0–2.5 or an IMR greater than 25 indicate CMD [4]. hMR is another Doppler-based valuable index, calculated as the ratio of distal coronary pressure to average peak velocity during hyperemia, with values greater than 2.5 indicating CMD [33].

Additionally, continuous thermodilution techniques allow for the quantification of true coronary blood flow (Q) and microvascular resistance ($R\mu$) without the need for hyperemic agents like adenosine. These techniques are considered to have lower variability compared to traditional methods and provide a more accurate assessment of microvascular function.

Coronary vasospasm testing: The diagnosis of coronary microvascular vasospasm, which may contribute to CMD, involves provocative testing with intracoronary acetylcholine (ACh). This test evaluates endothelium-dependent microvascular function by inducing vasoconstriction or spasms in the microvasculature. A positive test, defined by the reproduction of angina symptoms and ischemic electrocardiographic changes in the absence of significant epicardial spasms, confirms microvascular spasms as contributors to CMD.

Coronary angiography: While coronary angiography is primarily used to exclude obstructive CAD, it can also provide indirect evidence of CMD. For instance, the "coronary slow-flow phenomenon", characterized by the delayed opacification of coronary vasculature, suggests increased microvascular resistance and/or microvascular spasm—a hallmark feature of CMD [2,14,15].

3.2. Non-Invasive Diagnostic Techniques

Non-invasive diagnostic techniques are increasingly utilized in the evaluation of CMD, particularly in patients for whom invasive testing is not feasible. These methods generally evaluate the surrogate markers of coronary microvascular function and include modalities such as TTDE, PET, and CMR.

TTDE measures coronary flow velocity reserve (CFVR), analogous to CFR, by assessing maximal diastolic flow in the left anterior descending artery at rest and during pharmacologic stress. TTDE is advantageous due to its low cost, lack of radiation exposure, and widespread availability. However, it is technically challenging and requires significant operator expertise.

PET is considered the reference standard for the non-invasive assessment of myocardial blood flow (MBF) and CFR. It allows for a comprehensive evaluation of all coronary territories simultaneously by quantifying MBF at rest and during maximal hyperemia. Nevertheless, PET's use is limited by high costs, limited availability, and radiation exposure.

CMR offers high-resolution imaging capable of assessing myocardial perfusion and distinguishing between epicardial and microvascular dysfunction. CMR stress perfusion, typically performed with adenosine, allows for the quantification of myocardial perfusion reserve (MPR) and has shown promising results in detecting CMD. Additionally, late gadolinium enhancement on CMR has prognostic value, as it is associated with a higher risk of adverse cardiovascular events.

Dynamic Myocardial Perfusion Computed Tomography (CT) is an emerging technique that combines both the anatomical and functional assessments of coronary arteries. When combined with CTCA for the exclusion of obstructive CAD, dynamic myocardial perfusion CT offers promising diagnostic capabilities, though at the cost of increased radiation exposure [1,2,14].

4. Circulating Biomarkers in Coronary Microvascular Dysfunction

In the study of CMD, the identification of reliable biomarkers is crucial for understanding the underlying pathophysiological mechanisms and improving diagnosis and management. A comprehensive understanding of endothelial markers, particularly those associated with inflammation and oxidative stress, provides valuable insights into the pathophysiology of CMD. The key indicators of endothelial dysfunction, such as the von Willebrand factor (VWF), angiotensin-converting enzyme (ACE), and soluble adhesion molecules including sVCAM-1, sICAM-1, and E-selectin, not only reflect vascular injury but also underline the inflammatory and oxidative processes contributing to microvascular impairment [34].

Building on this framework, other biomarkers have also garnered attention due to their potential roles in CMD. Among these, the brain-derived neurotrophic factor (BDNF) has been studied for its critical function in maintaining vascular homeostasis and endothelial integrity [35]. Beyond these effects, BDNF has been implicated in modulating adverse cardiac remodeling [36] and promoting inflammation and atherosclerosis [37].

Additionally, heat shock proteins (HSPs), such as HSP60 and HSP72, have emerged as important mediators of microvascular and myocardial damage, particularly in patients with idiopathic left ventricular dysfunction [38]. Notably, HSP60 plays a pivotal role in the pathogenesis of cardiovascular diseases through its multifaceted involvement in mitochondrial function, protein folding, cardiac remodeling and failure, cardiomyocyte regulation, the promotion of atherosclerosis, and the activation of inflammatory pathways [39].

Although numerous biomarkers have been analyzed in the literature, this section focuses on biomarkers that have been studied specifically in patients where CMD was assessed using invasive diagnostic techniques, ensuring a high level of accuracy in evaluating the role of these biomarkers (Table 1).

4.1. Traditional Biomarkers of Heart Injury

To date, the role of troponin (Tn) as a biomarker in CMD remains not fully established. Several studies have explored its potential significance in CMD. For instance, research by Takashio et al. showed elevated plasma troponin T (TnT) levels in individuals with CMD compared to healthy controls [40]. Similarly, Fujii et al. observed that patients undergoing elective percutaneous coronary angioplasty (PTCA) exhibited higher post-PTCA values of IMR in those with elevated troponin I (TnI), pointing to significant microvascular dysfunction and evaluating the potential protective effect of statins [41].

In line with this, another study confirmed a similar association between post-PTCA values and plasma creatine kinase MB (CK-MB) levels [42], while a different study involving 55 PTCA patients demonstrated a link between post-PTCA CFR values and cTnT and CK levels [43].

Conversely, a study of 19 patients found no significant correlation between cTnI levels and CFR and a moderate association with IMR [44].

4.2. *Biomarkers of Inflammation*

The role of inflammation in CMD is increasingly recognized as a central mechanism in the pathogenesis and progression of the disease. Elevated levels of circulating biomarkers such as CRP have been consistently associated with microvascular dysfunction in various cardiovascular conditions [45]. CRP, a well-known inflammatory biomarker, is elevated in patients with cardiac syndrome X (CSX) [46,47] and has been shown to serve as a marker of disease activity, suggesting that inflammation plays a significant role in CMD pathophysiology [2,48].

Only a few studies have correlated circulating biomarkers with objective evidence of microvascular dysfunction using invasive tests. For instance, Ong et al. conducted ACh testing in patients with angina and unobstructed coronary arteries to evaluate coronary spasms. Notably, elevated levels of high-sensitivity CRP (hs-CRP) and soluble CD40 ligand (sCD40L) were significantly associated with coronary spasm, hence indicating an inflammatory component in CMD [49]. Furthermore, Teragawa et al. also used intracoronary ACh infusion to assess endothelial function and found that elevated CRP levels were associated with reduced coronary blood flow response [50].

Additional studies have examined the relationship between hs-CRP levels and coronary vasospastic angina [51], as well as the association between inflammation and microvascular resistance in patients with CSX. Using CFR and the IMR, Long et al. demonstrated that elevated hs-CRP levels were linked to impaired CFR and higher IMR, further supporting the role of inflammation in CMD [52].

4.3. *Biomarkers of Oxidative Stress*

Oxidative stress, defined as an imbalance between the production of reactive oxygen species (ROS) and the body's antioxidant defenses, is a major contributor to cardiovascular disease [53], CSX [54], plaque instability [55], and INOCA [56]. This imbalance leads to reduced NO bioavailability and promotes vascular damage.

An important antioxidant, glutathione, has been linked to CMD, with lower levels indicating greater oxidative stress and impaired microvascular function. Studies have demonstrated that glutathione is an independent predictor of CMD, reinforcing its potential as a biomarker of oxidative stress in this condition. Notably, the only study to evaluate coronary microvascular function invasively found that lower glutathione levels independently predicted impaired microvascular function and increased plaque necrotic core [57].

4.4. *Biomarkers of Endothelial Dysfunction*

Endothelial dysfunction is a key feature of CMD, characterized by impaired NO synthesis and increased vasoconstriction.

ET-1, the most potent endogenous vasoconstrictor, is a critical biomarker of endothelial dysfunction. James Theuerle et al. investigated the relationship between plasma levels of ET-1 and adrenomedullin (ADM) and their association with CMD and cardiovascular outcomes. The study found that elevated ET-1 levels were linked to higher microvascular resistance, while higher levels of ADM were associated with improved coronary flow and a reduced risk of major adverse cardiovascular events (MACEs) [58].

Moreover, lower circulating MOTS-c levels are associated with impaired coronary endothelial function in patients without significant coronary lesions [59].

4.5. Miscellaneous Biomarkers

Several other biomarkers have been studied in relation to CMD, even though they do not fit neatly into the aforementioned categories.

Homocysteine, a sulfur-containing amino acid, has been linked to CMD, with elevated levels correlating with impaired microvascular function [60].

Mangiacapra et al. found a significant association between elevated cholesterol levels, particularly LDL cholesterol, and coronary microvascular dysfunction [61].

Pacheco et al. found that higher cBIN1 levels were linked to CMD, suggesting a potential progression from CMD to HFpEF. These findings indicate that cardiomyocyte dysfunction may play a role in the pathophysiology of CMD in INOCA [62].

Soluble urokinase-type plasminogen activator receptor (suPAR) is another potential biomarker that has been associated with coronary microvascular dysfunction. In patients with non-obstructive coronary artery disease, plasma suPAR levels were found to be an independent predictor of coronary microvascular function [63].

Moreover, thrombogenicity has been associated with CMD in patients with acute myocardial infarction (AMI) [64].

Girum Mekonnen et al. from the NHLBI-sponsored Women's Ischemia Syndrome Evaluation—Coronary Vascular Dysfunction (WISE-CVD) study explored the relationship between CMD and circulating progenitor cells (CPCs) in women with ischemic symptoms but without obstructive coronary artery disease (CAD). In this study, CMD was evaluated through CFR using intracoronary adenosine. The results revealed that a lower CFR, indicating impaired microvascular function, was associated with higher levels of CPCs (CD34+, CD34+/CD133+, and CD34+/CXCR4+), suggesting that chronic ischemia in CMD may stimulate CPC mobilization [65].

Muroya et al. explored the relationship between omega-3 polyunsaturated fatty acids (PUFAs) and CMD in patients with stable coronary artery disease. The study assessed coronary microvascular function using the hMRI and found that lower levels of the eicosapentaenoic acid (EPA)/arachidonic acid (AA) ratio were associated with higher hMRI, indicating worse CMD [66].

Akhiyat et al. found that patients with CMD had significantly lower circulating levels of α-Klotho, a protein involved in vascular health and aging. The reduced α-Klotho levels were associated with CMD [67].

Serotonergic neurons in the central nervous system and enterochromaffin cells in the gastrointestinal tract are the primary producers of serotonin, also known as 5-hydroxytryptamine (5-HT), which platelets from the bloodstream subsequently collect [68]. It has been determined that plasma serotonin levels may serve as a biomarker for CMD in patients with nonobstructive coronary arteries and probable angina. Odaka et al. included 198 individuals who performed an acetylcholine provocation test to evaluate CMD—which is defined as myocardial lactate production without obvious epicardial spasms. A serotonin concentration above 9.55 nmol/L is a powerful independent predictor of the presence of CMD according to the findings, which also indicate a substantial correlation between elevated plasma serotonin levels and CMD [69].

Other studies explored the role of microRNAs (miRNAs) in endothelial and coronary dysfunction. MicroRNAs (miRNAs) are short, highly conserved RNA molecules that regulate gene expression through mRNA silencing, either by transcript degradation or translational repression [70]. These investigations aim to identify miRNA markers associated with both macrovascular and microvascular dysfunction, potentially offering new insights into the diagnosis and management of CMD [71,72]. Notably, the increased expression of miR-224-5p in circulating extracellular vesicles is associated with reduced CFR, indicating endothelial dysfunction. This miRNA was found to originate predominantly from hepatic cells, and it is linked to a higher expression of endothelial ICAM-1, which may contribute to vascular inflammation [73].

Even high levels of uric acid are associated with impaired coronary microvascular function, making it a relevant mostly gender-related marker in the study of CMD [74]. A study by Megha Prasad demonstrated that elevated uric acid levels are associated with inflammation and CMD in postmenopausal women. The research highlighted that higher serum uric acid was correlated with impaired endothelial function, as measured by acetylcholine-mediated coronary blood flow, and increased inflammatory markers, including hs-CRP and neutrophil counts. Uric acid levels have been found to be elevated in women with CMD, suggesting a potential gender-specific biomarker for CMD [75].

According to a pilot study by Keeley et al., women with coronary microvascular dysfunction have lower levels of omega-3 fatty acid-derived mediators, maresin 1 and resolvin D1, suggesting that the pathophysiology of CMD may be influenced by the impaired production of these mediators [76]. Furthermore, Ito et al. in patients without obstructive coronary artery disease (CAD), found a significant correlation between CMD and circulating malondialdehyde-modified low-density lipoprotein (MDA-LDL) levels [77].

Table 1. Circulating biomarkers and invasive measures of coronary microcirculation.

Authors	Year	Biomarkers Evaluated	N. Subjects	Main Findings	Invasive Measures
Takashio et al. [40]	2013	cTnT, BNP, hs-CRP, IL-6, TNF-α	90 patients 47 controls	cTnT levels were higher in patients with CMD and increased LVEDP	CFR
Kitabata et al. [42]	2013	CK-MB	24 patients	CK-MB levels were higher in patients with a higher MVRI	MVRI
Herrmann et al. [43]	2001	CK, cTnT	55 patients	rCVR after successful coronary stenting was independently associated with elevation of cTnT and CK	CFVR
Park et al. [44]	2014	cTnI	19 patients	cTnI was correlated with IMR but not with CFR values	CFR, IMR
Ong et al. [49]	2015	sCD40L, hs-CRP	62 patients	Patients with Ach-induced coronary spasm and no significant coronary stenoses had elevated levels of sCD40L and hs-CRP	ACh provocation test
Teragawa et al. [50]	2004	CRP	46 patients	Increased CRP levels had reduced CBF response to acetylcholine	ACh provocation test
Hung et al. [51]	2005	hs-CRP	428 patients	hs-CRP was associated with coronary vasospastic angina	Ergonovine provocation test
Long et al. [52]	2017	hs-CRP	20 patients 20 controls	hs-CRP was positively correlated with IMR	CFR, IMR
Dhawan et al. [57]	2011	Cystine/Glutathione, hs-CRP	47 patients	Lower plasma glutathione levels and higher cystine/glutathione ratios were associated with impaired coronary microvascular function and greater plaque necrotic core content	CFVR, hMR, IVUS
Theuerle et al. [58]	2019	ET-1, ADM	32 patients	Elevated ET-1 levels were associated with increased IMR. ADM levels were correlated with CFR.	IMR, CFR

Table 1. *Cont.*

Authors	Year	Biomarkers Evaluated	N. Subjects	Main Findings	Invasive Measures
Qin et al. [59]	2018	MOTS-c	20 patients 20 controls	MOTS-c levels were positively correlated with both microvascular and epicardial endothelial function	ACh provocation test
Ahmad et al. [60]	2020	Homocysteine	1418 patients	Elevated homocysteine levels positively correlate with CMD and early atherosclerosis	ACh provocation test
Mangiacapra et al. [61]	2012	Total cholesterol, LDL-c, HDL-c, TG	95 patients	Significant correlation between IMR and total cholesterol and LDL-c	IMR
Pacheco et al. [62]	2021	cBIN1	95 patients 50 controls	Higher cBIN1 score was associated with vasoconstriction to Ach	CFR, ACh provocation test
Mekonnen et al. [63]	2015	suPAR	66 patients	Higher suPAR levels correlated with lower CFR	CFR
Kang et al. [64]	2021	P-FCS, D-dimer	116 patients	P-FCS significantly increased the risk of CMD post-procedurally	IMR
Mekonnen et al. [65]	2016	CPCs	123 patients	Lower CFR was associated with higher levels of CPCs cells	CFR
Muroya et al. [66]	2018	EPA/AA, DGLA	108 patients	Lower levels of the EPA/AA were significantly correlated with increased hMVRI	hMVRI
Akhiyat et al. [67]	2024	α-Klotho	98 patients	Patients with CMD had significantly lower circulating α-Klotho levels	CFR
Odaka et al. [69]	2016	Serotonin	198 patients	Plasma serotonin concentration was found to be significantly higher in patients with CMD	ACh provocation test
James et al. [73]	2022	miR-224-5p	120 patients	miR-224-5p was found to be significantly elevated in the low-CFR group	CFR
Kuwahata et al. [74]	2010	UA	194 patients	Elevated UA levels were associated with impaired CBF	ACh and papaverin provocation test
Prasad et al. [75]	2017	UA	229 patients	Elevated UA levels were associated with impaired CBF	ACh provocation test
Keeley et al. [76]	2022	Resolvins, Maresin 1, EPA, DHA and 18-HEPE	31 patients 12 controls	Women with CMD had significantly lower levels of resolvin D1 and maresin 1 but higher levels of DHA and 18-HEPE compared to the control group	CFR
Ito et al. [77]	2024	MDA-LDL	95 patients	Elevated levels of MDA-LDL were significantly associated with CMD in patients without CAD	CFR, IMR

Abbreviations: 18-HEPE, 18-hydroxy-eicosapentaenoic acid; ACh, acetylcholine; ADM, adrenomedullin; BNP, B-type natriuretic peptide; CAD, coronary artery disease; CBF, coronary blood flow; cBIN1, cardiac bridging integrator 1; CFR, coronary flow reserve; CFVR, coronary flow velocity reserve; CK, creatine kinase; CK-MB, creatine kinase myocardial band; CMD, coronary microvascular dysfunction; CPCs, circulating progenitor cells; CRP, C-reactive protein; cystine/glutathione, ratio of oxidized cystine to reduced glutathione; cTnI, cardiac troponin I; cTnT, cardiac troponin T; DGLA, dihomo-gamma-linolenic acid; DHA, docosahexaenoic acid; D-dimer, fibrin degradation product; EPA, eicosapentaenoic acid; EPA/AA, eicosapentaenoic acid/arachidonic acid ratio; ET-1, endothelin-1; hMVRI, hyperemic microvascular resistance index; hs-CRP, high-sensitivity C-reactive protein; HDL-c, high-density lipoprotein cholesterol; IL-6, interleukin-6; IMR, index of microcirculatory resistance; LDL-c, low-density lipoprotein cholesterol; LVEDP, left ventricular end-diastolic pressure; MDA-LDL, malondialdehyde-modified low-density lipoprotein; miR, microRNA; MOTS-c, mitochondrial open reading frame of 12S rRNA-C; MVRI, microvascular resistance index; P-FCS, platelet-fibrin clot strength; rCVR, relative coronary vascular resistance; sCD40L, soluble CD40 ligand; suPAR, soluble urokinase-type plasminogen activator receptor; TG, triglycerides; TNF-α, tumor necrosis factor alpha; total cholesterol, total cholesterol; UA, uric acid; α-Klotho, alpha-klotho.

5. Therapeutic Implications of Biomarkers in Coronary Microvascular Disease

The management of CMD remains a significant clinical challenge due to the lack of large-scale randomized trials confirming effective treatment strategies. Current approaches emphasize addressing cardiovascular risk factors such as smoking, hypertension, diabetes, and obesity while tailoring therapies to individual phenotypic presentations. Animal models, from rodents to large mammals, are essential for exploring CMD's role in ischemia and metabolic disorders. While small models facilitate genetic studies, large models like swine better mimic human physiology, enabling the evaluation of targeted therapies under controlled conditions and ensuring translational relevance to clinical applications [78].

Inflammation, as indicated by elevated CRP and hs-CRP levels, underscores the importance of anti-inflammatory strategies. Statins, beyond their cholesterol-lowering effects, reduce vascular inflammation, enhance nitric oxide bioavailability, and improve endothelial function. These benefits have been observed not only in patients with obstructive coronary artery disease but also in those with CMD and non-obstructive CAD [79,80]. Similarly, angiotensin-converting enzyme (ACE) inhibitors and angiotensin II receptor blockers (ARBs) exert vasoprotective effects by improving microcirculatory function, reducing oxidative stress, and enhancing CFR [81,82].

Beta-blockers remain highly effective in reducing chest pain episodes in CMD by lowering myocardial oxygen demands and inducing endothelium-dependent vasodilation. Third-generation beta-blockers, such as nebivolol, offer additional benefits by improving endothelial function and exercise tolerance [83].

Nitrates, particularly short-acting formulations, provide symptomatic relief in CMD by inducing vasodilation [84]. Emerging therapies, such as phosphodiesterase type-5 inhibitors (e.g., sildenafil), have demonstrated potential in improving microvascular function by enhancing nitric oxide signaling [85]. Similarly, the inhibition of Rho-kinase pathways with agents like fasudil shows promise in preventing microvascular spasms and improving myocardial perfusion [86,87].

Ranolazine, widely used in CMD, reduces myocardial oxygen demand and alleviates ischemic symptoms [88], while other anti-anginal agents, including ivabradine, nicorandil, and trimetazidine, also target oxygen consumption and have demonstrated benefits in alleviating angina symptoms [89]. Additional promising therapeutic roles may be played by agents such as SGLT2 inhibitors, GLP-1 receptor agonists, and novel lipid-lowering therapies, for which specific studies are still needed.

Additionally, endothelial dysfunction, as evidenced by biomarkers like ADM and ET-1, underscores the therapeutic potential of interventions aimed at improving endothelial function. Notably, Zibotentan, a selective endothelin-A receptor antagonist, has emerged as a promising therapeutic agent targeting ET-1. By inhibiting ET-1 activity, Zibotentan may reduce vascular tone and improve microvascular function, offering a novel approach to managing CMD and its associated complications [90].

The integration of biomarkers into CMD management not only advances the understanding of underlying mechanisms but also enables a more precise and individualized therapeutic approach. Combining traditional therapies with emerging pharmacological strategies and lifestyle modifications offers a comprehensive framework for improving outcomes in CMD patients. Ongoing research into innovative treatments, including those targeting inflammation, endothelial dysfunction, and oxidative stress, will be pivotal in addressing unmet clinical needs in CMD management.

6. Conclusions

CMD represents a complex and multifactorial condition with significant implications for cardiovascular health. As this review has highlighted, CMD is underpinned by intricate interactions between functional and structural abnormalities within the coronary microcirculation, driven by endothelial dysfunction, inflammation, oxidative stress, and autonomic dysregulation. The challenges in diagnosing CMD, particularly given its subtle and non-obstructive nature, underscore the importance of invasive diagnostic methods such as coronary reactivity testing, which remains the gold standard. However, the advent of circulating biomarkers offers a promising adjunct to these invasive techniques, providing insights into the underlying pathophysiology and opening avenues for more personalized therapeutic strategies. By targeting specific biomarkers associated with inflammation, oxidative stress, and endothelial dysfunction, novel therapies can be developed to mitigate the impact of CMD, ultimately improving patient outcomes.

Author Contributions: Authors have made substantial contributions to all the following: the conception and design of the study, R.Q. and A.P; the acquisition of data, or the analysis and interpretation of data, G.M. and G.L.; drafting the article or revising it critically for important intellectual content, L.R.R. and A.P., supervision, C.A.M.S., F.F.G., A.C. and C.I. with approval of this version to be submitted. All authors have read and agreed to the published version of the manuscript.

Funding: Giovanni Martino, MD, received a research grant from the CardioPath program of the Federico II University of Naples, Italy.

Institutional Review Board Statement: Not applicable.

Informed Consent Statement: Not applicable.

Data Availability Statement: Not applicable.

Conflicts of Interest: The authors declare no conflicts of interest.

References

1. Crea, F.; Montone, R.A.; Rinaldi, R. Pathophysiology of Coronary Microvascular Dysfunction. *Circ. J.* **2022**, *86*, 1319–1328. [CrossRef]
2. Del Buono, M.G.; Montone, R.A.; Camilli, M.; Carbone, S.; Narula, J.; Lavie, C.J.; Niccoli, G.; Crea, F. Coronary Microvascular Dysfunction across the Spectrum of Cardiovascular Diseases: JACC State-of-the-Art Review. *J. Am. Coll. Cardiol.* **2021**, *78*, 1352–1371. [CrossRef]
3. Vrints, C.; Andreotti, F.; Koskinas, K.C.; Rossello, X.; Adamo, M.; Ainslie, J.; Banning, A.P.; Budaj, A.; Buechel, R.R.; Chiariello, G.A.; et al. 2024 ESC Guidelines for the management of chronic coronary syndromes. *Eur. Heart J.* **2024**, *45*, 3415–3537. [PubMed]
4. Kunadian, V.; Chieffo, A.; Camici, P.G.; Berry, C.; Escaned, J.; Maas, A.H.E.M.; Prescott, E.; Karam, N.; Appelman, Y.; Fraccaro, C.; et al. An EAPCI Expert Consensus Document on Ischaemia with Non-Obstructive Coronary Arteries in Collaboration with European Society of Cardiology Working Group on Coronary Pathophysiology & Microcirculation. *Eur. Heart J.* **2020**, *41*, 3504–3520. [CrossRef]
5. Tamis-Holland, J.E.; Jneid, H.; Reynolds, H.R.; Agewall, S.; Brilakis, E.S.; Brown, T.M.; Lerman, A.; Cushman, M.; Kumbhani, D.J.; Arslanian-Engoren, C.; et al. Contemporary diagnosis and management of patients with myocardial infarction in the absence of obstructive coronary artery disease: A scientific statement from the American Heart Association. *Circulation* **2019**, *139*, e891–e908. [CrossRef]
6. Severino, P.; D'Amato, A.; Prosperi, S.; Myftari, V.; Colombo, L.; Tomarelli, E.; Piccialuti, A.; Di Pietro, G.; Birtolo, L.I.; Maestrini, V.; et al. Myocardial Infarction with Non-Obstructive Coronary Arteries (MINOCA): Focus on Coronary Microvascular Dysfunction and Genetic Susceptibility. *J. Clin. Med.* **2023**, *12*, 3586. [CrossRef]
7. Indolfi, C.; Porto, I.; Muscoli, S.; Fedele, F.; Mancone, M. La patologia microvascolare coronarica: Meccanismi e contesti clinici. *G. Ital. Cardiol.* **2022**, *23*, 397–407. [CrossRef]
8. La Vecchia, G.; Fumarulo, I.; Caffè, A.; Chiatto, M.; Montone, R.A.; Aspromonte, N. Microvascular Dysfunction across the Spectrum of Heart Failure Pathology: Pathophysiology, Clinical Features and Therapeutic Implications. *Int. J. Mol. Sci.* **2024**, *25*, 7628. [CrossRef]

9. Padro, T.; Manfrini, O.; Bugiardini, R.; Canty, J.; Cenko, E.; De Luca, G.; Duncker, D.J.; Eringa, E.C.; Koller, A.; Tousoulis, D.; et al. ESC Working Group on Coronary Pathophysiology and Microcirculation position paper on 'coronary microvascular dysfunction in cardiovascular disease'. *Cardiovasc. Res.* **2020**, *116*, 741–755. [CrossRef] [PubMed]
10. Belmonte, M.; Foà, A.; Paolisso, P.; Bergamaschi, L.; Gallinoro, E.; Polimeni, A.; Scarsini, R.; Muscoli, S.; Amicone, S.; De Vita, A.; et al. Coronary microvascular dysfunction beyond the spectrum of chronic coronary syndromes. *Prog. Cardiovasc. Dis.* **2024**, *87*, 73–82. [CrossRef]
11. Camici, P.G.; Crea, F. Coronary microvascular dysfunction. *N. Engl. J. Med.* **2007**, *356*, 830–840. [CrossRef] [PubMed]
12. Indolfi, C.; Polimeni, A.; Mongiardo, A.; De Rosa, S.; Spaccarotella, C. Old unsolved problems: When and how to treat silent ischaemia. *Eur. Heart J. Suppl.* **2020**, *22*, L82–L85. [CrossRef]
13. Guo, Z.; Yang, Z.; Song, Z.; Li, Z.; Xiao, Y.; Zhang, Y.; Wen, T.; Pan, G.; Xu, H.; Sheng, X.; et al. Inflammation and coronary microvascular disease: Relationship, mechanism and treatment. *Front. Cardiovasc. Med.* **2024**, *11*, 1280734. [CrossRef] [PubMed]
14. Rocco, E.; Grimaldi, M.C.; Maino, A.; Cappannoli, L.; Pedicino, D.; Liuzzo, G.; Biasucci, L.M. Advances and Challenges in Biomarkers Use for Coronary Microvascular Dysfunction: From Bench to Clinical Practice. *J. Clin. Med.* **2022**, *11*, 2055. [CrossRef]
15. Scarsini, R.; Campo, G.; Di Serafino, L.; Zanon, S.; Rubino, F.; Monizzi, G.; Biscaglia, S.; Ancona, M.; Polimeni, A.; Niccoli, G.; et al. #FullPhysiology: A systematic step-by-step guide to implement intracoronary physiology in daily practice. *Minerva Cardiol. Angiol.* **2023**, *71*, 504–514. [CrossRef] [PubMed]
16. Montone, R.A.; Meucci, M.C.; De Vita, A.; Lanza, G.A.; Niccoli, G. Coronary provocative tests in the catheterization laboratory: Pathophysiological bases, methodological considerations and clinical implications. *Atherosclerosis* **2021**, *318*, 14–21. [CrossRef] [PubMed]
17. Djordjevic Dikic, A.; Dedic, S.; Jovanovic, I.; Boskovic, N.; Giga, V.; Nedeljkovic, I.; Tesic, M.; Aleksandric, S.; Cortigiani, L.; Ciampi, Q.; et al. Noninvasive evaluation of dynamic microvascular dysfunction in ischemia and no obstructive coronary artery disease patients with suspected vasospasm. *J. Cardiovasc. Med.* **2024**, *25*, 123–131.
18. Vijayan, S.; Barmby, D.S.; Pearson, I.R.; Davies, A.G.; Wheatcroft, S.B.; Sivananthan, M. Assessing coronary blood flow physiology in the cardiac catheterisation laboratory. *Curr. Cardiol. Rev.* **2017**, *13*, 232–243. [CrossRef] [PubMed]
19. Nowroozpoor, A.; Gutterman, D.; Safdar, B. Is microvascular dysfunction a systemic disorder with common biomarkers found in the heart, brain, and kidneys?—A scoping review. *Microvasc. Res.* **2021**, *134*, 104123. [CrossRef] [PubMed]
20. Sato, T.; Hanna, P.; Mori, S. Innervation of the coronary arteries and its role in controlling microvascular resistance. *J. Cardiol.* **2024**, *84*, 1–13. [CrossRef]
21. Crea, F.; Camici, P.G.; Merz, C.N.B. Coronary microvascular dysfunction: An update. *Eur. Heart J.* **2014**, *35*, 1101–1111. [CrossRef]
22. Chakrala, T.; Prakash, R.; Valdes, C.; Pepine, C.J.; Keeley, E.C. Circulating Biomarkers in Coronary Microvascular Dysfunction. *J. Am. Heart Assoc.* **2023**, *12*, e029341. [CrossRef] [PubMed]
23. Camici, P.G.; Tschöpe, C.; Di Carli, M.F.; Rimoldi, O.; Van Linthout, S. Coronary microvascular dysfunction in hypertrophy and heart failure. *Cardiovasc. Res.* **2020**, *116*, 806–816. [CrossRef] [PubMed]
24. Nunes, K.P.; Webb, R.C. New insights into RhoA/Rho-kinase signaling: A key regulator of vascular contraction. *Small GTPases* **2021**, *12*, 458–469. [CrossRef] [PubMed]
25. Masi, S.; Rizzoni, D.; Taddei, S.; Widmer, R.J.; Montezano, A.C.; Lüscher, T.F.; Schiffrin, E.L.; Touyz, R.M.; Paneni, F.; Lerman, A.; et al. Assessment and pathophysiology of microvascular disease: Recent progress and clinical implications. *Eur. Heart J.* **2021**, *42*, 2590–2604. [CrossRef] [PubMed]
26. Demola, P.; Beneduce, A.; Masiero, G.; Serino, F.; Baldi, E.; Polimeni, A.; Attisano, T.; Contarini, M.; Castiglioni, B.; De Marco, F.; et al. Diagnosi e trattamento dei pazienti affetti da ischemia/infarto miocardico in assenza di coronaropatia ostruttiva nei laboratori di emodinamica italiani: Risultati dell'indagine nazionale SICI-GISE promossa dal Comitato GISE Young. *G. Ital. Cardiol.* **2023**, *24*, 42S–52S.
27. Polimeni, A.; Campo, G. Editorial: Coronary epicardial and microvascular hemodynamics. *Front. Cardiovasc. Med.* **2022**, *9*, 969928. [CrossRef]
28. De Rosa, S.; Polimeni, A.; De Velli, G.; Conte, M.; Sorrentino, S.; Spaccarotella, C.; Mongiardo, A.; Sabatino, J.; Contarini, M.; Indolfi, C. Reliability of Instantaneous Wave-Free Ratio (iFR) for the Evaluation of Left Main Coronary Artery Lesions. *J. Clin. Med.* **2019**, *8*, 1143. [CrossRef] [PubMed]
29. De Rosa, S.; Polimeni, A.; Petraco, R.; Davies, J.E.; Indolfi, C. Diagnostic Performance of the Instantaneous Wave-Free Ratio: Comparison With Fractional Flow Reserve. *Circ. Cardiovasc. Interv.* **2018**, *11*, e004613. [CrossRef] [PubMed]
30. Niccoli, G.; Morrone, D.; De Rosa, S.; Montone, R.A.; Polimeni, A.; Aimo, A.; Mancone, M.; Muscoli, S.; Pedrinelli, R.; Indolfi, C.; et al. The central role of invasive functional coronary assessment for patients with ischemic heart disease. *Int. J. Cardiol.* **2021**, *331*, 17–25. [CrossRef]
31. Ong, P.; Camici, P.G.; Beltrame, J.F.; Crea, F.; Shimokawa, H.; Sechtem, U.; Kaski, J.C.; Merz, C.N.B.; Coronary Vasomotion Disorders International Study Group (COVADIS). International standardization of diagnostic criteria for microvascular angina. *Int. J. Cardiol.* **2018**, *250*, 16–20. [CrossRef] [PubMed]

32. Ong, P.; Safdar, B.; Seitz, A.; Hubert, A.; Beltrame, J.F.; Prescott, E. Diagnosis of coronary microvascular dysfunction in the clinic. *Cardiovasc. Res.* **2020**, *116*, 841–855. [CrossRef] [PubMed]
33. Pompei, G.; Ganzorig, N.; Kotanidis, C.P.; Alkhalil, M.; Collet, C.; Sinha, A.; Perera, D.; Beltrame, J.; Kunadian, V. Novel diagnostic approaches and management of coronary microvascular dysfunction. *Am. J. Prev. Cardiol.* **2024**, *19*, 100712. [CrossRef]
34. Goncharov, N.V.; Nadeev, A.D.; Jenkins, R.O.; Avdonin, P.V. Markers and Biomarkers of Endothelium: When Something Is Rotten in the State. *Oxid. Med. Cell. Longev.* **2017**, *2017*, 9759735. [CrossRef] [PubMed]
35. Cao, L.; Zhang, L.; Chen, S.; Yuan, Z.; Liu, S.; Shen, X.; Zheng, X.; Qi, X.; Lee, K.K.H.; Chan, J.Y.H.; et al. BDNF-mediated migration of cardiac microvascular endothelial cells is impaired during ageing. *J. Cell. Mol. Med.* **2012**, *16*, 1518–1526. [CrossRef]
36. Bahls, M.; Könemann, S.; Markus, M.R.P.; Wenzel, K.; Friedrich, N.; Nauck, M.; Völzke, H.; Steveling, A.; Janowitz, D.; Grabe, H.-J.; et al. Brain-derived neurotrophic factor is related with adverse cardiac remodeling and high NTproBNP. *Sci. Rep.* **2019**, *9*, 15421. [CrossRef] [PubMed]
37. Kytikova, O.Y.; Novgorodtseva, T.P.; Denisenko, Y.K.; Antonyuk, M.V.; Gvozdenko, T.A.; Atamas, O.V. Brain-derived neurotrophic factor and coronary artery disease. *Russ. Open Med. J.* **2022**, *11*, e0202. [CrossRef]
38. Giannessi, D.; Colotti, C.; Maltinti, M.; Del Ry, S.; Prontera, C.; Turchi, S.; L'Abbate, A.; Neglia, D. Circulating heat shock proteins and inflammatory markers in patients with idiopathic left ventricular dysfunction: Their relationships with myocardial and microvascular impairment. *Cell Stress Chaperones* **2007**, *12*, 265–274. [CrossRef] [PubMed]
39. Duan, Y.; Tang, H.; Mitchell-Silbaugh, K.; Fang, X.; Han, Z.; Ouyang, K. Heat Shock Protein 60 in Cardiovascular Physiology and Diseases. *Front. Mol. Biosci.* **2020**, *7*, 73. [CrossRef] [PubMed]
40. Takashio, S.; Yamamuro, M.; Izumiya, Y.; Sugiyama, S.; Kojima, S.; Yamamoto, E.; Tsujita, K.; Tanaka, T.; Tayama, S.; Kaikita, K.; et al. Coronary Microvascular Dysfunction and Diastolic Load Correlate with Cardiac Troponin T Release Measured by a Highly Sensitive Assay in Patients with Nonischemic Heart Failure. *J. Am. Coll. Cardiol.* **2013**, *62*, 632–640. [CrossRef] [PubMed]
41. Fujii, K.; Kawasaki, D.; Oka, K.; Akahori, H.; Iwasaku, T.; Fukunaga, M.; Eguchi, A.; Sawada, H.; Masutani, M.; Lee-Kawabata, M.; et al. The Impact of Pravastatin Pre-Treatment on Periprocedural Microcirculatory Damage in Patients Undergoing Percutaneous Coronary Intervention. *JACC Cardiovasc. Interv.* **2011**, *4*, 513–520. [CrossRef] [PubMed]
42. Kitabata, H.; Kubo, T.; Ishibashi, K.; Komukai, K.; Tanimoto, T.; Ino, Y.; Kashiwagi, M.; Ozaki, Y.; Shiono, Y.; Shimamura, K.; et al. Prognostic Value of Microvascular Resistance Index Immediately After Primary Percutaneous Coronary Intervention on Left Ventricular Remodeling in Patients with Reperfused Anterior Acute ST-Segment Elevation Myocardial Infarction. *JACC Cardiovasc. Interv.* **2013**, *6*, 1046–1054. [CrossRef] [PubMed]
43. Herrmann, J.; Haude, M.; Lerman, A.; Schulz, R.; Volbracht, L.; Ge, J.; Schmermund, A.; Wieneke, H.; von Birgelen, C.; Eggebrecht, H.; et al. Abnormal Coronary Flow Velocity Reserve After Coronary Intervention Is Associated with Cardiac Marker Elevation. *Circulation* **2001**, *103*, 2339–2345. [CrossRef]
44. Park, K.; Kim, M.; Cho, Y.-R.; Park, J.-S.; Park, T.-H.; Kim, M.H.; Kim, Y.-D. Association Between Cardiac Troponin Level and Coronary Flow Reserve in Patients Without Coronary Artery Disease: Insight from a Thermodilution Technique Using an Intracoronary Pressure Wire. *Korean Circ. J.* **2014**, *44*, 141–147. [CrossRef]
45. Ridker, P.M.; Cook, N. Clinical Usefulness of Very High and Very Low Levels of C-Reactive Protein Across the Full Range of Framingham Risk Scores. *Circulation* **2004**, *109*, 1955–1959. [CrossRef]
46. Li, J.J.; Zhu, C.G.; Nan, J.L.; Li, J.; Li, Z.C.; Zeng, H.S.; Gao, Z.; Qin, X.W.; Zhang, C.Y. Elevated circulating inflammatory markers in female patients with cardiac syndrome X. *Cytokine* **2007**, *40*, 172–176. [CrossRef]
47. Recio-Mayoral, A.; Rimoldi, O.E.; Camici, P.G.; Kaski, J.C. Inflammation and microvascular dysfunction in cardiac syndrome X patients without conventional risk factors for coronary artery disease. *J. Am. Coll. Cardiol. Imaging* **2013**, *6*, 660–667. [CrossRef] [PubMed]
48. Vaccarino, V.; Khan, D.; Votaw, J.; Faber, T.; Veledar, E.; Jones, D.P.; Goldberg, J.; Raggi, P.; Quyyumi, A.A.; Bremner, J.D. Inflammation Is Related to Coronary Flow Reserve Detected by Positron Emission Tomography in Asymptomatic Male Twins. *J. Am. Coll. Cardiol.* **2011**, *57*, 1271–1279. [CrossRef] [PubMed]
49. Ong, P.; Carro, A.; Athanasiadis, A.; Borgulya, G.; Schäufele, T.; Ratge, D.; Gaze, D.; Sechtem, U.; Kaski, J.C. Acetylcholine-induced coronary spasm in patients with unobstructed coronary arteries is associated with elevated concentrations of soluble CD40 ligand and high-sensitivity C-reactive protein. *Coron. Artery Dis.* **2015**, *26*, 126–132. [CrossRef] [PubMed]
50. Teragawa, H.; Fukuda, Y.; Matsuda, K.; Ueda, K.; Higashi, Y.; Oshima, T.; Yoshizumi, M.; Chayama, K. Relation between C-reactive protein concentrations and coronary microvascular endothelial function. *Heart* **2004**, *90*, 750–754. [CrossRef]
51. Hung, M.-J.; Cherng, W.-J.; Yang, N.-I.; Cheng, C.-W.; Li, L.-F. Relation of high-sensitivity C-reactive protein level with coronary vasospastic angina pectoris in patients without hemodynamically significant coronary artery disease. *Am. J. Cardiol.* **2005**, *96*, 1484–1490. [CrossRef] [PubMed]
52. Long, M.; Huang, Z.; Zhuang, X.; Huang, Z.; Guo, Y.; Liao, X.; Luo, C. Association of inflammation and endothelial dysfunction with coronary microvascular resistance in patients with cardiac syndrome X. *Arq. Bras. Cardiol.* **2017**, *109*, 397–403. [CrossRef] [PubMed]

53. Vichova, T.; Motovska, Z. Oxidative Stress: Predictive Marker for Coronary Artery Disease. *Exp. Clin. Cardiol.* **2013**, *18*, e88–e91.
54. Sahin, D.Y.; Elbasan, Z.; Gür, M.; Türkoglu, C.; Özaltun, B.; Sümbül, Z.; Çaylı, M. Relationship between Oxidative Stress Markers and Cardiac Syndrome X. *J. Clin. Exp. Investig.* **2012**, *3*, 174–180. [CrossRef]
55. Nishihira, K.; Yamashita, A.; Imamura, T.; Hatakeyama, K.; Sato, Y.; Nakamura, H.; Yodoi, J.; Ogawa, H.; Kitamura, K.; Asada, Y. Thioredoxin in Coronary Culprit Lesions: Possible Relationship to Oxidative Stress and Intraplaque Hemorrhage. *Atherosclerosis* **2008**, *201*, 360–367. [CrossRef] [PubMed]
56. Raad, M.; AlBadri, A.; Wei, J.; Mehta, P.K.; Maughan, J.; Gadh, A.; Thomson, L.; Jones, D.P.; Quyyumi, A.A.; Pepine, C.J.; et al. Oxidative Stress Is Associated with Diastolic Dysfunction in Women with Ischemia with No Obstructive Coronary Artery Disease. *J. Am. Heart Assoc.* **2020**, *9*, e015602. [CrossRef]
57. Dhawan, S.S.; Eshtehardi, P.; McDaniel, M.C.; Fike, L.V.; Jones, D.P.; Quyyumi, A.A.; Samady, H. The role of plasma aminothiols in the prediction of coronary microvascular dysfunction and plaque vulnerability. *Atherosclerosis* **2011**, *219*, 266–272. [CrossRef]
58. Theuerle, J.; Farouque, O.; Vasanthakumar, S.; Patel, S.K.; Burrell, L.M.; Clark, D.J.; Al-Fiadh, A.H. Plasma endothelin-1 and adrenomedullin are associated with coronary artery function and cardiovascular outcomes in humans. *Int. J. Cardiol.* **2019**, *291*, 168–172. [CrossRef]
59. Qin, Q.; Delrio, S.; Wan, J.; Widmer, R.J.; Cohen, P.; Lerman, L.O.; Lerman, A. Downregulation of circulating MOTS-c levels in patients with coronary endothelial dysfunction. *Int. J. Cardiol.* **2018**, *254*, 23–27. [CrossRef] [PubMed]
60. Ahmad, A.; Corban, M.T.; Toya, T.; Sara, J.D.; Lerman, B.; Park, J.Y.; Lerman, L.O.; Lerman, A. Coronary microvascular endothelial dysfunction in patients with angina and nonobstructive coronary artery disease is associated with elevated serum homocysteine levels. *J. Am. Heart Assoc.* **2020**, *9*, e017746. [CrossRef]
61. Mangiacapra, F.; De Bruyne, B.; Peace, A.J.; Melikian, N.; Wijns, W.; Barbato, E. High cholesterol levels are associated with coronary microvascular dysfunction. *J. Cardiovasc. Med.* **2012**, *13*, 439–442. [CrossRef]
62. Pacheco, C.; Wei, J.; Shufelt, C.; Hitzeman, T.C.; Cook-Wiens, G.; Pepine, C.J.; Handberg, E.; Anderson, R.D.; Petersen, J.; Hong, T.; et al. Association of coronary microvascular dysfunction and cardiac bridge integrator 1, a cardiomyocyte dysfunction biomarker. *Clin. Cardiol.* **2021**, *44*, 1586–1593. [CrossRef] [PubMed]
63. Mekonnen, G.; Corban, M.T.; Hung, O.Y.; Eshtehardi, P.; Eapen, D.J.; Al-Kassem, H.; Rasoul-Arzrumly, E.; Gogas, B.D.; McDaniel, M.C.; Pielak, T.; et al. Plasma soluble urokinase-type plasminogen activator receptor level is independently associated with coronary microvascular function in patients with non-obstructive coronary artery disease. *Atherosclerosis* **2015**, *239*, 55–60. [CrossRef]
64. Kang, M.G.; Koo, B.-K.; Tantry, U.S.; Kim, K.; Ahn, J.-H.; Park, H.W.; Park, J.R.; Hwang, S.-J.; Hwang, J.-Y.; Gurbel, P.A.; et al. Association between thrombogenicity indices and coronary microvascular dysfunction in patients with acute myocardial infarction. *JACC Basic Transl. Sci.* **2021**, *6*, 749–761. [CrossRef] [PubMed]
65. Mekonnen, G.; Hayek, S.S.; Mehta, P.K.; Li, Q.; Mahar, E.; Mou, L.; Kenkre, T.S.; Petersen, J.W.; Azarbal, B.; Samuels, B.; et al. Circulating progenitor cells and coronary microvascular dysfunction: Results from the NHLBI-sponsored Women's Ischemia Syndrome Evaluation—Coronary Vascular Dysfunction Study (WISE-CVD). *Atherosclerosis* **2016**, *253*, 111–117. [CrossRef]
66. Muroya, T.; Kawano, H.; Koga, S.; Ikeda, S.; Yamamoto, F.; Miwa, T.; Kohno, Y.; Maemura, K. Lower circulating omega-3 polyunsaturated fatty acids are associated with coronary microvascular dysfunction evaluated by hyperemic microvascular resistance in patients with stable coronary artery disease. *Int. Heart J.* **2018**, *59*, 1194–1201. [CrossRef] [PubMed]
67. Akhiyat, N.; Ozcan, I.; Gulati, R.; Prasad, A.; Tchkonia, T.; Kirkland, J.L.; Lewis, B.; Lerman, L.O.; Lerman, A. Patients with Coronary Microvascular Dysfunction Have Less Circulating α-Klotho. *J. Am. Heart Assoc.* **2024**, *13*, e031972. [CrossRef]
68. McFadden, E.P.; Clarke, J.G.; Davies, G.J.; Kaski, J.C.; Haider, A.W.; Maseri, A. Effect of Intracoronary Serotonin on Coronary Vessels in Patients with Stable Angina and Patients with Variant Angina. *N. Engl. J. Med.* **1991**, *324*, 648–654. [CrossRef] [PubMed]
69. Odaka, Y.; Takahashi, J.; Tsuburaya, R.; Nishimiya, K.; Hao, K.; Matsumoto, Y.; Ito, K.; Sakata, Y.; Miyata, S.; Manita, D.; et al. Plasma concentration of serotonin is a novel biomarker for coronary microvascular dysfunction in patients with suspected angina and unobstructive coronary arteries. *Eur. Heart J.* **2017**, *38*, 489–496. [CrossRef] [PubMed]
70. Iaconetti, C.; Gareri, C.; Polimeni, A.; Indolfi, C. Non-Coding RNAs: The "Dark Matter" of Cardiovascular Pathophysiology. *Int. J. Mol. Sci.* **2013**, *14*, 19987–20018. [CrossRef]
71. Matute-Blanco, L.; Fernández-Rodríguez, D.; Casanova-Sandoval, J.; Belmonte, T.; Benítez, I.D.; Rivera, K.; Garcia-Guimaraes, M.; Cortés Villar, C.; Peral Disdier, V.; Millán Segovia, R.; et al. Study protocol for the epigenetic characterization of angor pectoris according to the affected coronary compartment: Global and comprehensive assessment of the relationship between invasive coronary physiology and microRNAs. *PLoS ONE* **2023**, *18*, e0283097. [CrossRef] [PubMed]
72. Ferrone, M.; Ciccarelli, M.; Varzideh, F.; Kansakar, U.; Guerra, G.; Cerasuolo, F.A.; Buonaiuto, A.; Fiordelisi, A.; Venga, E.; Esposito, M.; et al. Non-Coding RNAs: The "Dark Matter" of Cardiovascular Pathophysiology. *Cardiovasc. Diabetol.* **2024**, *23*, 268. [CrossRef] [PubMed]
73. James, K.; Bryl-Gorecka, P.; Olde, B.; Gidlof, O.; Torngren, K.; Erlinge, D. Increased expression of miR-224-5p in circulating extracellular vesicles of patients with reduced coronary flow reserve. *BMC Cardiovasc. Disord.* **2022**, *22*, 321. [CrossRef] [PubMed]

74. Kuwahata, S.; Hamasaki, S.; Ishida, S.; Kataoka, T.; Yoshikawa, A.; Orihara, K.; Ogawa, M.; Oketani, N.; Saihara, K.; Okui, H.; et al. Effect of uric acid on coronary microvascular endothelial function in women: Association with eGFR and ADMA. *J. Atheroscler. Thromb.* **2010**, *17*, 259–269. [CrossRef] [PubMed]
75. Prasad, M.; Matteson, E.L.; Herrmann, J.; Gulati, R.; Rihal, C.S.; Lerman, L.O.; Lerman, A. Uric Acid Is Associated with Inflammation, Coronary Microvascular Dysfunction, and Adverse Outcomes in Postmenopausal Women. *Hypertension* **2017**, *69*, 236–242. [CrossRef] [PubMed]
76. Keeley, E.C.; Li, H.J.; Cogle, C.R.; Handberg, E.M.; Merz, C.N.B.; Pepine, C.J. Specialized Proresolving Mediators in Symptomatic Women with Coronary Microvascular Dysfunction (from the Women's Ischemia Trial to Reduce Events in Nonobstructive CAD [WARRIOR] Trial). *Am. J. Cardiol.* **2022**, *162*, 1–5. [CrossRef]
77. Ito, T.; Yokoi, M.; Kitada, S.; Kawada, Y.; Mizoguchi, T.; Kikuchi, S.; Goto, T.; Seo, Y. Increased circulating levels of malondialdehyde-modified low-density lipoprotein in patients with coronary microvascular dysfunction. *Am. J. Cardiol.* **2021**, *158*, 103–109. [CrossRef] [PubMed]
78. Sorop, O.; van de Wouw, J.; Chandler, S.; Ohanyan, V.; Tune, J.D.; Chilian, W.M.; Merkus, D.; Bender, S.B.; Duncker, D.J. Experimental animal models of coronary microvascular dysfunction. *Cardiovasc. Res.* **2020**, *116*, 756–770. [CrossRef] [PubMed]
79. Kayikcioglu, M. Benefits of statin treatment in cardiac syndrome-X. *Eur. Heart J.* **2003**, *24*, 1999–2005. [CrossRef] [PubMed]
80. Ridker, P.M.; MacFadyen, J.; Cressman, M.; Glynn, R.J. Efficacy of rosuvastatin among men and women with moderate chronic kidney disease and elevated high-sensitivity C-reactive protein: A secondary analysis from the JUPITER trial. *J. Am. Coll. Cardiol.* **2010**, *55*, 1266–1273. [CrossRef]
81. Pauly, D.F.; Johnson, B.D.; Anderson, R.D.; Handberg, E.M.; Smith, K.M.; Cooper-DeHoff, R.M.; Sopko, G.; Sharaf, B.M.; Kelsey, S.F.; Bairey Merz, C.N.; et al. In women with symptoms of cardiac ischemia, nonobstructive coronary arteries, and microvascular dysfunction, angiotensin-converting enzyme inhibition is associated with improved microvascular function: A double-blind randomized study from the National Heart, Lung and Blood Institute Women's Ischemia Syndrome Evaluation (WISE). *Am. Heart J.* **2011**, *162*, 678–684.
82. Pizzi, C.; Manfrini, O.; Fontana, F.; Bugiardini, R. Angiotensin-converting enzyme inhibitors and 3-hydroxy-3-methylglutaryl coenzyme A reductase in cardiac Syndrome X: Role of superoxide dismutase activity. *Circulation* **2004**, *109*, 53–58. [CrossRef]
83. Kalinowski, L.; Dobrucki, L.W.; Szczepanska-Konkel, M.; Jankowski, M.; Martyniec, L.; Angielski, S.; Malinski, T. Third-generation beta-blockers stimulate nitric oxide release from endothelial cells through ATP efflux: A novel mechanism for antihypertensive action. *Circulation* **2003**, *107*, 2747–2752. [CrossRef] [PubMed]
84. Bugiardini, R.; Borghi, A.; Pozzati, A.; Ottani, F.; Morgagni, G.L.; Puddu, P. The paradox of nitrates in patients with angina pectoris and angiographically normal coronary arteries. *Am. J. Cardiol.* **1993**, *72*, 343–347. [CrossRef] [PubMed]
85. Denardo, S.J.; Wen, X.; Handberg, E.M.; Bairey Merz, C.N.; Sopko, G.S.; Cooper-Dehoff, R.M.; Pepine, C.J. Effect of phosphodiesterase type 5 inhibition on microvascular coronary dysfunction in women: A Women's Ischemia Syndrome Evaluation (WISE) ancillary study. *Clin. Cardiol.* **2011**, *34*, 483–487. [CrossRef] [PubMed]
86. Mohri, M.; Shimokawa, H. Rho-kinase inhibitors for ischemic heart disease. *Nihon Rinsho* **2003**, *61* (Suppl. 5), 874–879.
87. Shimokawa, H.; Sunamura, S.; Satoh, K. RhoA/Rho-Kinase in the Cardiovascular System. *Circ. Res.* **2016**, *118*, 352–366. [CrossRef] [PubMed]
88. Bairey Merz, C.N.; Handberg, E.M.; Shufelt, C.L.; Mehta, P.; Minissian, M.B.; Wei, J.; Thomson, L.E.J.; Berman, D.S.; Shaw, L.J.; Petersen, J.W.; et al. A randomized, placebo-controlled trial of late Na current inhibition (ranolazine) in coronary microvascular dysfunction (CMD): Impact on angina and myocardial perfusion reserve. *Eur. Heart J.* **2016**, *37*, 1504–1513. [CrossRef]
89. Villano, A.; Di Franco, A.; Nerla, R.; Sestito, A.; Tarzia, P.; Lamendola, P.; Di Monaco, A.; Sarullo, F.M.; Lanza, G.A.; Crea, F. Effects of ivabradine and ranolazine in patients with microvascular angina pectoris. *Am. J. Cardiol.* **2013**, *112*, 8–13. [CrossRef]
90. Heerspink, H.J.L.; Greasley, P.J.; Ahlström, C.; Althage, M.; Dwyer, J.P.; Law, G.; Wijkmark, E.; Lin, M.; Mercier, A.K.; Sunnåker, M.; et al. Efficacy and safety of zibotentan and dapagliflozin in patients with chronic kidney disease: Study design and baseline characteristics of the ZENITH-CKD trial. *Nephrol. Dial. Transplant.* **2024**, *39*, 414–425. [CrossRef]

biomolecules

MDPI

Review

The Role of IL-6 in Ischemic Stroke

Hanna Pawluk [1,*], Alina Woźniak [1,*], Agnieszka Tafelska-Kaczmarek [2], Agnieszka Kosinska [3], Mateusz Pawluk [1], Krzysztof Sergot [4], Renata Grochowalska [5] and Renata Kołodziejska [1]

1 Department of Medical Biology and Biochemistry, Faculty of Medicine, Collegium Medicum in Bydgoszcz, Nicolaus Copernicus University in Toruń, Karłowicza 24, 85-092 Bydgoszcz, Poland; pawluk.mateusz23@gmail.com (M.P.); renatakol@poczta.fm (R.K.)
2 Department of Organic Chemistry, Faculty of Chemistry, Nicolaus Copernicus University, Gagarina 7, 87-100 Toruń, Poland; tafel@umk.pl
3 Centre for Languages & International Education, University College London, 26 Bedford Way, London WC1H 0AP, UK; a.kosinska@ucl.ac.uk
4 Laboratory of Laser Molecular Spectroscopy, Institute of Applied Radiation Chemistry, Faculty of Chemistry, Lodz University of Technology, Wroblewskiego 15, 93-590 Lodz, Poland; krzysztofsergot@gmail.com
5 Laboratory of Cell Biochemistry and Biology, Department of Biotechnology, Institute of Biological Sciences, Faculty of Biological Sciences, University of Zielona Góra, Prof. Szafran 1, 65-516 Zielona Góra, Poland; r.grochowalska@wnb.uz.zgora.pl
* Correspondence: hannapawluk1@wp.pl (H.P.); alina-wozniak@wp.pl (A.W.)

Abstract: The pathophysiology of a stroke is a complex process involving oxidative stress and inflammation. As a result of the actions of reactive oxygen species (ROS), not only does vascular damage occur, but the brain tissue is also damaged. It is a dynamic process, induced by a cellular–molecular immune response, focused on the development of an immediate reaction. During ischemia, inflammatory mediators are released, among which IL-6 plays a particularly important role in the acute phase of a stroke. Recently, a lot of attention has been devoted to this pleiotropic pro-inflammatory cytokine, which enhances the migration of leukocytes and is controlled by chemokines and the expression of adhesion handlers. The impact of IL-6 on the severity of neurological treatment and on patient prognosis in AIS is of interest to many researchers. More and more data indicate that it may be a reliable prognostic factor in strokes.

Keywords: ischemic stroke; interleukin 6; biomarker

1. Introduction

A stroke is one of the most common diseases in the world, affecting people over the age of 40 and being a leading cause of disability and even death. According to data from the World Health Organization, one in six people worldwide will experience a stroke during their lifetime. The ischemic stroke is a type of stroke that affects about 85–90% of patients, and it can be caused by cerebral microangiopathy, arteriosclerosis, embolism, or blood clotting disorders [1,2].

So far, it has been proven that ongoing inflammation and oxidative stress play a significant role in the etiology of a stroke [3]. As a result of the obstruction of the cerebral artery, there is a lack of oxygen, glucose, and lipids, ultimately leading to energy failure. ATP depletion disrupts ion gradients, causing cellular depolarization, calcium influx, and excessive glutamate release. The activation of the *N*-methyl-D-aspartate (NMDA) receptors leads to further calcium influx and the expression of destructive enzymes [3].

Oxidative stress is one of the causes of the pathogenesis of ischemic neuronal damage, in which calcium levels mediate the process. Disruptions in calcium homeostasis and its increase in cells enhance the production of reactive oxygen species (ROS). These can also be generated by damaged mitochondria or through the activation of cellular receptors that transmit signals from inflammatory mediators [3]. In patients with acute ischemic stroke (AIS), ROS generated during the ischemia and reperfusion phases lead to brain damage. When the redox potential balance in the cell is disturbed, it causes the peroxidation of polyunsaturated fatty acids and leads to other intracellular damages, ultimately resulting in apoptosis [4–6].

Other mechanisms also contribute to brain damage following a stroke, such as neuroinflammation and disturbances in hormonal balance [7,8]. The inflammatory response, both at the periphery and within the central nervous system (CNS), involves various types of cells and molecular pathways, in which cytokines play key roles [8].

Although cytokines play an important role in the pathophysiology of stroke, the loss of balance between pro-inflammatory and anti-inflammatory cytokines after a stroke affects not only the extent of brain damage but also the recovery process. Importantly, cytokines, as pleiotropic biomolecules, participate in both innate and adaptive immune responses. As the stroke injury progresses, various signaling pathways are activated, modified, or interlinked. Pathways associated with the S1P receptor, particularly S1P1 and S1P3, are involved in modulating various aspects of neuroinflammatory responses. They participate in immune cell trafficking, blood–brain barrier (BBB) integrity, and cytokine production [8].

An important role in the innate immune response is played by the intracellular multiprotein complex, the NLRP3 inflammasome, which detects cellular damage and mediates the production of pro-inflammatory factors. The activation of NLRP3 occurs in stages, with the second stage involving the active inflammasome converting pro-caspase-1 into its active form, which triggers the pro-inflammatory cellular response by cleaving the precursors of interleukin-1 beta (IL-1β) and interleukin-18 (IL-18) into their mature forms [9–11].

Other signaling cascades include the mitogen-activated protein kinase (MAPK), Janus kinase (JAK)/signal transducer and activator of transcription proteins (STAT3), transforming growth factor β (TGF-β)/Smad protein, high mobility group box 1 protein (HMGB1)/toll-like receptors (TLR), and nuclear factor kappa-light-chain-enhancer of activated B cells (NF-κB). All changes in the expression, organization, activation, or stimulation of these signaling pathways can occur at any stage of the developing inflammatory condition. There may be an increased secretion of cytokines that promote the inflammatory response and influence the course and progression of a stroke [8].

The released cytokines are recognized by specific membrane receptors, which transmit signals to the cell nucleus, where there is an upregulation of genes encoding acute phase proteins (APP). This particularly applies to pro-inflammatory cytokines responsible for inducing, mainly in the liver, the fibrinogen and c-reactive protein (CRP), a standard inflammation marker associated with stroke [1]. It has been shown that the increase in the acute phase immune response factor, stimulated by interleukin 6 (IL-6), can rise not only during the acute phase of stroke but also in response to a broad spectrum of systemic inflammatory conditions [12].

Many studies suggest that anti-inflammatory action can significantly reduce cardiovascular risk. However, there is limited information, and when available, the findings are inconclusive regarding the ongoing inflammatory process in stroke involving this cytokine. Data from several human studies suggest a potentially causal role of IL-6, a pro-inflammatory cytokine, in vascular disease [13,14], which may make it a promising therapeutic target.

In addition, the use of an anti-inflammatory treatment targeting IL-6 may seem like an important tool in the treatment of strokes. Cohort studies have managed to show strong associations between circulating IL-6 levels and the risk of coronary artery disease [15], but there is limited evidence regarding its associations with ischemic stroke [16,17].

There are also reports suggesting that it could be a reliable prognostic factor in the acute phase of AIS and a biomarker in the development of clinical complications resulting from acute ischemia [2]. We decided to use pooled data from the published literature through a systematic review and meta-analysis to investigate the relationship between circulating IL-6 levels and the risk of ischemic stroke in humans. Additionally, we aimed to examine the correlation between serum IL-6 levels, the severity of neurological deficits, and patient prognosis after AIS, depending on stroke volume.

2. The Role and Properties of Cytokine IL-6

IL-6 is a cytokine involved not only in inflammatory and infectious responses but also in the regulation of metabolic, regenerative, and neuronal processes. The regenerative or anti-inflammatory effects of this cytokine are associated with classical signaling, while trans signaling is more likely involved in pro-inflammatory processes of IL-6 [18].

IL-6 can be secreted by both immune cells (T lymphocytes, B lymphocytes, macrophages, and microglia) and non-immune cells (muscle cells, adipocytes, fibroblasts, endothelial cells, and neurons) [19–21].

Microglia and astrocytes produce IL-6 in response to brain injury, which may contribute to both chronic inflammation and regeneration, clearance of debris, or tissue repair [22,23]. Astrocytes secreting IL-6 promote the polarization of Th1 to Th2, mediate an immunosuppressive microenvironment, and contribute to neurogenesis, angiogenesis, and neuronal differentiation [24].

Endothelial cells secreting IL-6 increase vascular permeability and promote the early transcriptional activation of genes related to angiogenesis through stimulation of STAT3 phosphorylation [25]. At the same time, the IL-6 receptor (IL-6R) activates angiogenesis-related pathways such as phosphoinositide 3-kinase (PI3K)/serine/threonine kinase (AKT) and JAK/STAT [26]. By increasing angiogenesis, IL-6 enables neuronal regeneration and recovery of function during a stroke. It also facilitates post-traumatic healing in the CNS by repairing endothelial cells and promoting revascularization of brain blood flow in the period after ischemic stroke. However, excessive activation of angiogenesis can lead to brain edema [27].

IL-6 may support neuronal survival after ischemic injury [28]. Secreted by immune system cells (macrophages, T lymphocytes), it activates the inflammatory response and intensifies stroke-related changes, which may exacerbate brain damage [29].

The IL-6 molecular weight ranges from 20 to 30 kDa, depending on the cellular source. It is a type of multi-effector cytokine and, as a pleiotropic cytokine, plays an important role in host defenses [30].

IL-6 affects immune responses and regulates hematopoiesis in an autocrine or paracrine manner. It plays an important role in the growth, differentiation, regeneration, and degradation of nerve cells in both the peripheral and CNS.

IL-6 is not only responsible for the activation and stimulation of neutrophils and monocytes but also enhances endothelial cells to secrete adhesive molecules and inflammatory mediators and strengthens the local inflammatory response [31].

It has also been found that IL-6 is an important factor in the differentiation and growth of hematopoietic cells as well as B and T cells, endothelial cells, and osteoclasts. It can induce the production of the fibroblast growth factor, stimulate macrophage colony formation, promote tumor necrosis factor (TNF) production, and participate in the proliferation of smooth muscle cells [32]. On the other hand, TNF-α can stimulate brain pericytes to synthesize IL-6 through the activation of NF-κB. The soluble form (sTNF-α) acts at the systemic level, enhancing the action of macrophages and the expression of cytokines IL-6 and IL-1 [33]. Additionally, IL-6 facilitates the post-traumatic healing of CNS and endothelial cell repair [27,34]. This cytokine increases the survival of neurons and reduces their excitotoxic damage caused by injury or protects neurons from apoptosis [35].

It has been shown that the high-sensitivity C-reactive protein, IL-6, IL-1 receptor antagonist (IL-1Ra) [36,37], lipoprotein-associated phospholipase A2 (Lp-PLA2-A) [15], and chitinase-3-like protein 1 (YKL-40) [38,39] are associated not only with the initial but also with subsequent strokes. IL-6 is involved in the etiology of atherosclerosis and the stability of atherosclerotic plaques, which may influence the development of a subsequent stroke [40,41].

Furthermore, atherosclerosis of large arteries and the obstruction of small vessels share several common factors that may contribute to the development of atherosclerosis, which in turn is a cause of certain subtypes of strokes [42,43].

It has been demonstrated that the correlations between IL-6 and YKL-40 are more pronounced in patients with large artery atherosclerosis and small vessel occlusion. These markers are associated with recurrent stroke, complex vascular events, and poor functional outcomes [44,45].

3. IL-6 as a Mediator of Inflammatory Response in Stroke

In ischemic stroke, inflammation and oxidative stress play a key role in neuronal cell apoptosis. IL-6 is crucial in this complex, multi-stage pathophysiological process due to its synthesis in CNS and its ability to modulate inflammation through dual signaling pathways [46–48]. Depending on the presence of IL-6R or the membrane-associated signal transducer gp130, IL-6 signaling can lead to pro-inflammatory and anti-inflammatory responses [48].

IL-6 utilizes two unique signaling pathways: classical and trans signaling. In the classical pathway, IL-6 binds to IL-6R on the cell surface, forming a hexameric complex with two IL-6 molecules, two IL-6Rα subunits, and two gp130 subunits. This pathway primarily occurs in specific immune and non-immune cells such as T lymphocytes, hepatocytes, and neutrophils, and has an anti-inflammatory effect [49,50].

In contrast, IL-6 also activates pro-inflammatory trans signaling where it binds to the soluble isoform of IL-6R (sIL-6R), which lacks the transmembrane domain. This complex, IL-6: sIL-6R, can bind and activate gp130, triggering a broader cellular response. Natural inhibitors like soluble gp130 (sgp130) limit this trans signaling [49,51].

Unlike classical signaling, trans signaling does not require membrane-bound IL-6Rα for the indicator initiation, expanding the range of target cells to almost all cells in the body [52]. The soluble IL-6:IL-6Rα complex can initiate IL-6 signaling in cells containing the signal transducer gp130, which is present in neurons, neuronal cells, endothelial cells, and oligodendrocytes [52]. Both mechanisms contribute to neuroinflammation after stroke, but trans signaling is considered the dominant pathway in the brain [49,53].

While trans signaling aids in damage repair after stroke, it can also trigger harmful recurring inflammatory responses. It is linked to ischemic stroke risk due to its strong pro-inflammatory and pro-atherogenic effects [54].

Both signaling pathways activate intracellular kinases, including JAK (and to a lesser degree, TYK), which further activate proteins like the transcription factors STAT, the RAS-RAF-MAPK pathway, and PI3K (Figure 1) [30,32,55].

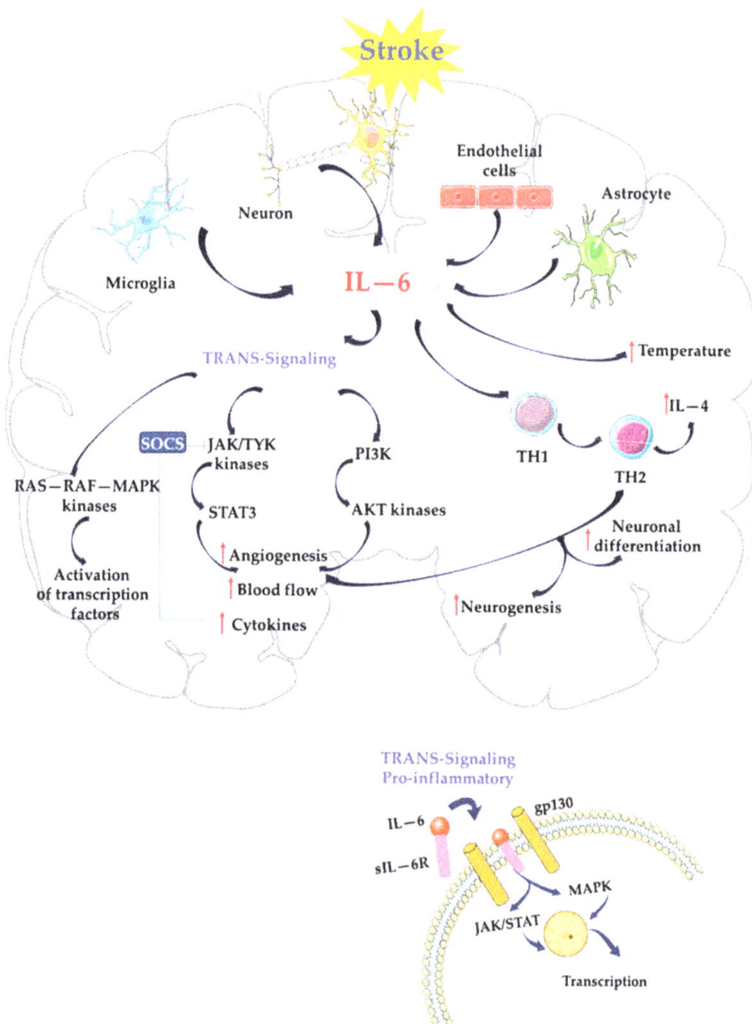

Figure 1. IL-6 trans signaling. The IL-6/sIL-6R/gp130 complex can activate intracellular tyrosine kinases that activate STAT transcription factors or the RAS-RAF-MAPK or PI3K pathway. Trans signaling can lead to the expression of a pro-inflammatory state via increased JAK/STAT signaling due to SOCS inhibition. STAT3 phosphorylation can enhance the transcription of genes related to angiogenesis and intense cerebral blood flow. AKT—protein kinase B; IL-4—interleukin 4; IL-6—interleukin 6; JAK—Janus kinase; pg130—signal transducing receptor; PI3K—phosphatidylinositol 3-kinase; RAS—RAF—MAPK—small guanosine triphosphatases—serine/threonine-specific protein—mitogen-activated protein kinases; sIL-6R—soluble isoform of IL-6 receptor; SOCS—proteins called suppressors of cytokine signaling; STAT3—signal transducer and activator of transcription 3; TH1, TH2—lymphocytes; TYK—tyrosine kinase. This figure was created using Servier Medical Art (available at https://smart.servier.com/) (accessed on 12 September 2024).

Trans signaling notably boosts JAK/STAT signaling, surpassing MAPK, by inhibiting suppressors of cytokine signaling (SOCS) and affecting NFκB. SOCS3 is a negative regulator of STAT3, which affects the reduction in IL-6 signal transduction and interferes with immune cell migration [49,56].

The increased activity of JAK/STAT caused by the lack of feedback regulation promotes cytokine expression through chronic activation of microglia and astrocytes. Excessive microglial activation leads to a phenotypic shift to pro-inflammatory M1, which is manifested by the increased expression of inflammatory markers and pro-inflammatory cytokines, triggering nerve inflammation [56].

Astrocytes can release pro-inflammatory cytokines and chemokines that recruit microglia, collectively contributing to the damage of BBB [56,57].

Disruption of BBB integrity allows the infiltration of peripheral immune cells, which release pro-inflammatory mediators. Persistent nerve inflammation causes oxidative stress, synaptic dysfunction, and neuronal death [56,58].

IL-6R also plays a role in angiogenesis after stroke by activating the JAK/STAT and PI3K/AKT pathways. IL-6 promotes STAT3 phosphorylation and the transcription of angiogenesis-related genes, enhancing neovascularization and cerebral blood flow in the post-ischemic period [26,30,59].

Astrocyte-derived IL-6 further contributes to angiogenesis, neurogenesis, and neuronal differentiation by promoting the polarization of TH1 lymphocytes into TH2, facilitating recovery after stroke [24,60–62]. Astrocytes, through IL-6 regulation, support neurovascular regeneration and functional recovery post-stroke [24].

4. IL-6 and Cognitive Functions in Stroke

Almost 1/3 of stroke patients develop cognitive impairments. Post-stroke cognitive impairment (PSCI) is a common complication after a stroke and directly affects functioning and quality of life, including the ability to work and maintain interpersonal relationships. In patients with PSCI, there is a decline in short- and long-term memory, reasoning, coordination of movements, and task planning. Around 20–30% of these patients develop dementia [63,64]. Post-stroke cognitive impairments contribute to the growing health, social, and economic burden [65].

It is not always easy to diagnose and assess the degree of PSCI based on clinical assessment, imaging, and neuropsychological tests, which is why inflammatory mediators and brain damage markers can help diagnose and predict cognitive function impairment [63,64]. It has been shown that memory deficits after acute brain ischemia are associated with increased levels of pro-inflammatory cytokines (IL-1, IL-6 and TNF-α) in the brain and neurodegeneration [65,66].

It is known that IL-1, IL-8, TNF-α, and IL-6 can be predictors of cognitive decline in late middle age [67–69].

Higher levels of IL-6 in serum are associated with a greater decrease in psychomotor speed [70]. It seems that elevated IL-6 levels may cause microvascular changes, leading to reduced neuronal propagation and impaired processing speed [67]. IL-6 can directly activate endothelial cells, which secrete several types of cytokines and chemokines, initiating a cytokine storm. A complication of the cytokine storm is the dysregulation of endothelial cells, characterized by abnormal coagulation and vascular leakage [71].

Furthermore, the dysregulation of IL-6 along with other pro-inflammatory cytokines, such as IL-1 and TNF-α, can lead to nerve inflammation, which is associated with neurodegenerative diseases such as Alzheimer's disease (AD), Parkinson's disease (PD), depression, and multiple sclerosis (MS) [72].

IL-6 stimulates the synthesis of the beta-amyloid precursor protein (Aβ) and exacerbates tau pathology in AD, enhancing neuronal damage. It contributes to cognitive decline by disrupting the integrity of BBB, impairing neuronal survival, and synaptic plasticity [28,49,70].

However, not all studies confirm the link between IL-6 levels and cognitive decline [73].

The risk of cognitive impairment with a stroke increases 5 to 8 times [65,66]. The cause of this is inflammation, which plays a role in the pathogenesis of post-stroke cognitive impairment. IL-6, along with other inflammatory cytokines such as TNF-α and IL-1β, may play an important role in cognitive impairments following a stroke, as they are key regulators of the acute phase response to inflammation and tissue damage. The levels of pro-inflammatory cytokines, including IL-6, are elevated in patients with PSCI [65,66,74–80].

Therefore, IL-6 may be responsible for cognitive disturbances after a stroke, as it is one of the main regulators of the acute-phase response to inflammation and tissue damage, and its concentration is elevated in patients with PSCI [44,45,49–55].

During ischemia, IL-6 activates microglia, autophagy, and endoplasmic reticulum (ER) stress, which leads to hippocampal neuron death and cognitive decline [81].

IL-6 is also one of the factors influencing demyelination and myelin regeneration following damage to the white matter integrity after a stroke. An increase in IL-6 levels in neurons at risk of ischemia may be associated with neuroprotection; however, excessive production of IL-6 in the brain leads to neurodegenerative changes [82].

It has been found that elevated IL-6 levels released by activated microglia in the central nervous system are associated with early neurological damage [83]. High concentrations of IL-6 may therefore be an important factor promoting the development of cognitive impairments and may be linked to their prevalence [79]. According to the Montreal Cognitive Assessment (MoCA), increased IL-6 levels after 18 months were associated with cognitive dysfunction in a 36-month follow-up of patients hospitalized for acute stroke [75]. Similarly, an increase in interleukin-6 levels was associated with a decline in cognitive functions according to MoCA, assessed between 1 year and 3 months, in patients with AIS or transient ischemic attack (TIA) [76]. The risk of cognitive deterioration after stroke increased by as much as 95% for the highest IL-6 levels compared to the lowest [76].

Elevated IL-6 levels may also be associated with psychological complications, the occurrence of anxiety or depression, which negatively affect rehabilitation outcomes, the effectiveness of neurotrophic agents, and even mortality in patients with AIS [78,79]. IL-6, along with IL-1β, TNF-α, and interleukin-17A (IL-17A), can regulate the enzyme 2,3-indoleamine-2,3-dioxygenase, thereby modulating serotonin, which is associated with the pathogenesis of psychological complications [78]. It has also been observed that with increasing IL-6 levels in plasma, the risk of post-stroke delirium rises, which is characterized by acute, fluctuating changes in attention, awareness, and cognition. Hence, both immunodepression and systemic inflammation, reflected by increased IL-6 levels, are associated with poor cognitive outcomes [77]. Chronic inflammation, in turn, can activate neurotoxic pathways, and damaged neurons may exacerbate neurological disturbances by producing chemokines, activating microglia, and astrocytes [84].

In general, pro-inflammatory cytokines, including IL-6, cross the blood–brain barrier, promoting communication between the central nervous system and the periphery, and causing brain damage. Additionally, hyperperfusion caused by a stroke leads to endothelial damage, which promotes nerve inflammation, disrupts the blood–brain barrier, and causes neurodegeneration [67,85].

However, the relationship between IL-6 and cognitive decline after a stroke is not fully established [86,87]. In a study examining the relationship between inflammatory markers and baseline cognitive status, as well as subsequent cognitive decline, it was found that interleukins IL-8 and IL-12, rather than IL-6, are independent predictors of cognitive deterioration [83]. Other studies have found that the levels of Th17-related cytokines, including IL-6, in serum, had minimal predictive value for the recovery of cognitive functions during subacute hospital rehabilitation after a stroke [88]. Central nervous system inflammation may likely mask the relationship between systemic inflammation and cognitive functions [83].

Despite the ambiguous clinical studies, it seems reasonable to assess the levels of inflammatory biomarkers, including IL-6, in patients with PSCI. These studies, combined with neuroimaging and neuropsychological evaluation, may help in the early detection of PSCI and in preventive actions against post-stroke cognitive impairments [65]. The measurement of circulating IL-6 could be used as a screening test to identify AIS patients with cognitive impairment [89]. This test can be extended to assess other pro-inflammatory cytokines, such as TNF-α, IL-8, and IL-17A, which positively correlate with anxiety, depression, or cognitive impairment at various time points after the onset of AIS [71,75].

5. Circulating IL-6 and the Risk in AIS

Interleukin-6 plays an important role in modulating the inflammatory response of the nervous system in stroke. It influences the promotion of homeostasis in response to immune system stimulation through the classical path. The homeostatic, anti-inflammatory route affects nerve regeneration and tissue repair [46,53]. On the other hand, the trans pathway, due to the activation of pg130 on neuroinflammatory cells, can intensify inflammation and worsen stroke outcomes [46].

IL-6 plays a role in the acute phase of stroke as a mediator of the inflammatory process, a reliable prognostic factor, and a neurotrophic factor in the later stages of brain ischemia development [35,90].

IL-6 is a key marker of inflammation, and its concentration rises dynamically in serum within a few hours of AIS and can remain elevated for up to 3 months after the stroke [2,16,91–107].

In population studies in humans, it has been shown that higher levels of IL-6 are associated with stroke risk factors and with a higher risk of AIS occurrence [12,16,30,45,108–118]. IL-6 levels can be useful as an analytical biomarker for stroke recurrence after AIS [108], and its concentration reflects ethnic differences [16,119]. Jiang et al. found that the risk of stroke recurrence rises by 8% for every 1 pg/mL increase in serum IL-6 levels, both in unadjusted and adjusted analyses [108]. A meta-analysis conducted by McCabe et al. demonstrated that inflammatory biomarkers' levels, including IL-6, were independently associated with vascular recurrence after stroke. For every threefold increase in IL-6 levels, there was a 20% to 25% increase in the risk of major cardiovascular events [110]. Multivessel disease (PolyVD) combined with elevated IL-6 levels ($\geq$2.64 pg/mL) facilitates the risk of stroke recurrence in patients with AIS or TIA after one year of observation [120]. Additionally, IL-6 promoter polymorphisms predict recurrent young strokes. During one year of observation of patients with different IL-6 promoter genotypes, it was found that the frequency of recurrent strokes in young individuals with moderate internal carotid artery stenosis increased for the IL-6 G allele promoter [109].

Elevated levels of IL-6 are associated with unsuccessful reperfusion in patients with AIS and large vessel occlusion (LVO) treated with mechanical thrombectomy (MT) [121]. On the other hand, decreased IL-6 levels at admission were associated with the first-pass effect (FPE), meaning complete or nearly complete reperfusion achieved after a single thrombectomy. IL-6 thus enables the differentiation between FPE and non-FPE, with the optimal threshold for its level being 3.0 pg/mL [122].

Interleukin-6 may have a major impact on the prognosis in AIS [2,91–93,95,96,99,103,115,120,123–129]. Elevated levels of this marker are correlated with the severity of neurological damage both clinically (National Institute of Health Stroke Scale (NIHSS); Scandinavian

Stroke Scale (SSS); European Stroke Scale (ESS); Canadian Neurological Scale (CNS)) and radiologically, with diagnostic subtypes (according to the Trial of Org 10172 in Acute Stroke Treatment (TOAST) classification or Oxfordshire Community Stroke Project Classification (OCSP)), activation of the acute phase immune response, and poor prognosis [2,91–93,95,96, 99,103–105,115,116,120,123–129].

It has been shown that an increase in IL-6 levels in the serum within the first 24 h after a stroke is associated with worsened functional status and poorer neurological outcomes in patients [2,91,95,130], as well as with the volume of damage [131–133].

A similar correlation has been established for IL-6 in the cerebrospinal fluid of stroke patients. Increased IL-6 levels in the cerebrospinal fluid at 6 h post-stroke have been shown to correlate with the degree of neurological deficit, assessed on days 1 and 7 post-stroke using the NIHSS, as well as with functional deficits measured by the Barthel index (BI) [134].

Recent studies also suggest that elevated levels of IL-6 and TNF-α assessed on days 1 and 7 significantly correlate with greater stroke severity and worse functional outcomes, as indicated by higher NIHSS and mRS scores. These findings align with studies indicating strong associations between inflammatory cytokines and post-stroke disability, further emphasizing their importance as prognostic indicators [2,135].

Higher IL-6 levels are associated with in-hospital neurological deterioration (ND) [103,136]. IL-6 may be an independent prognostic factor for ND in patients with AIS with large artery atherosclerosis (LAA) and small vessel occlusion (SVO) [103]. IL-6 levels may be independently associated with infarct volume, indicating brain infarction extension without the need for neuroimaging techniques [124]. It can help diagnose ischemic stroke and predict its severity by increasing the sensitivity/specificity of another non-invasive biomarker, miR-221 [123]. Elevated IL-6 levels after a stroke may be associated with mortality, adverse short- and long-term outcomes, and functional status, assessed using the modified Rankin Scale (mRS) at 3 months and 1 year [2,45,91,96,103,106,107,114,116,123–126,137]. An increase in IL-6 concentration was associated with an increased risk of stroke recurrence and disability within 90 days, with stroke recurrence mediating less than 20% of the association between IL-6 and functional outcome [114]. Therefore, the level of this biomarker can be an independent predictor of poor post-stroke outcomes [107,116,120,125] and is associated with a lower likelihood of a favorable outcome (mRS 0–2) [127]. It may also be a good candidate as a biomarker for individuals at risk of stroke-associated infection (SAI), enabling early identification of those at risk of infection, which consequently reduces the risk of premature death and hospitalization, especially in older adults [138]. A relationship was also shown between selected polymorphisms in IL-6 genes and outcomes after AIS and adverse effects (ADE). Individuals with the CC or GG genotype of the IL-6-174 G/C polymorphism had significantly lower NIHSS scores and poorer prognosis [139].

IL-6 plays a role in the pathogenesis of stroke, especially the atherosclerotic stroke of large arteries, and is associated with disease progression and functional outcomes within 3 months of the stroke [44]. There is also a relationship between the functional genetic variant of the interleukin 6 receptor and the risk and outcome of LAA stroke [140]. A causal relationship has been found between IL-6 levels and the severity of stenosis, susceptibility to atherosclerotic plaques, and long-term progression of the atherosclerotic plaque in the carotid artery [31,141].

Along with the increased IL-6 concentration, there is an increase in other inflammatory biomarkers, including C-reactive protein, fibrinogen, IL-1 receptor antagonist, and TNF-α. IL-6 can induce further phases of inflammatory markers and influence the prognosis of patients after AIS [2,35]. Low production of pro-inflammatory cytokines may be associated with relatively good outcomes if immunodeficiency is not accompanied by systemic inflammation linked to high circulating IL-6 levels. A multi-marker assessment based on cytokines may, therefore, be significant in predicting stroke outcomes by enhancing the prognostic value of the clinical model [142].

However, some reports do not confirm such correlations. According to Bustamante et al., based on a meta-analysis of 24 studies, it was found that IL-6 levels are associated with poor outcomes after stroke, but its predictive value was moderate, suggesting that IL-6′s practical application in clinical practice is unlikely [130]. It was also found that increased IL-6 expression indicates an inverse correlation with stroke severity, infarct size [143], short-term prognosis [144], and post-hospital improvement, assessed between discharge and 3 months [136]. Cytokines can contribute to both neurotoxicity and neuroprotection, locally as well as systemically, which is related to the ischemic damage timeline [46,118,143].

The level of IL-6 is directly associated with both fever and the strength of the acute phase response, so excluding patients with fever or infectious diseases may explain the protective effect of IL-6 on specific ethnic populations [118].

In summary, in Table 1, the concentrations of IL-6 and its impact on the risk of ischemic stroke, the degree of neurological damage, and functional outcomes after AIS are presented.

Table 1. The influence of IL-6 concentration on the risk of ischemic stroke, the degree of neurological damage, and functional outcomes after AIS.

Collection Time	Average Concentration	Control	Results	Conclusions	Reference
from 5 to 7 days	Peak plasma IL-6 30.5 pg/mL (n = 37)	-	Peak plasma IL-6 concentration above the median sample (30.5 pg/mL) at 5 to 7 days correlated significantly with cerebral infarct volume on computed tomography (r = 0.62, p = 0.006), mRS at 3 months (r = 0.62, p = 0.002), mRS at 12 months (r = 0.77, p < 0.001), stroke severity (r = 0.66, p < 0.001), Barthel index (BI) at 3 months (r = −0.74, p < 0.001), BI at 12 months (r = -0.77, p < 0.001) and was associated with increased mortality at 12 months (p = 0.0011).	Peak plasma IL-6 concentration in the first week of ischemic stroke significantly correlated with infarct volume, stroke severity, and clinical outcome at 3 months and mortality at 12 months.	[126]
at admission	6.4 ± 2.6 pg/mL (n = 11)	2.8 ± 1.5 pg/mL (n = 9) p = 0.002	Significant correlation of IL-6 with larger stroke volume (p = 0.012, r = 0.72,) and poorer prognosis at 1 year, as measured by ESS (p = 0.014, r = −0.71) and BI (p = 0.006, r = −0.80).	In acute ischemic stroke, IL-6 can be an early indicator of prognosis.	[92]
24–72 h	10.5 (6–27.25) pg/mL (n = 60)	9 (2.90–18) pg/mL (n = 123) p = 0.002	No differences in IL-6 levels at 24–72 h after stroke onset in non-lacunar versus lacunar strokes.	Higher IL-6 values were demonstrated compared to healthy individuals.	[94]
<20 h	9.7 (7.9–15.7) pg/mL (n = 50)	-	The level IL-6 was significantly higher in patients with better outcome (median 12.6 (7.6–21.4)) when compared with those with worse outcome (median 8.492.4–13.70 p < 0.0001). There was a significant inverse correlation between IL-6 and TNF-α levels in the entire cohort of patients (r = 0.78, p = 0.002)	Serum IL-6 concentration was inversely correlated with final neurological impairment and infarct size (p < 0.001). In patients with ischemic brain injury, IL-6 is associated with neuroprotection but not neurotoxicity.	[143]
50 h	13.78 ± 19.46 pg/mL (n = 60)	4.38 ± 15.88 pg/mL (n = 30) p = 0.002	The levels of serum IL-6 were significantly higher in stroke patients. IL-6 showed negative correlation with protein S (r = –0.504, p = 0.000) and CNS scores (r = –0.451, p = 0.000).	IL-6 was correlated with the worse clinical outcome.	[105]

Table 1. *Cont.*

Collection Time	Average Concentration	Control	Results	Conclusions	Reference
at admission	8 (4–21) pg/mL (n = 68)	2 (2–4) pg/mL (n = 71) p = 0.035	IL-6 was higher in stroke patients than in controls. Higher IL-6 concentration was significantly associated with greater neurological deficit (NIHSS ≤ 16 (n = 30), 7 (4–11) pg/mL, NIHSS 17–22 (n = 24), 7 (5–16) pg/mL; NIHSS ≥ 23 (n = 17), 22 (10–71) pg/mL; p = 0.007) at admission and greater degree of patient disability assessed with BI (BI ≤ 29 (n = 28), 13 (6–28) pg/mL; BI 30–59 (n = 15), 8 (5–20) pg/mL; BI ≥ 60 (n = 28), 6 (4–12) pg/mL; p = 0.030) and mRS (mRS = 0–2 (n = 28), 6 (4–12) pg/mL; mRS = 3–4 (n = 14), 8 (5–11) pg/mL; mRS = 5–6 (n = 29), 13 (6–36) pg/mL; p = 0.019).	IL-6 levels measured 12 h after the onset of cerebral ischemia were higher in stroke patients compared to the control group, and this increase was associated with more severe stroke and worse stroke outcomes.	[104]
within <36 h of symptom onset	5.14 ± 4.98 pg/mL (mean ± SD) [3.18 (1.97–6.22) median (25–75%)] (n = 113)	-	The proportion of patients with severe stroke (NIHSS > 10) at admission rose in a graded fashion from 0% in patients with IL-6 in the bottom tertile (≤2.26 pg/mL) to 46% in those in the top tertile (>4.63 pg/mL). Similar trends were observed at day 1–2, day 4–5, and at discharge (p < 0.01). Disability during follow-up (mRS ≥ 2) rose from 13% for IL-6 in the bottom tertile to 48% in the top tertile (p = 0.01). Association with poor physical function (Stroke Impact Scale (SIS ≤ 85)) revealed a comparable trend rising from 32% at the bottom tertile to 63% at the top tertile (p = 0.04).	High plasma IL-6 levels are associated with stroke severity and poorer functional outcomes.	[129]
at admission	10 (6–28) pg/mL (n = 107)	8 (3.1–12) pg/mL (n = 102) p < 0.001	Significant association between IL-6 demographic variables (age), diagnostic subtype (lacunar (5 (2–8) pg/mL) or cardioembolic (12 (6.5–18) pg/mL; p = 0.003), and SSS score (b1= −0.068; p < 0.0001).	Significant association between IL-6 and severity of neurological deficit at admission, diagnostic subtype.	[93]
24 h	26.5 +/−2.3 pg/mL (range 6.4–161.3) (n = 46)	3.9 +/−1.5 pg/mL (range 2.3–5.9) (n = 98; p < 0.0001	The concentration of IL-6 in the blood of patients with acute ischemic stroke was significantly higher than in healthy controls.	There is an association between inflammation and circulating IL-6 levels in patients with acute hypoxic–ischemic injury, therefore IL-6 might be used as a warning signal, the concentration of which increases at an early inflammatory state.	[98]

Table 1. *Cont.*

Collection Time	Average Concentration	Control	Results	Conclusions	Reference
24 h	<6.47 pg/mL (n = 200) lower NIHSS ≥6.47 pg/mL (n = 50) higher NIHSS	–	Plasma IL-6 concentration correlated with a more severe stroke as demonstrated by NIHSS and mRS (r = 0.318, $p < 0.001$ and r = 0.302, $p < 0.001$, respectively). IL-6 concentration was found to be the optimal predictor, with a cut point of 6.47, χ^2 (I, n= 250) = 20.5, $p < 0{\cdot}001$. The value of IL-6 above 6.47 during the acute phase predicted a subsequent lack of survival (p= 0.006, odds ratio 8.0).	IL-6 shows potential as an early signal of survival after acute ischemic stroke and suggests a clear cut-off point for high-risk patients who might benefit from closer clinical surveillance and/or therapeutic interventions.	[127]
2–6 h, 6 h, 24 h, 3 days, 7 days	Figure 1a in [99] ng/L (n = 69)	–	IL-6 showed different time courses depending on stroke outcome. The IL-6 levels were significantly increased in the patients with severe strokes compared to the mild/moderate stroke group as early as 6 h after stroke onset. The levels stayed elevated only until 24 h after symptom onset (6 h p = 0.022, 12 h p = 0.011, 24 h p = 0.006). IL-6 at 12 h, 24 h, and 3 days was independently related to the NIHSS. In the first days after stroke, IL-6 significantly correlated with CRP and MMP-9 at different time points, while only at 6 h after stroke onset, IL-6 correlated with MCP-1.	The data indicate significant differences in the early time course of several potential markers for the complex network of inflammation and brain damage after ischemic stroke, depending on the outcome.	[99]
72 h 6 months	No infection 1.4 ± 0.3 ng/mL (n = 37) Infection 1.6 ± 0.3 ng/mL (n = 45)	–	Plasma IL-6 at baseline was associated with decreased survival at 2 years (OR = 9.2, [95% CI 1.0, 85.1], p = 0.031) and this association was independent of SAI, severity of stroke, age and sex.	IL-6 is associated with SAI, after adjusting for recognized risk factors, and IL-6 is associated with mortality in the first two years post stroke.	[138]
at admission	4.6 (0.9 to 13.0) pg/mL (n = 175)	–	Strong association between IL-6 and poor stroke outcomes (favorable outcome (n = 111), 2.2 (0.6 to 5.5) *versus* poor outcome (n = 64), 12.1 (6.7 to 48.0); $p < 0.001$). The logistic regression analysis of individual blood markers after adjusting for age and initial NIHSS score revealed that plasma logIL-6 (adjusted OR: 1.75, 95% CI: 1.25 to 2.25, p = 0.001) reached a statistical significance. The statistical significance of logIL-6 remained after further adjustment for 72 h infarct volume and Afib (logIL-6, OR: 1.74, 95% CI: 1.24 to 2.44, p = 0.003), time of stroke onset (logIL-6, OR: 1.74, 95% CI: 1.24 to 2.44, p = 0.001) and previous statin use (logIL-6, OR: 1.81, 95% CI: 1.27 to 2.57, p = 0.001).	Circulating IL-6 levels have clinical value in predicting stroke outcomes.	[96]

Table 1. *Cont.*

Collection Time	Average Concentration	Control	Results	Conclusions	Reference
1 day	42.92 ± 72.2 pg/mL	–	Statistically significant correlation between IL-6 and NIHSS and mRS of patients from the moment of admission to the end of the observation period ($p < 0.001$, r = 0.6). Statistically significant correlation between IL-6 and infarct size in brain magnetic resonance imaging (MRI) ($p < 0.001$, r = 0.7).	IL-6 contributes to determination of severity of ischemic stroke.	[91]
5 days	56.91 ± 82.63 pg/mL				
The samples were drawn on the day of randomization, but before the initiation of antiplatelet therapy	Age: <50 1.8 pg/mL (1.4–3.4) Age: 50–65 2.4 pg/mL (1.6–3.7) Age:>65–75 2.2 pg/mL (1.6–3.6) Age:>75 2.5 pg/mL (1.7–3.5) (n = 1244)	–	In unadjusted analyses, IL-6 was significantly related to the risk of having a recurrent ischemic stroke (adjusted HR per SD, 1.1; 95% confidence interval [CI], 1.0–1.2; $p < 0.01$. IL-6 was also associated with an increase in risk of major vascular events (adjusted HR per SD, 1.1; 95% CI, 1.02–1.19; $p = 0.008$). In categorical analyses, there was evidence that the treatment effect of dual antiplatelet therapy depended on IL-6 levels (p for interaction = 0.04 for major vascular events).	Among recent lacunar stroke patients, IL-6 concentration predict risk of recurrent vascular events, and they are associated with the effect of antiplatelet therapies.	[118]
48 h	15.82 ± 16.64 pg/mL (n = 42)	6.64 ± 2.5 (n = 34) $p = 0.000035$	Patients with early ischemic stroke had significantly higher serum IL-6 levels than controls.	Ischemic stroke is associated with higher serum IL-6 levels.	[102]
at admission	21.91 ±30.75 pg/mL (n = 33)	1.04 ± 1.12 pg/mL (n = 60)	Baseline levels of IL-6 and stroke severity: The higher baseline IL-6 level showed a significant positive correlation with the NIHSS score at day 1($r^2 = 0.18$; $p = 0.01$), on day 7 following admission ($r^2 = 0.20$; $p = 0.01$) and with the infarct volume on DWI ($r^2 = 0.17$; $p = 0.01$). IL-6 was found to correlate with worse long-term outcomes at both 1 month ($r^2 = 0.19$; $p = 0.01$) and 3 months ($r^2 = 0.15$; $p = 0.04$), as measured by MRS. The baseline IL-6 levels was independent predictors of short term outcome as assessed by NIHSS at day 7 ($p = 0.04$); also for long term outcome as measured by MRS at one month ($p = 0.02$) and three months ($p = 0.04$). Baseline levels of IL-6 were higher ($p = 0.04$) in those patients who died than those who survived.	Serum IL-6 concentrations were significantly higher in patients with ischemic stroke than in the healthy control group. A correlation was observed between the clinical and radiological severity of stroke and the level of IL-6, and the poor outcome could be predicted (the higher baseline IL-6 may predict mortality in ischemic stroke). This suggests that the IL-6 serum may be a potentially useful inflammatory biomarker for the prognosis of acute ischemic stroke.	[95]
27 days	12.80 ±21.18 pg/mL (n = 33)				
1 month	6.18 ±14.21 pg/mL (n = 33)				
3 months	2.75 ± 2.54 pg/mL (n = 33)				

Table 1. *Cont.*

Collection Time	Average Concentration	Control	Results	Conclusions	Reference
24 h	2.61 (1.60–4.73) pg/mL ($n = 7053$)	–	Stroke recurrence was observed in 458 (6.5%) patients, and functional disability was seen in 1708 (24.2%) patients at the 90-day follow-up. Per stand deviation (4.26 pg/mL) increase in IL-6 concentration was associated with a significantly 19% increased risk of stroke recurrence (OR, 1.19; 95% CI, 1.09–1.29) and a significantly 22% increased risk of disability (OR, 1.22; 95% CI, 1.15–1.30) at 90 days. The mediated proportion of the association between IL-6 and functional disability by stroke recurrence was 18.72% (95% CI, 9.26–28.18%) in the adjusted model.	Stroke recurrence mediates less than 20% of the association between IL-6 and functional outcome at 90 days among patients with acute ischemic stroke.	[114]
–	Lacunar stroke 6.6 (7.5) pg/mL ($n = 52$)	–	Baseline IL-6 was significantly associated with risk of main outcome in them ultivariable analysis, after adjustment for age, sex, and small vessel disease score (1.4 (1.02–2.2); $p = 0.04$). IL-6 was significantly related to along-term risk of vascular events or death in patients with cerebral small vessel disease (CSVD).	It demonstrated the important prognostic role of Il-6 in persons with different clinical manifestations of CSVD. The strongest association occurred between IL-6 and recurrent stroke, other vascular events and death.	[115]
–	4.5 (3.1) ng/mL ($n = 557$)	cohort random sample 3.7 (2.6) ng/mL ($n = 951$) $p < 0.001$	IL-6 was associated with risk factors for stroke. After adjusting for different factors (age, sex, race) the hazard ratio (HR; 95% CI) for incident stroke for the highest versus lowest quartile of IL-6 was 2.4 (1.6–3.4). IL-6 mediated the black-white disparity in stroke risk, but mediation was mediated by associations of IL-6 with stroke risk factors.	IL-6 was strongly associated with stroke risk factors and a substantially increased risk of incident stroke, independent of racial disparity.	[16]

Table 1. *Cont.*

Collection Time	Average Concentration	Control	Results	Conclusions	Reference
at admission	Intra-hospital improvement [h]/extra-hospital improvement [i] 21.1 ± 15.8/22.4 ± 15.3 (NO n = 1176/1326) 21.3 ± 15.2/20.1 ± 15.3 (YES n = 3119/2527) p = 0.844/0.043	-	Th high levels of IL-6 on admission had a negative correlation with extrahospital improvement (β: −0.305, $p < 0.0001$), while high levels of IL6 at 24 hours correlates with a negative intrahospital improvement and a positive extrahospital improvement (β:−0.121, $p < 0.0001$ vs. β: 0.131, $p < 0.0001$).	High IL-6 level at 24 hours was associated with a worse intra-hospital improvement and with better extra-hospital improvement. The variations in IL6 level in the first 24 hours clearly showed a relationship between the molecular components of the ischemic cascade and the clinical outcome of patients.	[136]
24 h	62.2 ± 47.5/20.1 ± 25.0 (NO n = 1176/1326) 51.1 ± 46.3/96.5 ± 30.8 (YES n = 3119/2527) p = 0.006/<0.0001				
24 h	18.8 (4.0) pg/mL (n = 176)	6.7 (4.6) pg/mL (n = 176) $p < 0.001$	The results showed that the best predictors of IS were SBP, glucose, NOx, hydroperoxides, IL-6, and WBC (all positively associated), 25(OH)D (negatively associated), and sex (male) (x^2 = 159.64, df = 8, $p < 0.001$, Nagelkerke = 0.787) and that 89.4% of all cases were correctly classified with sensitivity of 86.2% and specificity of 93.0%. IL-6 is significantly higher in patients with an endpoint mRS ≥ 3 (27.5 (4.1) pg/mL) vs. those with an endpoint mRS < 3 (8.7(7.2) pg/mL; $p < 0.001$).	The cumulative effects of immune-inflammatory, metabolic, oxidative, and nitrosative stress (IMO&NS) biomarkers are associated with AIS and predict a poor outcome at 3-month follow-up.	[107]
<7 days	Figure 2a,b in [106]	-	Baseline IL-6 was associated with poor functional outcome (odds ratio [OR] per 1-unit increase 1.02, CI 1.01–1.04, $p = 0.002$). After adjustment for age, sex, qualifying event, baseline mRS, hypertension, anti-platelet and statins on discharge, the association remained unchanged (adjusted OR 1.02, CI 1.01–1.04, $p = 0.004$). A relationship between IL-6 and mortality after 1 year was observed. When the highest quartile was compared to the lowest quartile IL-6, the HR was 2.30 (CI 1.10–4.84, $p = 0.03$)	Baseline inflammatory cytokines independently predicted late recurrence, supporting a rationale for randomized trials of anti-inflammatory agents for prevention after stroke and suggesting that targeted therapy to high-risk patients with high baseline inflammation may be beneficial.	[106]

Table 1. *Cont.*

Collection Time	Average Concentration	Control	Results	Conclusions	Reference
at admission	[a] 1.3 [0.3–3.6] pg/mL	–	High IL-6 levels at 6, 24, and 48 h, in combination with a higher age, a pre-stroke mRS score > 2, a past history of hypertension or diabetes, current smoking, a higher baseline National Institute of Health Stroke Scale Score, the absence of associated intravenous thrombolysis, an intracranial internal carotid artery or a tandem occlusion and an increased infarct growth were associated with futile reperfusion.	IL-6 is a marker of futile reperfusion in the setting of MT.	[121]
	[b] 1.3 [0.7–3.4] pg/mL				
	[c] 2.0 [0.7–4.5] pg/mL				
6 h	[a] 3.1 [2.0–6.0] pg/mL				
	[b] 2.1 [1.1–4.4] pg/mL				
	[c] 3.3 [2.2–6.2] pg/mL				
24 h	[a] 5.0 [3.3–7.3] pg/mL				
	[b] 2.7 [1.7–5.5] pg/mL				
	[c] 4.3 [2.6–9.0] pg/mL				
48 h	[a] 5.2 [2.9–15.9] pg/mL				
	[b] 2.5 [1.2–5.2] pg/mL				
	[c] 5.3 [2.5–8.2] pg/mL				
3 months	[a] 1.2 [0.3–2.0] pg/mL				
	[b] 0.6 [0.3–1.1] pg/mL				
	[c] 1.0 [0.4–2.6] pg/mL				
24 h	(4.3; 8; 14.9) pg/mL (n = 332)	–	In linear regression analysis that was adjusted for etiology, IL-6 was found to be a biomarker independently associated with infarct volume (Atheromatous, (n = 59), 8.9 (4.9–14.6); Cardioembolic (n = 126), 10.9 (6.1–23.4); Small vessel (n = 63), 4.8 (2.8–9.7); Indeterminate, (n = 77), 8.1 (5.4–14.7); Unusual, (n = 7), 4.5 (1.9–10.0); $p < 0.001$).	IL-6 levels may be independently associated with infarct volume. The results may have clinical significance because IL-6 levels in blood may indicate the extension of brain infarction without the need to use neuroimaging techniques.	[124]
27–96 h	IL-6: Q1: ≤1.6 ng/L (n = 2896) Q2: 1.6–2.6 ng/L (n = 1872) Q3: 2.6–5.1 ng/mL (n = 1934) Q4: >5.1 ng/mL (n = 3051)	–	The highest quartiles of IL-6 (adjusted HR, 1.36; 95% CI 1.13–1.64; p = 0.001) were associated with increased risk of recurrent stroke; and the highest quartiles of IL-6 (adjusted OR 1.93; 95% CI 1.64–2.27; p < 0.0001), were correlated with increased risk of poor functional outcome.	In patients with ischemic stroke, an association was found between IL-6 and recurrent stroke, composite vascular events, and poor functional outcome.	[45]

Table 1. *Cont.*

Collection Time	Average Concentration	Control	Results	Conclusions	Reference
Oset, 24 h, 72 h	poor outcome (mRS $\geq$ 2)/ good outcome after 90 days 8.72 (IQR 3.04–26.89) pg/mL/ 3.72 (IQR, 1.88–8.58) pg/mL p = 0.012	–	The level of IL-6 was significantly increased in the NIHSS > 5 group at admission (p < 0.001) compared to the NIHSS $\leq$ 5 group. IL-6 was notably higher in the poor outcome group (p = 0.012). IL-6 was an independent predictor of risk for AIS patients with an adjusted OR of 1.152 (p = 0.028). Patients with a NIHSS score of less than 5 exhibited lower IL-6 levels, indicating that increased levels of IL-6 correlated with AIS severity after intravenous thrombolytic treatment (IVT).	IL-6 might be useful plasma marker to predict the prognosis in AIS patients at 90 days after IVT.	[116]
4.5 h	6.37 (4.35; 7.43) pg/mL (n = 125)	3.59(3.42; 4.04) (n = 28; p < 0.001)	Correlations were found between IL-6 serum concentration measured during the onset and on the 1st day of ischemic stroke and the NIHSS results assessed on admission (R = 0.43, p < 0.01 and R = 0.4, p < 0.01, respectively) and at discharge (R = 0.61, p < 0.01 and R = 0.52, p < 0.01, respectively. Moreover, an association was observed between IL-6 serum levels determined in <4.5 h and on the 1st day of ischemic stroke, and results evaluated according to the mRS scale on admission (R = 0.52, p < 0.01 and R = 0.44, p < 0.01, respectively), at discharge (R = 0.61, p < 0.01 and R = 0.47, p < 0.01, respectively), after 3 months (R = 0.68, p < 0.01 and R = 0.3, p < 0.01, respectively), and 1 year since the stroke (R = 0.73, p < 0.01 and R = 0.50, p < 0.01, respectively) and IL-6 concentration measured 7 days after discharge (R = 0.28, p = 0.04). The significant statistical connection between IL-6 levels during onset and TNF-α levels assessed in <4.5 h and after 24 h since the stroke (p < 0.01) were detected. The IL-6 (5.59 pg/mL vs. 7.29 pg/mL) with p = 0.016 assessed in time < 4.5 h was statistically lower in patients suffering from the LACI (lacunar cerebral infarct) ischemic stroke in comparison to the patients with PACI (partial anterior circulation infarct) stroke. In the case of IL-6 level measured on the 7th day (>4 pg/mL), an increase in mortality was observed.	IL-6 levels were higher compared to the control group and in the subgroup of patients with an unfavorable functional outcome (mRS: 3–6 points). A positive correlation was found between IL-6 and TNF-α levels in patients with AIS within <4.5 h and on the day of stroke. Higher IL-6 levels in the acute phase of stroke and on the first and seventh day were associated with a worse early and late prognosis in patients treated with intravenous thrombolysis. An association between IL-6 levels in subacute AIS and the severity of neurological deficit was shown. IL-6 may be a prognostic factor in thrombolytic treatment of AIS.	[2]
24 h	7.09 (6.48; 9.30) pg/L (n = 125)				
7 days	6.54 (4.44; 7.58) pg/mL (n = 125)				

Table 1. *Cont.*

Collection Time	Average Concentration	Control	Results	Conclusions	Reference
at admission	FPE <3.0 pg/mL (n = 43) Figure 2 in [122]	Non-FPE (n = 108) $p < 0.05$ Figure 2 in [122].	FPE was found to be associated with low IL-6 levels and platelet count, an older age, a lack of hypertension, a lack of tandem occlusion, a shorter thrombus length, and a reduced procedural time. Multivariate analysis showed that low IL-6 admission level was associated with FPE (OR 0.66, 95% CI 0.46 to 0.94). Optimal cut-off of IL-6 level for distinguishing FPE from non-FPE was 3.0 pg/mL (sensitivity 92.3%, specificity 42.3%).	A lower admission level of IL-6 is associated with FPE.	[122]
-	310.7 ± 110.6 pg/mL (155–522) (n = 45)	34.2 ± 14.3 pg/mL (13–62) (n = 45) $p < 0.0001$	The patient subgroups had significantly different levels of IL-6 as the severe cases had the highest level and mild ones had the lowest level ($p < 0.001$). There was a significant negative correlation between serum miR 221 with NIHSS score and IL-6 while there was a significant positive correlation with the grade of weakness (r = −0.75, $p < 0.0001$, r = −0.74, $p < 0.0001$ and r = 0.79, $p < 0.001$, respectively). The ROC curve with AUC of 0.99 for serum miR-221 combined with IL-6, showed both higher sensitivity and specificity of 96.4% and 89%, respectively.	Serum level of miR-221 may be considered a sensitive and specific marker for diagnosis and for assessing the severity of ischemic stroke especially if combined with IL-6. Such markers could be recommended as diagnostic tools and early predictive markers for the severity of the ischemic stroke disease.	[123]
<24 h	Table 1 in [137].	–	IL-6 were significantly correlated ($p < 0.05$) with disability assessed by mRS score. Higher intracranial levels of IL-6 were associated with a lower likelihood (mRS score at 90 days > 2) of favorable outcome (adjusted OR per SD increase, 0.51; 95% CI 0.27 to 0.95).	The correlation between IL-6 and mRS score outcome support an avenue for add-on and localized immune modulatory strategies in AIS.	[137]
-	LAA 3.7 (2.2–8.2) pg/mL; (n = 272) ^{d}/3.1(1.9–5.7); (n = 1467) ^{e} CE 6.0(3.0–12.1) pg/mL; (n = 85) ^{d}/4.3 (2.1–7.7); (n = 369) ^{e} SAO 2.5 (1.6–4.9) pg/mL; (n = 173) ^{d}/2.2 (1.4–3.7); (n = 1397) ^{e}	–	At 3 months poststroke, 1026 (14.8%) patients experienced worse outcomes. Compared with patients with good outcomes, patients with poorer outcomes had significantly higher levels of inflammatory biomarkers, including IL-6 ($p < 0.05$). The highest quartile of IL-6 (1.43 [1.16–1.76]) was associated with an increased risk of worse outcomes at 3 months.	Significant associations have been found between elevated levels of inflammatory markers, including IL-6, and increased risk of poorer outcomes in patients with acute ischemic stroke. Targeting the inflammatory response, through novel anti-inflammatory therapies, may hold promise for improving neurological outcomes in stroke patients.	[113]

Table 1. *Cont.*

Collection Time	Average Concentration	Control	Results	Conclusions	Reference
24 h	Mean ± S 3.8 ± 4.0 pg/mL Median (IQR) 2.4(1.5–4.2) pg/mL; (n = 3910) [f] Mean ± S 5.7 ± 5.7 pg/mL Median (IQR) 2.8(1.9–8.1) pg/mL; (n = 121) [g]	–	Patients with ND had higher levels of IL-6 (median, 2.4 vs. 2.8 pg/mL) than patients without ND. IL-6 levels (adjusted OR, 1.43 [95% CI, 1.24–1.66]) remained independent predictor of in-hospital ND. The above relationship was also observed in long-term poor outcomes, including unfavorable functional outcomes (defined as mRS 2–5) and death.	Elevated IL-6 was an independent predictor for ND in minor AIS patients with LAA and SVO subtypes. IL-6 had the most notable incremental predictive value of in-hospital ND in addition to conventional predictors, both collectively and in the LAA subgroup.	[103]

[a] No or partial reperfusion (n = 31); [b] successful reperfusion—effective (n = 84); [c] successful reperfusion—futile (n = 46);[d] worse outcome (worse outcomes at 3 months were defined as $\Delta mRS_{3m\text{-}discharge} \geq 1$ ($\Delta mRS_{3m\text{-}discharge} = mRS_{3m\text{-}mRSdischarge}$); [e] good outcome; [f] no ND, neurological deterioration; [g] ND, neurological deterioration; [h] positive value of the NIHSS on admission minus the value of the NIHSS at discharge. [i] positive value of the NIHSS scale at discharge minus the value of the NIHSS scale at three months.

6. Modulation of IL-6

IL-6 represents a complex therapeutic target, as it plays both a toxic, inflammatory role and a regenerative role in the pathophysiology of stroke. Preclinical studies related to the inhibition or amplification of IL-6 are ambiguous [37–42]. It has been shown that IL-6 is a pro-inflammatory interleukin, which is why it is believed that reducing its expression is beneficial for neurological recovery. However, some of the literature reports suggest neurological benefits from upregulating this cytokine [32]. A detailed study of IL-6, focusing on the timing of its administration and the dosing regimen after stroke, may reveal the optimal strategy for achieving therapeutic effects. Many researchers have shown that manipulating its levels can impact the prognosis of patients [34–36] and improve post-stroke recovery care [13,30].

Controlling its concentration at different time intervals could be significant in post-stroke therapies. Experimental and preclinical studies have shown that inflammatory therapies, which reduce the number of serious adverse cardiovascular events (MACE) in coronary artery disease or stroke, can be used in stroke treatment. Establishing a specific relationship between inflammatory markers and the risk of recurrence of ischemic events may be useful in stroke-related anti-inflammatory therapies [145].

This "inflammatory hypothesis" was the basis for several clinical studies using different drugs in patients with coronary artery disease, and colchicine has shown promising therapeutic potential so far. In the CONVINCE study (Colchicine for Prevention of Vascular Inflammation in non-cardio Embolic Stroke) evaluating the impact of colchicine on stroke recurrence, its role in reducing both the inflammation caused by atherosclerosis and in stabilizing atherosclerotic plaques was demonstrated. Colchicine inhibits the function of neutrophils and reduces their ability to migrate to the site of inflammation, thus shortening the duration of the inflammatory response. Reduction in the production of inflammatory cytokines such as IL-1, IL-6, and TNF-α is also a result of the anti-inflammatory effects of colchicine. Therefore, daily use of colchicine may provide benefits in preventing recurrent strokes and major vascular events compared to usual care [45,146].

In another large-scale study, CANTOS (Canakinumab Anti-inflammatory Thrombosis Outcome Study), it was observed that the monoclonal antibody (canakinumab) had a beneficial effect on the final cardiovascular outcome, including stroke, in patients who had an unsuccessful myocardial infarction [147].

However, the monoclonal antibody was not able to show effectiveness in every stroke due to limited power and non-atherosclerotic causes of stroke. These observations suggest that monoclonal antibodies targeting IL-6 may be a promising tool in preventing ischemic stroke in patients with carotid artery atherosclerosis. The threshold of 2.0 pg/mL may help identify individuals who would benefit from anti-IL-6 drugs to prevent stroke [19,31].

So far, it has been shown that in stroke patients, higher levels of IL-6 in plasma correlate with increased stroke severity and poor long-term prognosis. Treatment with an immunosuppressive drug approved by the Food and Drug Administration (FDA) could block IL-6 receptors to inhibit its pro-inflammatory action. In an experimental study, researchers demonstrated that following the use of an immunosuppressant, young male mice with ischemic stroke showed a reduction in the infarct size. This led to the conclusion that controlling cytokine signaling, influenced by temporal and environmental changes, is crucial for improving post-stroke patient outcomes [148].

Depending on the timeline of ischemia, a direct injection of IL-6 into the brain after ischemia can reduce ischemic brain damage [35]. IL-6 has been found to play an anti-inflammatory role in both local and systemic acute inflammatory responses by regulating the levels of pro-inflammatory but not anti-inflammatory cytokines. This is supported by other literature reports, indicating that IL-6 is a promising tool in reducing the risk of stroke [12,35,45,149].

Endogenous IL-6 may play a key role in preventing neuronal damage during the acute phase of brain ischemia by activating the signaling protein and STAT3. The administration of a monoclonal antibody against the mouse IL-6 receptor (IL-6RA) 6 h after MCAO resulted in a significant reduction in the amount of phosphorylated STAT3 in the peri-infarct area of the cerebral cortex. After 24 h from MCAO, blocking IL-6 signaling led to an increase in the number of apoptotic cells in the peri-infarct area and an enlargement of the infarct size, as well as adversely affected neurological functions [150].

Another stroke therapy utilizing the potential of IL-6 could be modulation of the trans signaling pathway, enhancing trans signaling at an early stage or suppressing it at a later stage, to avoid a neuroinflammatory response [34,53,54].

It is known that in trans signaling, sIL6R, by forming a binary complex with IL-6, can activate gp130, triggering a pro-inflammatory cascade. However, if IL-6: sIL6R forms a ternary complex with soluble sgp130, the pro-inflammatory cascade is inhibited. The formation of the IL-6:sIL6R:sgp130 complex prevents the binding of gp130 and IL-6 trans signaling [22,53].

The ratio of the pro-inflammatory binary complex to the three-component complex (B/T) could be a new biomarker for assessing the increased risk of ischemic stroke, particularly of atherosclerotic origin, as the IL-6 trans signaling pathway is primarily associated with atherosclerotic cerebrovascular disease [54,151].

As a result of assessing the B/T ratio, it is possible to measure IL-6 signaling at different time points after ischemic injury. This could directly influence the achievement of the desired therapeutic effect by individualizing stroke therapy to support the body's natural responses to reduce ongoing inflammation [152].

Knowledge of an elevated B/T ratio could enable clinicians to decide on the implementation of therapies using IL-6 trans signaling antagonists, such as spg130, which could lead to optimal modulation of the inflammatory signaling pathway [53].

Therefore, consideration could be given to introducing cell-based therapies into stroke treatment that can respond to local environments and increase or decrease IL-6 signaling. Stem cell-based therapies have a high therapeutic potential for regulating IL-6 signaling. This approach is currently receiving more attention in preclinical strategies, but such therapy with stem cells has not yet been implemented in stroke patient treatment [46].

Stem cells can be obtained at various stages of differentiation, and in the human body, they can be transformed into different cell types. They can show effective therapeutic results in the treatment of brain stroke or other diseases. After ischemic injury, they can be used to enhance any signaling pathway that would be most beneficial in treatment, considering the timeline of hypoxia [46]. It has been shown that treatment with embryonic stem cells and umbilical cord-derived stem cells (hESC-MSC-CM) leads to neurogenesis, angiogenesis, and protection of the brain tissue from ischemic damage. Inflammation and apoptosis are inhibited [152].

Additionally, pre-conditioning with IL-6 increases the efficacy of cell transplantation therapy in ischemic stroke by protecting neural stem cells (NSCs) from reperfusion injury. Conditioned NSCs promote angiogenesis, reduce infarct size, and improve behavioral outcomes [153].

Similarly, in an experimental study on mice with middle cerebral artery occlusion (dMCAO) receiving mesenchymal stem cell (HUMSC) transplants, which were protected by the caspase-1 inhibitor (VX765), their anti-apoptotic effect was demonstrated. This led to the activation of autophagy through the AMPK/mTOR signaling pathway both in vivo and in vitro [154].

Transplantation of human bone marrow-derived stem cells (BMSC) in ischemic stroke leads to an increase in T regulatory cells (Tregs), active mediators of immunomodulation, enabling neuroprotection, which is partly due to the anti-inflammatory properties of the cells in vitro. Regulatory T cells exert neuroprotective effects in stroke by inhibiting both inflammation and the activation of effector T lymphocytes [155].

Culturing oligodendrocyte precursor cells (OPC) with BMSC and Tregs show greater myelination potential and an increased production of IL-6 and basic fibroblast growth factor (FGF-β) compared to OPC cultured with BMSC alone. This indicates the potential of OPCs in an ischemic environment by modulating optimal IL-6 signaling pathways [46,156].

In further experimental animal studies, the injection of human neural stem cells into the hippocampus of mice reduced inflammation in ischemia–reperfusion brain injury by reducing pro-inflammatory cytokines and attenuating damage to BBB [157].

Similarly, the administration of adipose-derived mesenchymal stem cells (ADMSC) reduced neurological severity, limited the infarct area, and decreased cell apoptosis. They altered IL-6 signaling to restore BBB integrity and reduced the infiltration of peripheral inflammatory cells [158].

It has been demonstrated that the administration of NSC and NSC-1 cells 24 h after stroke led to improvements in motor and behavioral neurological functions, as well as a reduction in infarct size, which persisted for the following 7 days after the ischemic event. The authors concluded that the therapeutic effect was influenced by the increased levels of fibroblast growth factor 2 (bFGF) and IL-6 secreted by NSC-01. These results suggest that IL-6 and bFGF mediate the mechanism responsible for the improvement observed after in vitro MSC treatment [159].

7. Summary

In this study, we aimed to present the functional role of interleukin IL-6 and its association with ischemic stroke. Based on the available literature, it has been established that hypoxia can lead to various types of cellular damage and induce an immune response in which IL-6 amplifies the local inflammatory response. IL-6 activates the JAK/STAT pathway to promote the expression of pro-cytokines. The study attempted to demonstrate that IL-6 could serve as a reliable prognostic factor in the acute phase of AIS, as well as a biomarker for the development of clinical complications resulting from acute ischemia, or to assess the severity of neurological deficits and the prognosis of patients after AIS.

However, developing effective anti-inflammatory methods for preventing stroke remains an important, yet still unresolved clinical issue.

Understanding the temporal profile and cellular sources of IL-6, as well as its potential transport to the ischemic region in the early phase after a stroke, is crucial for comprehending how this cytokine interacts with potentially salvageable neurons in the ischemic penumbra. However, selecting the appropriate therapy requires consideration of the patient diversity in stroke cases. Monitoring IL-6 levels across various time points in different patients will enable more precise targeting of anti-IL-6 therapies, minimizing systemic side effects. Individual factors such as age, comorbidities, and genetic background can also affect the body's response to IL-6 therapy and treatment. Moreover, inhibiting IL-6 signaling pathways must be approached cautiously. Rather than completely blocking IL-6, it may be more beneficial to focus on therapies that specifically target trans-signaling.

It is also important to consider the possibility of determining a specific relationship between this inflammatory marker and the risk of recurrence of ischemic events, so that it could be used in anti-inflammatory therapies for stroke patients. However, further research is needed to better understand which IL-6 signaling pathway is utilized at different time points after stroke to effectively support patient recovery.

Author Contributions: Conceptualization, H.P. and R.K.; writing—original draft preparation, H.P., A.T.-K., A.K., M.P., K.S., R.G. and R.K.; writing—review and editing, H.P., A.W., A.T.-K., A.K., M.P., K.S., R.G. and R.K.; visualization, R.K. and H.P.; supervision, H.P. and R.K.; project administration, H.P. and R.K.; funding acquisition, A.W. All authors have read and agreed to the published version of the manuscript.

Funding: This research received no external funding.

Institutional Review Board Statement: Not applicable.

Informed Consent Statement: Not applicable.

Data Availability Statement: Not applicable.

Conflicts of Interest: The authors declare no conflicts of interest. Prognostic value of blood interleukin-6 in the prediction of functional outcome after stroke: a systematic review and meta-analysis.

References

1. Pawluk, H.; Grześk, G.; Kołodziejska, R.; Kozakiewicz, M.; Woźniak, A.; Grzechowiak, E.; Szumny, M.; Sobolewski, P.; Bieniaszewski, L.; Kozera, G. Effectof IL-6 and hsCRP serum levels on functional prognosis in stroke patients undergoing IV-thrombolysis: Retrospective analysis. *Clin. Interv. Aging* **2020**, *6*, 1295–1303. [CrossRef]
2. Pawluk, H.; Kołodziejska, R.; Grześk, G.; Kozakiewicz, M.; Woźniak, A.; Pawluk, M.; Kosinska, A.; Grześk, M.; Wojtasik, J.; Kozera, G. Selected mediators of inflammation in patients with acute ischemic stroke. *Int. J. Mol. Sci.* **2022**, *23*, 10614. [CrossRef] [PubMed]
3. Pawluk, H.; Kołodziejska, R.; Grześk, G.; Woźniak, A.; Kozakiewicz, M.; Kosinska, A.; Pawluk, M.; Grzechowiak, E.; Wojtasik, J.; Kozera, G. Increased oxidative stress markers in Acute Ischemic Stroke patients treated with thrombolytics. *Int. J. Mol. Sci.* **2022**, *23*, 15625. [CrossRef]
4. Žitňanová, I.; Šiarnik, P.; Kollár, B.; Chomová, M.; Pazderová, P.; Andrezálová, L.; Ježovičová, M.; Koňariková, K.; Laubertová, L.; Krivošíková, Z.; et al. Oxidative stress markers and their dynamic changes in patients after acute ischemic stroke oxidative. *Oxid. Med. Cell. Longev.* **2016**, *2016*, 9761697. [CrossRef]
5. Rodrigo, R.; Fernandez-Gajardo, R.; Gutiérrez, R.; Matamala, J.M.; Carrasco, R.; Miranda-Merchak, A.; Feuerhake, W. Oxidative stress and pathophysiology of ischemic stroke: Novel therapeutic opportunities. *CNS Neurol. Disord.-Drug Targets* **2013**, *12*, 698–714. [CrossRef]
6. Zheng, Z.; Yenari, M.A. Post-ischemic inflammation: Molecular mechanisms and therapeutic implications. *Neurol. Res.* **2004**, *26*, 884–892. [CrossRef]
7. Shi, K.; Tian, D.C.; Li, Z.G.; Ducruet, A.F.; Lawton, M.T.; Shi, F.D. Global brain inflammation in stroke. *Lancet Neurol.* **2019**, *18*, 1058–1066. [CrossRef]
8. Kumari, S.; Dhapola, R.; Sharma, P.; Naga, P.; Medhi, B.; Hari, D.; Reddy, K. The impact of cytokines in neuroinflammation-mediated stroke. *Cytokine Growth Factor Rev.* **2024**, *78*, 105–119. [CrossRef]
9. Ozaki, E.; Campbell, M.; Doyle, S.L. Targeting the NLRP3 inflammasome in chronic inflammatory diseases: Current perspectives. *J. Inflamm. Res.* **2015**, *8*, 15–27. [CrossRef]
10. Yang, Y.; Wang, H.; Kouadir, M.; Song, H.; Shi, F. Recent advances in the mechanisms of NLRP3 inflammasome activation and its inhibitors. *Cell Death Dis.* **2019**, *10*, 128–139. [CrossRef]
11. Yang, C.-S.; Kim, J.-J.; Kim, T.S.; Lee, P.Y.; Kim, S.Y.; Lee, H.-M.; Shin, D.-M.; Nguyen, L.T.; Lee, M.-S.; Jin, H.S.; et al. Small heterodimer partner interacts with NLRP3 and negatively regulates activation of the NLRP3 inflammasome. *Nat. Commun.* **2015**, *6*, 6115–6126. [CrossRef] [PubMed]
12. Papadopoulos, A.; Palaiopanos, K.; Björkbacka, H.; Peters, A.; de Lemos, J.A.; Seshadri, S.; Dichgans, M.; Georgakis, M.K. Circulating interleukin-6 levels and incident Ischemic Stroke: A systematic review and meta-analysis of prospective studies. *Neurology* **2022**, *98*, e1002–e1012. [CrossRef] [PubMed]
13. Georgakis, M.K.; Malik, R.; Gill, D.; Franceschini, N.; Sudlow, C.L.M.; Dichgans, M. Interleukin-6 signaling effects on ischemic stroke and other cardiovascular outcomes: A mendelian randomization study. *Circ. Genom. Precis. Med.* **2020**, *13*, e002872. [CrossRef] [PubMed]
14. Georgakis, M.K.; Malik, R.; Li, X.; Gill, D.; Levin, M.G.; Vy, H.M.T.; Judy, R.; Ritchie, M.; Verma, S.S.; Nadkarni, G.N.; et al. Genetically downregulated interleukin-6 signaling is associated with a favorable cardiometabolic profile: A phenome-wide association study. *Circulation* **2021**, *143*, 1177–1180. [CrossRef]

15. Kaptoge, S.; Seshasai, S.R.; Gao, P.; Freitag, D.F.; Butterworth, A.S.; Borglykke, A.; Di Angelantonio, E.; Gudnason, V.; Rumley, A.; Lowe, G.D.O.; et al. Inflammatory cytokines and risk of coronary heart disease: New prospective study and updated meta-analysis. *Eur. Heart J.* **2014**, *35*, 578–589. [CrossRef]
16. Jenny, N.S.; Callas, P.W.; Judd, S.E.; McClure, L.A.; Kissela, B.; Zakai, N.A.; Cushman, M. Inflammatory cytokines and ischemic stroke risk: The REGARDS cohort. *Neurology* **2019**, *92*, e2375–e2384. [CrossRef]
17. Patterson, C.C.; Smith, A.E.; Yarnell, J.W.; Rumley, A.; Ben-Shlomo, Y.; Lowe, G.D. The associations of interleukin-6 (IL-6) and downstream inflammatory markers with risk of cardiovascular disease: The Caerphilly Study. *Atherosclerosis* **2010**, *209*, 551–557. [CrossRef]
18. Scheller, J.; Chalaris, A.; Schmidt-Arras, D.; Rose-John, S. The pro- and anti-inflammatory properties of the cytokine interleukin-6. *Biochim. Biophys. Acta* **2011**, *1813*, 878–888. [CrossRef]
19. Ridker, P.M.; Rane, M. Interleukin-6 signaling and anti-interleukin-6 therapeutics in cardiovascular disease. *Circ. Res.* **2021**, *128*, 1728–1746. [CrossRef]
20. Libby, P.; Rocha, V.Z. All roads lead to IL-6: A central hub of cardiometabolic signaling. *Int. J. Cardiol.* **2018**, *259*, 213–215. [CrossRef]
21. Loppnow, H.; Libby, P. Proliferating or interleukin 1-activated human vascular smooth muscle cells secrete copious interleukin 6. *J. Clin. Investig.* **1990**, *85*, 731–738. [CrossRef] [PubMed]
22. Rothaug, M.; Becker-Pauly, C.; Rose-John, S. The role of interleukin-6 signaling in nervous tissue. *Biochim. Biophys. Acta* **2016**, *1863*, 1218–1227. [CrossRef] [PubMed]
23. Rayasam, A.; Hsu, M.; Kijak, J.A.; Kissel, L.; Hernandez, G.; Sandor, M.; Fabry, Z. Immune responses in stroke: How the immune system contributes to damage and healing after stroke and how this knowledge could be translated to better cures? *Immunology* **2018**, *154*, 363–376. [CrossRef] [PubMed]
24. Yao, H.; Zhang, Y.; Shu, H.; Xie, B.; Tao, Y.; Yuan, Y.; Shang, Y.; Yuan, S.; Zhang, J. Hyperforin promotes post-stroke neuroangiogenesis via astrocytic IL-6-mediated negative immune regulation in the ischemic brain. *Front. Cell. Neurosci.* **2019**, *13*, 201. [CrossRef]
25. Fang, J.; Wang, Z.; Miao, C.-Y. Angiogenesis after ischemic stroke. *Acta Pharmacol. Sin.* **2023**, *44*, 1305–1321. [CrossRef]
26. Zhu, H.; Zhang, Y.; Zhong, Y.; Ye, Y.; Hu, X.; Gu, L.; Xiong, X. Inflammation-mediated angiogenesis in ischemic stroke. *Front. Cell. Neurosci.* **2021**, *15*, 652647. [CrossRef]
27. Swartz, K.R.; Liu, F.; Sewell, D.; Schochet, T.; Campbell, I.; Sandor, M.; Fabry, Z. Interleukin-6 promotes post-traumatic healing in the Central Nervous System. *Brain Res.* **2001**, *896*, 86–95. [CrossRef]
28. Kummer, K.K.; Zeidler, M.; Kalpachidou, T.; Kress, M. Role of IL-6 in the regulation of neuronal development, survival and function. *Cytokine* **2021**, *144*, 155582. [CrossRef]
29. Chiba, T.; Umegaki, K. Pivotal roles of monocytes/macrophages in stroke. *Mediat. Inflamm.* **2013**, *27*, 1–10. [CrossRef]
30. Zhu, H.; Hu, S.; Li, Y.; Sun, Y.; Xiong, X.; Hu, X.; Chen, J.; Qiu, S. Interleukins and Ischemic Stroke. *Front. Immunol.* **2022**, *13*, 828447. [CrossRef]
31. Kamtchum-Tatuene, J.; Saba, L.; Heldner, M.R.; Rundek, T.; Kakkos, S.K.; Chaturvedi, S.; Poorthuis, M.H.F.; de Borst, G.J.; Topakian, R.; Polak, J.F.; et al. Interleukin-6 predicts Carotid Plaque Severity, vulnerability, and progression. *Circ. Res.* **2022**, *131*, e22–e33. [CrossRef] [PubMed]
32. Boulanger, M.J.; Chow, D.; Brevnova, E.E.; Garcia, K.C. Hexameric Structure and Assembly of the Interleukin-6/IL-6 α-Receptor/gp130 Complex. *Science* **2003**, *300*, 2101–2104. [CrossRef] [PubMed]
33. Ramiro, L.; Simats, A.; García-Berrocoso, T.; Montaner, J. Inflammatory molecules might become both biomarkers and therapeutic targets for stroke management. *Ther. Adv. Neurol. Disord.* **2018**, *1*, 1–24. [CrossRef]
34. Monsour, M.; Croci, D.M.; Agazzi, S. The role of IL-6 in TBI and PTSD, a potential therapeutic target? *Clin. Neurol. Neurosurg.* **2022**, *218*, 107280. [CrossRef]
35. Suzuki, S.; Tanaka, K.; Suzuki, N. Ambivalent aspects of interleukin-6 in Cerebral Ischemia: Inflammatory versus neurotrophic aspects. *J. Cereb. Blood Flow Metab.* **2009**, *29*, 464–479. [CrossRef]
36. Schofer, N.; Ludwig, S.; Rübsamen, N.; Schnabel, R.; Lackner, K.J.; Ruprecht, H.J.; Bickel, C.; Landmesser, U.; Blankenberg, S.; Zeller, T. Prognostic impact of interleukin-1 receptor antagonist in patients with documented coronary artery disease. *Int. J. Cardiol.* **2018**, *257*, 24–29. [CrossRef]
37. Li, J.; Zhao, X.; Meng, X.; Lin, J.; Liu, L.; Wang, C.; Wang, A.; Wang, Y.; Wang, Y. High-sensitive C-reactive protein predicts recurrent stroke and poor functional outcome: Subanalysis of the clopidogrel in high-risk patients with acute nondisabling cerebrovascular events trial. *Stroke* **2016**, *47*, 2025–2030. [CrossRef]
38. Kjaergaard, A.D.; Bojesen, S.E.; Johansen, J.S.; Nordestgaard, B.G. Elevated plasma YKL-40 levels and ischemic stroke in the general population. *Ann. Neurol.* **2010**, *68*, 672–680. [CrossRef]

39. Park, H.Y.; Jun, C.D.; Jeon, S.J.; Choi, S.-S.; Kim, H.-R.; Choi, D.-B.; Kwak, S.; Lee, H.-S.; Cheong, J.S.; So, H.-S.; et al. Serum YKL-40 levels correlate with infarct volume, stroke severity, and functional outcome in acute ischemic stroke patients. *PLoS ONE* **2012**, *7*, e51722. [CrossRef]
40. Hartman, J.; Frishman, W.H. Inflammation and atherosclerosis: A review of the role of interleukin-6 in the development of atherosclerosis and the potential for targeted drug therapy. *Cardiol. Rev.* **2014**, *22*, 147–151. [CrossRef]
41. Schieffer, B.; Selle, T.; Hilfiker, A.; Hilfiker-Kleiner, D.; Grote, K.; Tietge, U.J.; Traut-Wein, C.; Luchtefeld, M.; Schmittkamp, C.; Heeneman, S.; et al. Impact of interleukin-6 on plaque development and morphology in experimental atherosclerosis. *Circulation* **2004**, *110*, 3493–3500. [CrossRef] [PubMed]
42. Jung, K.W.; Shon, Y.M.; Yang, D.W.; Kim, B.S.; Cho, A.H. Coexisting carotid atherosclerosis in patients with intracranial small- or large-vessel disease. *J. Clin. Neurol.* **2012**, *8*, 104–108. [CrossRef] [PubMed]
43. Li, J.; Lin, J.; Pan, Y.; Wang, M.; Meng, X.; Li, H.; Wang, Y.; Zhao, X.; Qin, H.; Liu, L.; et al. Residual inflammatory risk predicts poor prognosis in Acute Ischemic Stroke or Transient Ischemic Attack Patients. *Stroke* **2021**, *52*, 2827–2836. [CrossRef] [PubMed]
44. Li, C.; Wang, Y.; Mei, P.; Tung, T.-H.; Wu, G.; Wang, F.; Wang, E.; Ni, H.; Zhu, X.; He, Z.; et al. High Interleukin-6 levels are associated with Large-Artery Atherosclerotic Stroke. *Neurologist* **2023**, *28*, 277–280. [CrossRef]
45. Li, J.; Lin, J.; Pan, Y.; Wang, M.; Meng, X.; Li, H.; Wang, Y.; Zhao, X.; Qin, H.; Liu, L.; et al. Interleukin-6 and YKL-40 predicted recurrent stroke after ischemic stroke or TIA: Analysis of 6 inflammation biomarkers in a prospective cohort study. *J. Neuroinflamm.* **2022**, *19*, 131. [CrossRef]
46. Lockard, G.M.; Alayli, A.; Monsour, M.; Gordon, J.; Schimmel, S.; Elsayed, B.; Borlongan, C.V. Probing Interleukin-6 in stroke pathology and neural stem cell transplantation. *Int. J. Mol. Sci.* **2022**, *23*, 15453. [CrossRef]
47. Erta, M.; Quintana, A.; Hidalgo, J. Interleukin-6, a major cytokine in the central nervous system. *Int. J. Biol. Sci.* **2012**, *8*, 1254–1266. [CrossRef]
48. Rose-John, S. Interleukin-6 signalling in health and disease. *F1000Research* **2020**, *9*, F1000 Faculty Rev-1013. [CrossRef]
49. Hodes, G.E.; Menard, C.; Russo, S.J. Integrating Interleukin-6 into depression diagnosis and treatment. *Neurobiol. Stress* **2016**, *4*, 15–22. [CrossRef]
50. Wolf, J.; Rose-John, S.; Garbers, C. Interleukin-6 and its receptors: A highly regulated and dynamic system. *Cytokine* **2014**, *70*, 11–20. [CrossRef]
51. Schuett, H.; Oestreich, R.; Waetzig, G.H.; Annema, W.; Luchtefeld, M.; Hillmer, A.; Bavendiek, U.; von Felden, J.; Divchev, D.; Kempf, T.; et al. Transsignaling of interleukin-6 crucially contributes to atherosclerosis in mice. *Arterioscler. Thromb. Vasc. Biol.* **2012**, *32*, 281–920. [CrossRef] [PubMed]
52. Hou, T.; Tieu, B.C.; Ray, S.; Iii, A.R.; Cui, R.; Tilton, R.G.; Brasier, A.R. Roles of IL-6-gp130 signaling in vascular inflammation. *Curr. Cardiol. Rev.* **2008**, *4*, 179–192. [CrossRef] [PubMed]
53. Monsour, M.; Croci, D.M.; Agazzi, S.; Borlongan, C.V. Contemplating IL-6, a double-edged sword cytokine: Which side to use for stroke pathology? *CNS Neurosci. Ther.* **2023**, *29*, 493–497. [CrossRef] [PubMed]
54. Ziegler, L.; Wallen, H.; Aspberg, S.; de Faire, U.; Gigante, B. IL6 trans-signaling associates with ischemic stroke but not with atrial fibrillation. *BMC Neurol.* **2021**, *21*, 306. [CrossRef]
55. Rose-John, S.; Jenkins, B.J.; Garbers, C.; Moll, J.M.; Scheller, J. Targeting IL-6 trans-signalling: Past, present and future prospects. *Nat. Rev. Immunol.* **2023**, *23*, 666–681. [CrossRef]
56. Yan, M.; Sun, Z.; Zhang, S.; Yang, G.; Jiang, X.; Wang, G.; Li, R.; Wang, Q.; Tian, X. SOCS modulates JAK-STAT pathway as a novel target to mediate the occurrence of neuroinflammation: Molecular details and treatment options. *Brain Res. Bull.* **2024**, *213*, 110988. [CrossRef]
57. Hirano, T.; Ishihara, K.; Hibi, M. Roles of STAT3 in mediating the cell growth, differentiation and survival signals relayed through the IL-6 family of cytokine receptors. *Oncogene* **2000**, *19*, 2548–2556. [CrossRef]
58. Mallick, R.; Basak, S.; Chowdhury, P.; Bhowmik, P.; Das, R.K.; Banerjee, A.; Paul, S.; Pathak, S.; Duttaroy, A.K. Targeting cytokine-mediated inflammation in brain disorders: Developing new treatment strategies. *Pharmaceuticals* **2025**, *18*, 104. [CrossRef]
59. Chen, J.-Y.; Yu, Y.; Yuan, Y.; Zhang, Y.-J.; Fan, X.-P.; Yuan, S.-Y.; Zhang, J.-C.; Yao, S.-L. Enriched housing promotes post-stroke functional recovery through astrocytic HMGB1-IL-6-mediated angiogenesis. *Cell Death Discov.* **2017**, *3*, 17054. [CrossRef]
60. Gertz, K.; Kronenberg, G.; Kalin, R.E.; Baldinger, T.; Werner, C.; Balkaya, M.; Eom, G.D.; Hellmann-Regen, J.; Kröber, J.; Miller, K.R.; et al. Essential role of interleukin-6 in post-stroke angiogenesis. *Brain* **2012**, *135*, 1964–1980. [CrossRef]
61. Kong, L.; Li, W.; Chang, E.; Wang, W.; Shen, N.; Xu, X.; Wang, X.; Zhang, Y.; Wen Sun, W.; Hu, W.; et al. mtDNA-STING axis mediates microglial polarization via IRF3/NF-κB signaling after Ischemic Stroke. *Front. Immunol.* **2022**, *13*, 860977. [CrossRef]
62. He, T.; Yang, G.-Y.; Zhang, Z. Crosstalk of astrocytes and other cells during Ischemic Stroke. *Life* **2022**, *12*, 910. [CrossRef] [PubMed]
63. Pendlebury, S.T.; Rothwell, P.M. Prevalence, incidence, and factors associated with pre-stroke and post-stroke dementia: A systematic review and meta-analysis. *Lancet Neurol.* **2009**, *8*, 1006–1018. [CrossRef] [PubMed]

64. Kim, K.Y.; Shin, K.Y.; Chang, K.-A. Potential biomarkers for post-stroke cognitive impairment: A systematic review and meta-analysis. *Int. J. Mol. Sci.* **2022**, *23*, 602. [CrossRef]
65. Ma, Y.; Chen, Y.; Yang, T.; He, X.; Yang, Y.; Chen, J.; Han, L. Blood biomarkers for post-stroke cognitive impairment: A systematic review and meta-analysis. *J. Stroke Cerebrovasc. Dis.* **2024**, *33*, 107632. [CrossRef]
66. Kulesh, A.; Drobakha, V.; Kuklina, E.; Nekrasova, I.; Shestakov, V. Cytokine response, tract-specific fractional anisotropy, and brain morphometry in post-stroke cognitive impairment. *J. Stroke Cerebrovasc. Dis.* **2018**, *27*, 1752–1759. [CrossRef]
67. Singh-Manoux, A.; Dugravot, A.; Brunner, E.; Kumari, M.; Shipley, M.; Elbaz, A.; Kivimaki, M. Interleukin-6 and C-reactive protein as predictors of cognitive decline in late midlife. *Neurology* **2014**, *83*, 486–493. [CrossRef]
68. Fromme, S.E.; Joergens, S.; Schwarte, K.; Hohoff, C.; Dietrich, D.E.; Baune, B.T. The association between cytokines and cognitive function in patients with major depressive disorder and controls. *J. Affect. Disord.* **2025**, *373*, 374–382. [CrossRef]
69. Khan, H.; Naseem, T.; Kaushik, P.; Narang, J.; Khan, R.; Panwar, S.; Parvez, S. Decoding paradoxical links of cytokine markers in cognition: Cross talk between physiology, inflammaging, and Alzheimer's disease- related cognitive decline. *Ageing Res. Rev.* **2024**, *101*, 102535. [CrossRef]
70. Palta, P.; Xue, Q.L.; Deal, J.A.; Fried, L.P.; Walston, J.D.; Carlson, M.C. Interleukin-6 and C-reactive protein levels and 9-year cognitive decline in community-dwelling older women: The women's health and aging Study II. *J. Gerontol. A Biol. Sci. Med. Sci.* **2015**, *70*, 873–878. [CrossRef]
71. Kang, S.; Kishimoto, T. Interplay between interleukin-6 signaling and the vascular endothelium in cytokine storms. *Exp. Mol. Med.* **2021**, *53*, 1116–1123. [CrossRef] [PubMed]
72. Elendu, C.; Amaechi, D.C.; Elendu, T.C.; Ibhiedu, J.O.; Egbunu, E.O.; Ndam, A.R.; Ogala, F.; Ologunde, T.; Peterson, J.C.; Boluwatife, A.I.; et al. Stroke and cognitive impairment: Understanding the connection and managing symptoms. *Ann. Med. Surg.* **2023**, *85*, 6057–6066. [CrossRef]
73. Dik, M.G.; Jonker, C.; Hack, C.E.; Smit, J.H.; Comijs, H.C.; Eikelenboom, P. Serum inflammatory proteins and cognitive decline in older persons. *Neurology* **2005**, *64*, 1371–1377. [CrossRef]
74. Loga-Andrijić, N.; Petrović, N.T.; Filipović-Danić, S.; Marjanovic, S. The significance of interleukin-6 and tumor necrosis factor-alpha levels in cognitive impairment among first-ever acute ischaemic stroke patients. *Psychiatr. Danub.* **2021**, *33*, 37–42.
75. Sandvig, H.V.; Aam, S.; Alme, K.N.; Askim, T.; Beyer, M.K.; Ellekjær, H.; Ihle-Hansen, H.; Lydersen, S.; Mollnes, T.E.; Munthe-Kaas, R.; et al. Plasma inflammatory biomarkers are associated with poststroke cognitive impairment: The Nor-COAST study. *Stroke* **2023**, *54*, 1303–1311. [CrossRef]
76. Wang, Y.; Li, J.; Pan, Y.; Wang, M.; Lin, J.; Meng, X.; Liao, X.; Wang, Y. Interleukin-6 as predictor of one-year cognitive function after Ischemic Stroke or TIA. *Neuropsychiatr. Dis. Treat.* **2022**, *18*, 391–399. [CrossRef]
77. Kowalska, K.; Klimiec, E.; Weglarczyk, K.; Pera, J.; Slowik, A.; Siedlar, M.; Dziedzic, T. Reduced ex vivo release of pro-inflammatory cytokines and elevated plasma Interleukin-6 are inflammatory signatures of post-stroke delirium. *J. Neuroinflamm.* **2018**, *15*, 111. [CrossRef]
78. Shi, Q.; Li, R.; Qu, Z.; Lang, Y.; Sheng, G.; Ning, J.; Zhang, W. Longitudinal change of six common inflammatory cytokines and their relationship to anxiety, depression, and cognitive impairment in acute ischemic stroke patients. *Braz. J. Med. Biol. Res.* **2023**, *56*, e13025. [CrossRef]
79. Li, R.; Fan, W.; Li, D.; Liu, X. Correlation of common inflammatory cytokines with cognition impairment, anxiety, and depression in acute ischemic stroke patients. *Braz. J. Med. Biol. Res.* **2022**, *55*, e11517. [CrossRef]
80. Zhang, X.; Bi, X. Post-stroke cognitive impairment: A review focusing on molecular biomarkers. *J. Mol. Neurosci.* **2020**, *70*, 1244–1254. [CrossRef]
81. Zhou, S.; Zhong, Z.; Huang, P.; Xiang, B.; Li, X.; Dong, H.; Zhang, G.; Wu, Y.; Li, P. L-6/STAT3 induced neuron apoptosis in hypoxia by downregulating ATF6 expression. *Front. Physiol.* **2021**, *12*, 729925. [CrossRef] [PubMed]
82. Guo, Y.-S.; Yuan, M.; Han, Y.; Shen, X.-Y.; Gao, Z.-K.; Bi, X. Therapeutic potential of cytokines in demyelinating lesions after stroke. *J. Mol. Neurosci.* **2021**, *71*, 2035–2052. [CrossRef] [PubMed]
83. Fahmi, R.M.; Elsaid, A.F. Infarction size, interleukin-6, and their interaction are predictors of short-term stroke outcome in young Egyptian adults. *J. Stroke Cerebrovasc. Dis.* **2016**, *25*, 2475–2481. [CrossRef]
84. Zhang, L.; Yang, L. Anti-inflammatory effects of vinpocetine in atherosclerosis and ischemic stroke: A review of the literature. *Molecules* **2014**, *20*, 335–347. [CrossRef]
85. Banks, W.A.; Kastin, A.J.; Broadwell, R.D. Passage of cytokines across the blood-brain barrier. *Neuroimmunomodulation* **1995**, *2*, 241–248. [CrossRef]
86. Narasimhalu, K.; Lee, J.; Leong, Y.L.; Ma, L.; De Silva, D.A.; Wong, M.C.; Chang, H.M.; Chen, C. Inflammatory markers and their association with post stroke cognitive decline. *Int. J. Stroke* **2015**, *10*, 513–518. [CrossRef]
87. Grigolashvili, M.A.; Mustafina, R.M. The role of the inflammatory process in the development of post-stroke cognitive impairment. *Zhurnal Nevrol. Psikhiatrii Im. SS Korsakova* **2021**, *12*, 16–21. [CrossRef]

88. Zhang, Y.; Song, H.; Wang, J.; Xi, X.; Cefalo, P.; Wood, L.J.; Luo, X.; Wang, Q.M. Multiplex array analysis of serum cytokines offers minimal predictive value for cognitive function in the subacute phase after stroke. *Front. Neurol.* **2022**, *13*, 886018. [CrossRef]
89. Loga-Andrijić, N.; Petrović, N.T.; Filipović-Danić, S.; Marjanović, S.; Mitrović, V.; Loga-Zec, S. Analysis of serum inflammatory markers in Cognitive Impairment among Acute Ischaemic Stroke patients. *Psychiatr. Danub.* **2021**, *33*, 1287–1293.
90. Hall, C.; Nguyen, D.T.; Mendoza, K.; Tan, C.; Chauhan, A. Inhibition of IL-6 trans-signaling promotes post-stroke functional recovery in a sex and dose-dependent manner. *J Neuroinflamm.* **2025**, *22*, 52. [CrossRef]
91. Shaafi, S.; Sharifipour, E.; Rahmanifar, R.; Hejazi, S.S.; Andalib, S.; Nikanfar, M.; Baradarn, B.; Mehdizadeh, R. Interleukin-6, a reliable prognostic factor for ischemic stroke. *Iran. J. Neurol.* **2014**, *13*, 70–76. [PubMed]
92. Waje-Andreassen, U.; Kråkenes, J.; Ulvestad, E.; Thomassen, L.; Myhr, K.-M.; Aarseth, J.; Vedeler, C.A. IL-6: An early marker for outcome in acute ischemic stroke. *Acta Neurol. Scand.* **2005**, *111*, 360–365. [CrossRef] [PubMed]
93. Tuttolomondo, A.; Sciacca, R.D.; Raimondo, D.D.; Serio, A.; D'Aguanno, G.; La Placa, S.; Pecoraro, R.; Arnao, V.; Marino, L.; Monaco, S.; et al. Plasma levels of inflammatory and thrombotic/fibrinolytic markers in acute ischemic strokes. Relationship with TOAST subtype, outcome and infarct site. *J. Neuroimmunol.* **2009**, *215*, 84–89. [CrossRef] [PubMed]
94. Licata, A.; Tuttolomondo, S.; Corrao, D.; Di Raimondo, D.; Fernandez, P.; Caruso, C.; Avellone, G.; Pinto, A. Immunoinflammatory activation during the acute phase of lacunar and non-lacunar ischemic stroke: Association with time of onset and diabetic state. *Int. J. Immunopathol. Pharmacol.* **2006**, *19*, 639–646. [CrossRef]
95. Choudhary, S.; Chowdhur, D.; Mishra, T.K.; Agarwal, S. Temporal profile of serum levels of IL-6 in acute ischemic stroke and its relationship with stroke severity and outcome in Indian population. *Int. J. Integr. Med. Sci.* **2018**, *5*, 555–560. [CrossRef]
96. Park, S.Y.; Kim, J.; Kim, O.J.; Kim, J.-K.; Song, J.; Shin, D.-A.; Oh, S.-H. Predictive value of circulating interleukin-6 and heart-type fatty acid binding protein for three months clinical outcome in acute cerebral infarction: Multiple blood markers profiling study. *Crit. Care.* **2013**, *17*, R45. [CrossRef]
97. Whiteley, W.; Chong, W.L.; Sengupta, A.; Sandercock, P. Blood markers for the prognosis of ischemic stroke: A systematic review. *Stroke* **2009**, *40*, 380–389. [CrossRef]
98. Cojocaru, I.M.; Cojocaru, M.; Tănăsescu, R.; Iliescu, I.; Dumitrescu, L.; Silosi, I. Expression of IL-6 activity in patients with acute ischemic stroke. *Rom. J. Intern. Med.* **2009**, *47*, 393–396.
99. Worthmann, H.; Tryc, A.B.; Goldbecker, A. The temporal profile of inflammatory markers and mediators in blood after acute ischemic stroke differs depending on stroke outcome. *Cerebrovasc. Dis.* **2010**, *30*, 85–92. [CrossRef]
100. Nayak, A.R.; Kashyap, S.R.; Kabra, D.; Hemant, J.; Taori, G.M.; Daginawala, H.F. Time course of inflammatory cytokines in acute ischemic stroke patients and their relation to inter-alfa trypsin inhibitor heavy chain 4 and outcome. *Ann. Indian Acad. Neurol.* **2012**, *15*, 181–185. [CrossRef]
101. Puz, P.; Lasek-Bal, A.; Kazibutowska, Z. Evaluation of the serum concentration of selected inflammatory cytokines in patients with carotid artery stenosis. *Chir. Pol.* **2014**, *16*, 57–65.
102. Wytrykowska, A.; Prosba-Mackiewicz, M.; Nyka, W.M. IL-1β, TNF-α, and IL-6 levels in gingival fluid and serum of patients with ischemic stroke. *J. Oral Sci.* **2016**, *58*, 509–513. [CrossRef] [PubMed]
103. Yi, L.; Li, Z.-X.; Jiang, Y.-Y.; Jiang, Y.; Meng, X.; Li, H.; Zhao, X.-Q.; Wang, Y.-L.; Liu, L.-P.; Wang, Y.-J.; et al. Inflammatory marker profiles and in-hospital neurological deterioration in patients with acute minor ischemic stroke. *CNS Neurosci. Ther.* **2024**, *30*, e14648. [CrossRef] [PubMed]
104. Kes, V.B.; Simundic, A.-M.; Nikolac, N.; Topic, E.; Demarin, V. Pro-inflammatory and anti-inflammatory cytokines in acute ischemic stroke and their relation to early neurological deficit and stroke outcome. *Clin. Biochem.* **2008**, *41*, 1330–1334. [CrossRef]
105. Akyol, A.; Ozkul, A.; Yenisey, C.; Kiylioglu, N. The relationship between protein C, protein S and cytokines in acute ischemic stroke. *Neuroimmunomodulation* **2006**, *13*, 187–193. [CrossRef]
106. Coveney, S.; Murph, S.; Belton, O.; Cassidy, T.; Crowe, M.; Dolan, E.; de Gaetano, M.; Harbison, J.; Horgan, G.; Marnane, M.; et al. Inflammatory cytokines, high-sensitivity C-reactive protein, and risk of one-year vascular events, death, and poor functional outcome after stroke and transient ischemic attack. *Int. J. Stroke* **2021**, *17*, 163–171. [CrossRef]
107. Alfieri, D.F.; Lehmann, M.F.; Flauzino, T.; Martins de Araújo, M.C.; Pivoto, N.; Tirolla, R.M.; Colado Simão, A.; Maes, M.; Vissoci Reiche, E.M. Immune-inflammatory, metabolic, oxidative, and nitrosative stress biomarkers predict Acute Ischemic Stroke and Short-Term Outcome. *Neurotox. Res.* **2020**, *38*, 330–343. [CrossRef]
108. Jiang, Y.; Fan, T. IL-6 and stroke recurrence in ischemic stroke. *Biomark. Med.* **2024**, *18*, 739–747. [CrossRef]
109. Ou, M.; Liu, S.; Ma, X.; Xing, X.; He, W.; Gao, H. IL-6 promoter polymorphism increased risks of recurrent stroke in the young patients with moderate internal carotid artery stenosis. *J. Cell. Biochem.* **2018**, *119*, 2886–2890. [CrossRef]
110. McCabe, J.J.; Walsh, C.; Gorey, S.; Harris, K.; Hervella, P.; Iglesias-Rey, R.; Jern, C.; Li, L.; Miyamoto, N.; Montaner, J.; et al. C-Reactive protein, interleukin-6, and vascular recurrence according to stroke subtype: An individual participant data meta-analysis. *Neurology* **2024**, *102*, e208016. [CrossRef]

111. McCabe, J.J.; Walsh, C.; Gorey, S.; Arnold, M.; DeMarchis, G.M.; Harris, K.; Hervella, P.; Iglesias-Rey, R.; Jern, C.; Katan, M.; et al. Interleukin-6, C-Reactive protein, and recurrence after stroke: A time-course analysis of individual-participant data. *Stroke* **2024**, *55*, 2825–2834. [CrossRef] [PubMed]
112. Esenwa, C.C.; Elkind, M.S. Inflammatory risk factors, biomarkers and associated therapy in ischaemic stroke. *Nat. Rev. Neurol.* **2016**, *12*, 594–604. [CrossRef]
113. Jia, Y.; Zhang, K.; Shi, M.; Guo, D.; Yang, P.; Bu, X.; Chen, J.; Wang, A.; Xu, T.; He, J.; et al. Associations of rheumatoid factor, rheumatoid arthritis, and interleukin-6 inhibitor with prognosis of ischemic stroke: A prospective multicenter cohort study and Mendelian randomization analysis. *Transl. Stroke Res.* **2024**, *15*, 750–760. [CrossRef]
114. Gu, H.-Q.; Yang, K.-X.; Li, J.-J.; Lin, J.-X.; Jing, J.; Xiong, Y.-Y.; Zhao, X.-Q.; Wang, Y.-L.; Liu, L.-P.; Meng, X.; et al. Mediation effect of stroke recurrence in the association between post-stroke interleukin-6 and functional disability. *CNS Neurosci. Ther.* **2023**, *29*, 3579–3587. [CrossRef]
115. Staszewski, J.; Skrobowska, E.; Piusińska-Macoch, R.; Brodacki, B.; Stępień, A. IL-1α and IL-6 predict vascular events or death in patients with cerebral small vessel disease—Data from the SHEF-CSVD study. *Adv. Med. Sci.* **2019**, *64*, 258–266. [CrossRef]
116. Lu, H.; Li, S.; Zhong, X.; Huang, S.; Jiao, X.; He, G.; Jiang, B.; Liu, Y.; Gao, Z.; Wei, J.; et al. Immediate outcome prognostic value of plasma factors in patients with acute ischemic stroke after intravenous thrombolytic treatment. *BMC Neurol.* **2022**, *22*, 359. [CrossRef]
117. Yuan, S.; Lin, A.; He, Q.-Q.; Burgess, S.; Larsson, S.C. Circulating interleukins in relation to coronary artery disease, atrial fibrillation and ischemic stroke and its subtypes: A two-sample Mendelian randomization study. *Int. J. Cardiol.* **2020**, *313*, 99–104. [CrossRef]
118. Boehme, A.K.; McClure, L.A.; Zhang, Y.; Luna, J.M.; Del Brutto, O.H.; Benavente, O.R.; Elkind, M.S.V. Inflammatory markers and outcomes after lacunar stroke: Levels of inflammatory markers in treatment of stroke study. *Stroke* **2016**, *47*, 659–667. [CrossRef]
119. Mukaz, D.K.; Zakai, N.A.; Cruz-Flores, S.; McCullough, L.D.; Cushman, M. Identifying genetic and biological determinants of race-ethnic disparities in stroke in the United States. *Stroke* **2020**, *51*, 3417–3424. [CrossRef]
120. Tian, Y.; Jing, J.; Wang, H.; Wang, A.; Zhang, Y.; Jiang, Y.; Lin, J.; Zhao, X.; Li, H.; Wang, Y.; et al. Association of polyvascular disease and elevated interleukin-6 with outcomes in Acute Ischemic Stroke or Transient Ischemic attack. *Front. Neurol.* **2021**, *12*, 661779. [CrossRef]
121. Mechtouff, L.; Bochaton, T.; Paccalet, A.; Da Silva, C.C.; Buisson, M.; Amaz, C.; Derex, L.; Ong, E.; Berthezene, Y.; Eker, O.F.; et al. Association of interleukin-6 levels and futile reperfusion after mechanical thrombectomy. *Neurology* **2021**, *96*, e752–e757. [CrossRef] [PubMed]
122. Mechtouff, L.; Bochaton, T.; Paccalet, A.; Da Silva, C.C.; Buisson, M.; Amaz, C.; Derex, L.; Ong, E.; Berthezene, Y.; Dufay, N.; et al. A lower admission level of interleukin-6 is associated with first-pass effect in ischemic stroke patients. *J. Neurointerv. Surg.* **2022**, *14*, 248–251. [CrossRef] [PubMed]
123. Emam, W.A.; Ibrahim, N.E.; El-Senosy, F.M.; Elsheikh, A.A.; Elsayed, A.E.; Attar, R.E.; Elhadad, S.M.; Alhabibi, A.M.; Mohammed, A.R. Prognostic utility of MicroRNA-221 and interleukin-6 in cerebral ischemic stroke. *Egypt J. Immunol.* **2023**, *30*, 11–20. [PubMed]
124. Purroy, F.; Farré-Rodriguez, J.; Mauri-Capdevila, G.; Vicente-Pascual, M.; Farré, J. Basal IL-6 and S100b levels are associated with infarct volume. *Acta Neurol. Scand.* **2021**, *144*, 517–523. [CrossRef]
125. Jia, W.-L.; Jiang, Y.-Y.; Jiang, Y.; Meng, X.; Li, H.; Zhao, X.-Q.; Wang, Y.-L.; Wang, Y.-J.; Gu, H.-Q.; Li, Z.-X. Associations between admission levels of multiple biomarkers and subsequent worse outcomes in acute ischemic stroke patients. *J. Cereb. Blood Flow. Metab.* **2024**, *44*, 742–756. [CrossRef]
126. Smith, C.J.; Emsley, H.C.A.; Gavin, C.M.; Georgiou, R.F.; Vail, A.; Barberan, E.M.; del Zoppo, G.J.; Hallenbeck, J.M.; Rothwell, N.J.; Hopkins, S.J.; et al. Peak plasma interleukin-6 and other peripheral markers of inflammation in the first week of ischaemic stroke correlate with brain infarct volume, stroke severity and long-term outcome. *BMC Neurol.* **2004**, *4*, 2. [CrossRef]
127. Shenhar-Tsarfaty, S.; Ben, E.; Assayag, I.; Bova, L.; Shopin, M.; Fried, S.; Berliner, I.; Shapira, N.M. Bornstein Interleukin-6 as an early predictor for one-year survival following an ischaemic stroke/transient ischaemic attack. *Int. J. Stroke* **2010**, *5*, 16–20. [CrossRef]
128. Meisinger, C.; Freuer, D.; Schmitz, T.; Ertl, M.; Zickler, P.; Naumann, M.; Linseisen, J. Inflammation biomarkers in acute ischemic stroke according to different etiologies. *Eur. J. Neurol.* **2024**, *31*, e16006. [CrossRef]
129. Orion, D.; Schwammenthal, Y.; Reshef, T.; Schwartz, R.; Tsabari, R.; Merzeliak, O.; Chapman, J.; Mekori, Y.A.; Tanne, D. Interleukin-6 and soluble intercellular adhesion molecule-1 in acute brain ischaemia. *Eur. J. Neurol.* **2008**, *15*, 323–328. [CrossRef]
130. Bustamante, A.; Sobrino, T.; Giralt, D.; García-Berrocoso, T.; Llombart, V.; Ugarriza, I.; Espadaler, M.; Rodríguez, N.; Sudlow, C.; Castellanos, M.; et al. Prognostic value of blood interleukin-6 in the prediction of functional outcome after stroke: A systematic review and meta-analysis. *J. Neuroimmunol.* **2014**, *274*, 215–224. [CrossRef]
131. Martínez-Sánchez, P.; Gutiérrez-Fernández, M.; Fuentes, B.; Masjuán, J.; de Leciñana Cases, M.A.; Novillo-López, M.E.; Díez-Tejedor, E. Biochemical and inflammatory biomarkers in ischemic stroke: Translational study between humans and two experimental rat models. *J. Transl. Med.* **2014**, *3*, 220–229. [CrossRef]

132. Hotter, B.; Hoffmann, S.; Ulm, L.; Meisel, C.; Fiebach, J.B.; Meisel, A. IL-6 Plasma levels correlate with cerebral perfusion deficits and infarct sizes in stroke patients without associated infections. *Front. Neurol.* **2019**, *10*, 83. [CrossRef]
133. Rodríguez-Yáñez, M.; Castillo, J. Role of inflammatory markers in brain ischemia. *Curr. Opin. Neurol.* **2008**, *21*, 353–357. [CrossRef]
134. Beridze, M.; Sanikidze, T.; Shakarishvilil, R.; Intskirveli, N.; Bornstein, N.M. Selected acute phase CSF factors in ischemic stroke: Findings and prognostic value. *BMC Neurol.* **2011**, *11*, 41–48. [CrossRef]
135. Băcilă, C.I.; Vlădoiu, M.G.; Văleanu, M.; Mogą, D.F.C.; Pumnea, P.M. The role of IL-6 and TNF-alpha biomarkers in predicting disability outcomes in Acute Ischemic Stroke patients. *Life* **2025**, *15*, 47. [CrossRef]
136. Hervella, P.; Rodríguez-Castro, E.; Rodríguez-Yáñez, M.; Arias, S.; Santamaría-Cadavid, M.; López-Dequidt, I.; Estany-Gestal, A.; Maqueda, E.; López-Loureiro, I.; Sobrino, T.; et al. Intra- and extra-hospital improvement in ischemic stroke patients: Influence of reperfusion therapy and molecular mechanisms. *Sci. Rep.* **2020**, *10*, 3513. [CrossRef]
137. Dargazanli, C.; Blaquière, M.; Moynier, M.; de Bock, F.; Labreuche, J.; Schiphorst, A.T.; Derraz, I.; Radu, R.A.; Gascou, G.; Lefevre, P.H.; et al. Inflammation biomarkers in the intracranial blood are associated with outcome in patients with ischemic stroke. *J. Neurointerv. Surg.* **2024**, *17*, 159–166. [CrossRef]
138. Kwan, J.; Horsfield, G.; Timothy Bryant, T.; Gawne-Cain, M.; Durward, G.; Byrne, C.D.; Englyst, N.A. IL-6 is a predictive biomarker for stroke associated infection and future mortality in the elderly after an ischemic stroke. *Exp. Gerontol.* **2013**, *48*, 960–965. [CrossRef]
139. Dusanovic Pjevic, M.; Vojvodic, L.; Grk, M.; Todorovic, J.; Maksimovic, N.; Rasic, M.; Perovic, D.; Damnjanovic, T.; Trickovic, J.; Kacar, K.; et al. Association of IL-6 rs1800795, but not TNF-α rs1800629, and IL-1β rs16944 polymorphisms' genotypes with recovery of ischemic stroke patients following thrombolysis. *Neurol. Res.* **2024**, *46*, 157–164. [CrossRef]
140. Huang, X.; Ye, Q.; Zhu, Z.; Chen, W.; Chen, Y.; Li, J.; Sun, J.; Ye, Z. Association between a functional interleukin 6 receptor genetic variant and the risk and functional outcome of large artery atherosclerotic stroke. *Int. J. Neurosci.* **2020**, *130*, 355–362. [CrossRef]
141. Rosário, M.; Fonseca, A.C. Update on biomarkers associated with Large-Artery Atherosclerosis stroke. *Biomolecules* **2023**, *13*, 1251. [CrossRef] [PubMed]
142. Klimiec-Moskal, E.; Piechota, M.; Pera, J.; Weglarczyk, K.; Slowik, A.; Siedlar, M.; Dziedzic, T. The specific ex vivo released cytokine profile is associated with ischemic stroke outcome and improves its prediction. *J Neuroinflamm.* **2020**, *17*, 7. [CrossRef] [PubMed]
143. Sotgiu, S.; Zanda, B.; Marchetti, B.; Fois, M.L.; Arru, G.; Pes, G.M.; Salaris, F.S.; Arru, A.; Pirisi, A.; Rosati, G. Inflammatory biomarkers in blood of patients with acute brain ischemia. *Eur. J. Neurol.* **2006**, *13*, 505–513. [CrossRef]
144. Chen, Z.; He, Y.; Su, Y.; Sun, Y.; Zhang, Y.; Chen, H. Association of inflammatory and platelet volume markers with clinical outcome in patients with anterior circulation ischaemic stroke after endovascular thrombectomy. *Neurol. Res.* **2021**, *43*, 503–510. [CrossRef]
145. Bahorik, A.L.; Hoang, T.D.; Jacobs, D.R.; Levine, D.A.; Yaffe, K. Association of changes in C-reactive protein level trajectories through early adulthood with cognitive function at midlife the CARDIA study. *Neurology* **2024**, *103*, e209526. [CrossRef]
146. Kelly, P.J.; Murphy, S.; Coveney, S.; Purroy, F.; Lemmens, R.; Tsivgoulis, G.; Price, C. Anti-inflammatory approaches to ischaemic stroke prevention. *J. Neurol. Neurosurg. Psychiatry* **2018**, *89*, 211–218. [CrossRef]
147. Ridker, P.M.; Everett, B.M.; Thuren, T.; MacFadyen, J.G.; Chang, W.H.; Ballantyne, C.; Fonseca, F.; Nicolau, J.; Koenig, W.; Anker, S.D.; et al. Antiinflammatory therapy with canakinumab for atherosclerotic disease. *N. Engl. J. Med.* **2017**, *377*, 1119–1131. [CrossRef]
148. Hudobenko, J. Ischemic Stroke Damage is Reduced by Inhibition of IL-6 Signaling with Tocilizumab. Master's Thesis, The University of Texas MD Anderson Cancer Center, Houston, TX, USA, 2018.
149. Xing, Z.; Gauldie, J.; Cox, G.; Baumann, H.; Jordana, M.; Lei, X.F.; Achong, M.K. IL-6 is an antiinflammatory cytokine required for controlling local or systemic acute inflammatory responses. *J. Clin. Investig.* **1998**, *101*, 311–320. [CrossRef]
150. Yamashita, T.; Sawamoto, K.; Suzuki, S.; Suzuki, N.; Adachi, K.; Kawase, T.; Mihara, M.; Ohsugi, Y.; Abe, K.; Okano, H. Blockade of interleukin-6 signaling aggravates ischemic cerebral damage in mice: Possible involvement of Stat3 activation in the protection of neurons. *J. Neurochem.* **2005**, *94*, 459–468. [CrossRef]
151. Ziegler, L.; Frumento, P.; Wallen, H.; de Faire, U.; Gigante, B. The predictive role of interleukin 6 trans-signalling in middle-aged men and women at low-intermediate risk of cardiovascular events. *Eur. J. Prev. Cardiol.* **2020**, *27*, 122–129. [CrossRef]
152. Taei, A.A.; Dargahi, L.; Khodabakhsh, P.; Kadivar, M.; Farahmandfar, M. Hippocampal neuroprotection mediated by secretome of human mesenchymal stem cells against experimental stroke. *CNS Neurosci. Ther.* **2022**, *28*, 1425–1438. [CrossRef]
153. Sakata, H.; Narasimhan, P.; Niizuma, K.; Maier, C.M.; Wakai, T.; Chan, P.H. Interleukin 6-preconditioned neural stem cells reduce ischaemic injury in stroke mice. *Brain* **2012**, *135*, 3298–3310. [CrossRef] [PubMed]
154. Sun, Z.; Gu, L.; Wu, K.; Wang, K.; Ru, J.; Yang, S.; Wang, Z.; Zhuge, Q.; Huang, L.; Huang, S. VX-765 enhances autophagy of human umbilical cord mesenchymal stem cells against stroke-induced apoptosis and inflammatory responses via AMPK/mTOR signaling pathway. *CNS Neurosci. Ther.* **2020**, *26*, 952–961. [CrossRef] [PubMed]

155. Neal, E.G.; Acosta, S.A.; Kaneko, Y.; Ji, X.; Borlongan, C.V. Regulatory T-cells within bone marrow-derived stem cells actively confer immunomodulatory and neuroprotective effects against stroke. *J. Cereb. Blood Flow Metab.* **2019**, *39*, 1750–1758. [CrossRef]
156. Zarriello, S.; Neal, E.G.; Kaneko, Y.; Borlongan, C.V. T-regulatory cells confer increased myelination and stem cell activity after stroke-induced white matter injury. *J. Clin. Med.* **2019**, *8*, 537. [CrossRef]
157. Huang, L.; Wong, S.; Snyder, E.Y.; Hamblin, M.H.; Lee, J.-P. Human neural stem cells rapidly ameliorate symptomatic inflammation in early-stage ischemic-reperfusion cerebral injury. *Stem Cell Res. Ther.* **2014**, *5*, 129. [CrossRef]
158. Chi, L.; Huang, Y.; Mao, Y.; Wu, K.; Zhang, L.; Nan, G. Tail vein infusion of adipose-derived mesenchymal stem cell alleviated inflammatory response and improved Blood Brain Barrier condition by suppressing endoplasmic reticulum stress in a Middle Cerebral Artery Occlusion rat model. *Med. Sci. Monit.* **2018**, *24*, 3946–3957. [CrossRef]
159. Kaneko, Y.; Lee, J.Y.; Tajiri, N.; Tuazon, J.P.; Lippert, T.; Russo, E.; Yu, S.J.; Bonsack, B.; Corey, S.; Coats, A.B.; et al. Translating intracarotid artery transplantation of bone marrow-derived NCS-01 cells for ischemic stroke: Behavioral and histological readouts and mechanistic insights into stem cell therapy. *Stem Cells Transl. Med.* **2020**, *9*, 203–220. [CrossRef]

biomolecules

MDPI

Review

Human Blood-Derived lncRNAs in Autism Spectrum Disorder

Carmela Serpe, Paola De Sanctis, Marina Marini *, Silvia Canaider, Provvidenza Maria Abruzzo * and Cinzia Zucchini

Department of Medical and Surgical Sciences, University of Bologna, Via Massarenti 9, 40138 Bologna, Italy; carmela.serpe@unibo.it (C.S.); paola.desanctis@unibo.it (P.D.S.); silvia.canaider@unibo.it (S.C.); cinzia.zucchini@unibo.it (C.Z.)

* Correspondence: marina.marini@unibo.it (M.M.); provvidenza.abruzzo2@unibo.it (P.M.A.)

Abstract: Autism spectrum disorder (ASD) is a complex and heterogeneous neurodevelopmental disorder with a significant impact on public health. ASD diagnosis is based on clinical observation and typically occurs around three years of age. The identification of reliable ASD markers could facilitate early diagnosis and help pinpoint therapeutic targets for effective interventions. Long non-coding RNAs (lncRNAs), particularly those derived from blood, have been recently proposed as potential biomarkers in many pathological conditions, including neurological diseases. This manuscript summarizes original studies examining human dysregulated blood-derived lncRNAs as potential ASD biomarkers. LncRNAs are described by grouping them according to the selection strategy used by the authors: (i) lncRNAs involved in biological processes impaired in ASD or in pathological conditions sharing the disrupted signaling pathways of ASD; and (ii) lncRNAs identified through high-throughput analysis. The study highlights key priorities for future research: assessing the ability of lncRNAs to distinguish ASD from other neurological disorders, extending analyses to larger and younger cohorts to validate candidate biomarkers in early life, and integrating multiple data sources to establish validated biomarker networks for clinical application. This review indicates that research on blood-derived lncRNAs in ASD is still in its early stages.

Keywords: long non-coding RNA; lncRNA; autism spectrum disorder; ASD; biomarkers; non-coding RNA; peripheral blood

1. Introduction

Autism spectrum disorder (ASD) is a complex and heterogeneous neurodevelopmental disorder with a significant impact on public health [1]. It is characterized by deficits in social interaction and communication, along with repetitive and restrictive behaviors [2]. In addition to these core symptoms, ASD is frequently associated with various comorbidities, including intellectual disability, depressive disorders, sleep disturbances, and gastrointestinal and immunological abnormalities [3]. Moreover, ASD subjects often experience additional health concerns, such as epilepsy, electroencephalogram abnormalities, dysmorphic features, and atypical magnetic resonance imaging findings [4–6]. The prevalence of ASD has markedly increased worldwide in the last few decades. According to the CDC's Autism and the Developmental Disabilities Monitoring (ADDM) Network, 1 in 36 children aged 8 years in the United States is currently diagnosed with ASD, with a male-to-female ratio of approximately 4:1 [7]. The increase in ASD prevalence over time results from a complex interplay of factors, including improvements in diagnostic criteria and greater awareness of the disorder [8]. ASD is diagnosed by expert clinicians around

 https://doi.org/10.3390/biom15070937

the age of 3 years [9], using standardized diagnostic tools such as the Autism Diagnostic Observation Schedule (ADOS), the Childhood Autism Rating Scale (CARS), and the Autism Diagnostic Interview—Revised (ADI-R) to assess children's behavior and symptoms [10,11]. However, due to the highly heterogeneous nature of ASD, this approach often delays ASD diagnosis [12]. Although numerous biological, physiological, and medical alterations have been identified in ASD, none of these are used as a diagnostic marker to aid early diagnosis, treatment, and management [13,14]. The lack of reliable biomarkers is primarily due to the limited understanding of ASD etiology. To date, ASD is considered a multifactorial disorder caused by a complex interplay of genetic and environmental factors [15]. Several studies, including twin studies, have shown that genetic factors greatly contribute to ASD susceptibility [16]. Common and rare genetic alterations, including de novo and rare copy number variations (CNVs) and single nucleotide variations (SNVs), have been found in several genes associated with chromatin remodeling, neural development, synaptic function, and neuronal communication, significantly contributing to ASD risk [17,18]. However, DNA alterations account for only 35–40% of ASD cases, suggesting that other factors are required for ASD onset. Environmental factors, including prenatal and perinatal causes, and microbial–gut–brain axis alterations are implicated in ASD etiology [19]. These factors may directly affect the transcription of specific susceptibility genes through epigenetic modifications, such as DNA methylation, chromatin remodeling, and alterations in the expression of non-coding RNA molecules, like microRNAs (miRNAs) and long non-coding RNAs (lncRNAs).

The recent advent of next-generation sequencing (NGS) has revealed that 85% of the human genome is actively transcribed, with only 2–3% of these transcripts coding for proteins [20,21]. The huge amount of genome containing information for non-coding RNA (ncRNA) transcripts indicates that these molecules might have wide and diversified roles in biological processes. Based on their size, ncRNAs are classified into two main groups: short ncRNAs, which are less than 200 nucleotides long, and long non-coding RNAs (lncRNAs), which are longer than 200 nucleotides. Short ncRNAs include miRNAs, piwi-interacting RNAs (piRNAs), circular RNAs (circRNAs), and small nucleolar RNAs (snoRNAs). LncRNAs are the most abundant class of ncRNAs in the human genome [20]. They are mainly transcribed by RNA polymerase II (PolII) and share characteristics with coding RNA genes, such as 7-methylguanosine cap, polyA tail, and alternative splicing [21,22]. Most lncRNAs overlap with protein-coding loci and can be divided into sense, antisense, and intronic lncRNAs. Additionally, some lncRNAs are located in the intergenic regions of the genome (lincRNAs) [21,22]. LncRNAs exert their functions in both the nucleus and cytoplasm, where they regulate gene expression at multiple levels by interacting with RNA, DNA, and proteins [23,24]. For instance, lncRNAs function as epigenetic regulators by acting as scaffolds for histone-modifying complexes and DNA methyltransferases or demethylases, thus modulating chromatin architecture [22,25,26]. LncRNAs also regulate transcriptional and post-transcriptional processes by interacting directly with transcription factors and proteins involved in mRNA maturation and alternative splicing [26–29]. In addition, lncRNAs can interact with miRNAs by acting as miRNA sponges or competing endogenous RNAs (ceRNAs) [26,30,31]. Finally, lncRNAs are also implicated in translational and post-translational gene expression control [26,29,32,33]. This scenario highlights the complex role of these molecules in the regulation of several biological pathways. LncRNAs are implicated in a wide variety of physiological processes, including proliferation and differentiation, metabolism, signaling pathways, and apoptosis [34]. The abnormal expression of lncRNAs, observed in many diseases [35–39], has attracted the attention of researchers seeking novel biomarkers. LncRNAs, and in particular, blood-derived lncRNAs, have been recently proposed as potential biomarkers in many pathological conditions, including cancer, neurological diseases, and ASD [40–45]. Their high specificity, stability, and relative abundance as well as cost-effectiveness and accessibility make them attractive diagnostic and prognostic tools for clinical applications [40].

This is a narrative review summarizing original studies that have examined dysregulated blood-derived lncRNAs in ASD as possible biomarkers. For this purpose, a search was performed on PubMed up to 1 September 2024 using the search terms (lncRNA), (long non-coding RNA) in combination with the terms (autism), (autism spectrum disorder). Among the retrieved original articles, only those that measured lncRNA expression levels in peripheral blood, serum, or plasma of ASD subjects, compared to controls, were considered eligible. This screening, carried out by two independent reviewers through reading the title, abstract, and materials and methods sections, led to the selection of 18 articles. Additional information was retrieved from the analysis of the reference list of the selected articles (Figure 1). Finally, the relevance of the selected studies was assessed by reading the full text. The following sections summarize dysregulated lncRNAs in ASD identified in one or more studies, along with details on their diagnostic power.

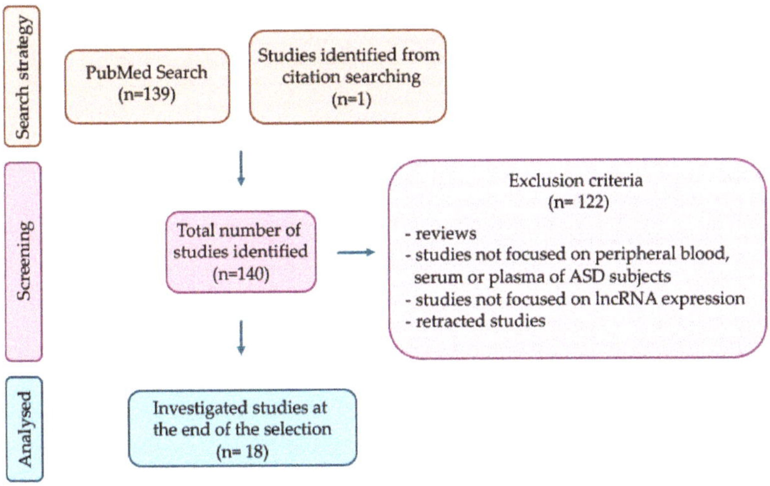

Figure 1. Study selection process. The flow diagram shows the number of investigated studies identified (n), according to the declared criteria.

2. Blood-Derived lncRNAs Dysregulated in Autism Spectrum Disorder (ASD)

The literature review revealed that authors employed two main approaches to identify dysregulated lncRNAs. A common strategy is based on the selective choice of lncRNAs according to their involvement in biological processes impaired in ASD or in pathological conditions that share the same disrupted signaling pathways with ASD. As an alternative approach, other authors have employed high-throughput analysis, which offers the advantage of detecting many dysregulated lncRNAs simultaneously, irrespective of their putative role in the disease. Therefore, in the following sections, blood-derived lncRNAs are described by grouping them according to the above-mentioned selection strategies.

2.1. LncRNAs Selected According to ASD-Related Pathways

Alterations in neurological processes during neurogenesis, including synaptic function, neuronal activity, neuronal cell adhesion, neurite growth, and synaptogenesis, are well-established contributors to ASD pathogenesis. Genes involved in these processes are among the risk loci associated with ASD [46]. Mounting evidence shows that lncRNAs are mainly expressed in the nervous system; therefore, it is not surprising that many lncRNAs have received attention as candidate biomarkers [47,48]. It is known that the Disrupted in schizophrenia 2 (DISC2) lncRNA, located on human chromosome 1q42 and transcribed antisense to exon 9 of the gene *DISC1* [49,50], plays a role in neuronal differentiation [51]. A deletion of chromosome 1q42, which includes *DISC2*, has been described in a 3-year-old male child affected by ASD [52]. However, the involvement of this lncRNA in ASD pathogenesis remains unclear, since contradictory results are reported. Tamizkar and colleagues [53] found an increased expression of DISC2 in the peripheral blood of ASD children compared to controls, while, more recently, Raahmani and colleagues [54] observed no significant difference in the expression levels of DISC2, suggesting that further investigation is needed. In addition to DISC2, Tamizkar and colleagues [53] observed alterations in the expression of three other lncRNAs: PRKAR2A antisense RNA 1 (PRKAR2A-AS1), LINC03091 long intergenic non-protein coding RNA 3091 (LINC03091 or LOC101928237), and LRRC2 antisense RNA 1 (LRRC2-AS1). In particular, the authors found an increase in the expression of PRKAR2A-AS1 and LOC101928237 and a decrease in LRRC2-AS1 levels in ASD children. Although the specific function of these lncRNAs in ASD is still unknown, the authors suggest that they may contribute to ASD pathogenesis and could be used as potential markers. Indeed, LOC101928237 and LRRC2-AS1 show a strong ability in distinguishing ASD from control children (AUC: 0.90 and 0.929, respectively) (Table 1). The authors also highlighted the need to validate these results in a larger cohort of subjects and to investigate these lncRNA signatures during the early stages of ASD pathogenesis [53].

The Small nucleolar RNA host gene 6 (SNHG6) lncRNA is related to the vitamin D receptor (VDR) pathway [55], which affects neurodevelopment, axon connectivity, and dopamine formation [56]. Ghafouri-Fard and collaborators [57] found a downregulation in SNHG6 expression in the blood of ASD subjects compared to the control group. A possible molecular mechanism comes from a previous study showing that SNHG6 can suppress miR-181c [58], which was found overexpressed in the amygdala of an ASD rat model [59]. Therefore, the authors highlighted the role of the SNHG6/miR-181c axis as a potential functional mechanism in the development of ASD. Moreover, based on Receiver Operating Characteristic Curve (ROC) analysis, which revealed high performance in discriminating ASD cases from controls, the authors proposed SNHG6 as a potential biomarker (AUC: 0.94) (Table 1).

Recently, Sane and colleagues [60] evaluated in blood samples of ASD children and healthy subjects the expression levels of five lncRNAs involved in the modulation of neurological response and in the differentiation of neuronal cells: the Long intergenic non-protein coding RNA, regulator of reprogramming (LincRNA-ROR), Long intergenic non-protein coding, p53-induced transcript (LINC-PINT), Long intergenic non-coding RNA p21 (LincRNA-p21), Prostate cancer-associated transcript 29 (PCAT-29), and Prostate cancer-associated transcript-1 (PCAT-1). The study revealed that all the examined lncRNAs were significantly downregulated in individuals with ASD compared to the healthy controls. ROC analysis indicated that LincRNA-ROR and PCAT-1 had better diagnostic power than PCAT-29, LINC-PINT, and LincRNA-p21 in distinguishing ASD from healthy subjects (Table 1) [60].

The Maternally expressed gene 3 (MEG3) lncRNA, encoded by an imprinted gene located on human chromosome 14q32.3, regulates AMPA receptor surface expression and is involved in neuronal synaptic plasticity [61]. Taheri and colleagues [62] found a significant upregulation of MEG3 lncRNA in the blood of ASD children compared to controls and suggested MEG3 as a putative biomarker, according to ROC analysis (AUC 0.792) (Table 1). Recent findings indicated that MEG3 is also involved in the apoptotic pathway in the hippocampal tissues of valproic acid (VPA)-treated rats; in this model, MEG3 is overexpressed and promotes cadherin2 (CDH2) expression via E1A binding protein p300 transcription factor (EP300), repressing neuronal viability [63] (Figure 2).

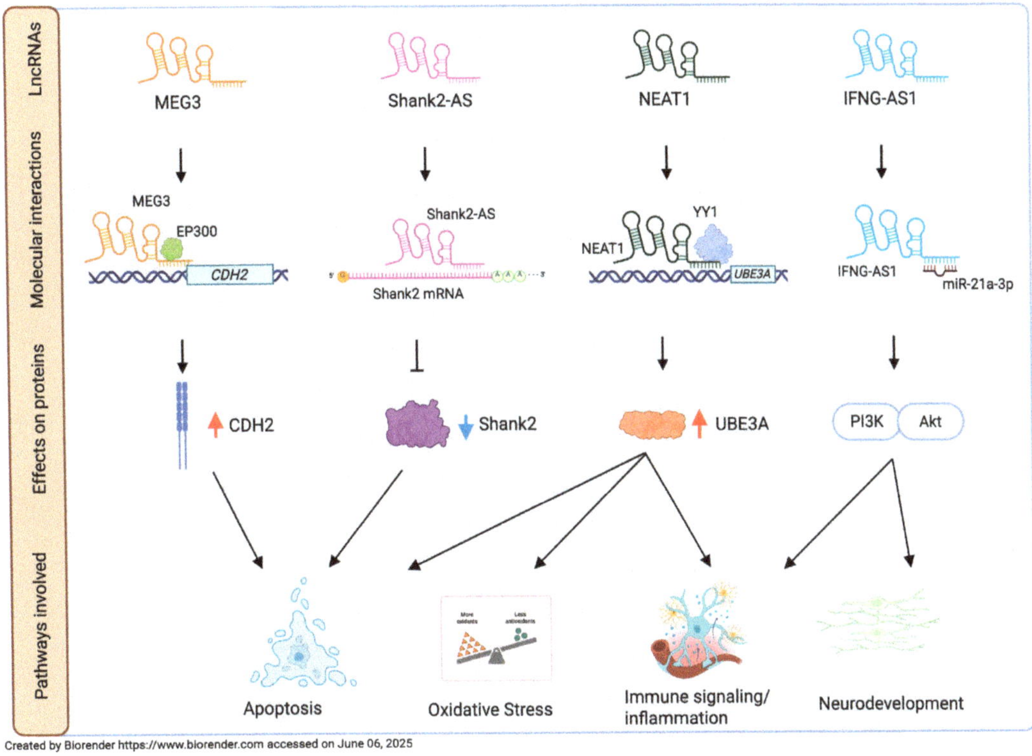

Figure 2. LncRNA-affected molecular mechanism in ASD pathogenesis. A schematic representation of molecular mechanisms observed for MEG3, NEAT1, Shank2-AS, and IFNG-AS1 in ASD cellular and animal models. CDH2, cadherin2; EP300, E1A binding protein p300 transcription factor; YY1, Yin-Yang 1 transcription factor; UBE3A, ubiquitin protein ligase E3A; PI3K/AKT, phosphoinositide 3-kinase/AKT serine/threonine kinase; ⊥ means inhibition; ↓ means promotion; red arrow means increased expression; blue arrow means decreased expression.

Apoptosis is one of the most explored biological pathways to discover new candidate markers for ASD. Apoptosis plays a crucial role in eliminating excess neurons and refining neural circuits during brain development. Its dysregulation affects neuronal density, number, organization, and composition, leading to impaired connectivity and synaptic dysfunctions [64]. Aberrant regulation of apoptotic pathways has been observed in both postmortem brain tissue and peripheral blood of subjects with ASD [65,66]. The antisense lncRNA of SH3 and multiple ankyrin repeat domains 2 (Shank2-AS) is reversely transcribed from the 10th intron of the *SHANK2* gene, a locus associated with ASD [67,68]. Its overexpression decreases neurite length and numbers, inhibits the proliferation of neuronal cells, and promotes their apoptosis by downregulating Shank2 protein expression by forming a double-stranded RNA [69] (Figure 2). Luo and colleagues [69] found an upregulation of Shank2-AS in lymphocytes from ASD peripheral blood. This result is consistent with the Shank2-AS overexpression previously observed by Wang and colleagues in a high-throughput study conducted on ASD leukocytes, which is described below [70]. These findings suggest that targeting lncRNAs located at ASD-causing loci may be a promising approach in the search for new putative biomarkers of the disease. Unfortunately, although the concordance of Shank2-AS dysregulation in two independent cohorts of subjects supports its potential role as a candidate biomarker, the authors did not provide data regarding its diagnostic utility.

The Nuclear paraspeckle assembly transcript 1 (NEAT1) lncRNA, located on chromosome 11q13.7, was originally investigated by Sayad and colleagues [71] for its regulatory role in apoptosis in colon cancer [72]. This apoptotic function was subsequently confirmed in many neurological diseases, including ASD [73,74]. In summary, NEAT1 promotes apoptosis, oxidative stress (OS) and inflammation in VPA-induced ASD rats by recruiting Yin-Yang 1 (YY1) transcription factor to regulate ubiquitin protein ligase E3A (UBE3A) expression [74] (Figure 2). A significant increase in NEAT1 levels was observed in the peripheral blood of ASD patients compared to controls. The authors proposed that NEAT1 might contribute to ASD pathogenesis via the miR-497/brain-derived neurotrophic factor (BDNF) pathway, similarly to what has been observed in diabetic retinopathy; this pathological condition is characterized by NEAT1 downregulation, which, in turn, decreases BDNF expression through the upregulation of miR-497 and leads to the death of Muller cells [75]. Conversely, increased NEAT1 is in line with the elevated peripheral BDNF levels in ASD subjects, thus supporting the involvement of miR-497/BDNF in ASD [71,76]. Although the precise mechanism is still unclear and warrants further investigation, NEAT1 has been proposed by Sayad and collaborators [71] as a potential diagnostic biomarker for ASD based on its AUC value (AUC: 0.759) (Table 1). The same research group also found a significant overexpression of Taurine upregulated gene 1 (TUG1) lncRNA in the peripheral blood of children with ASD compared to controls. Moreover, according to ROC analysis, TUG1 could acceptably distinguish ASD from healthy subjects (AUC: 0.733) (Table 1) [71]. Several authors have highlighted the apoptotic role of TUG1 exerted by its sponge activity in many neuronal dysfunctions [77–80]. Although Sayad and collaborators [71] did not explore the mechanism by which TUG1 would be implicated in the pathogenesis of ASD, they suggested that TUG1 might act as a sponge for miR9. This apparently contrasts with the finding that, in a knockdown cell model, the lncRNA TUG1 acts as a positive regulator of TBC/LysM-associated domain containing 1 (TLDC1) [81], a member of the TLDc family proteins that exerts a protective function against OS. Consistent with the findings reported by Sayad and colleagues [71], Zucchini and collaborators [82] found that TUG1 expression level was increased in the Peripheral Blood Mononuclear Cells (PBMCs) of ASD children, thus reinforcing the implication of TUG1 in this disorder (Table 1). Moreover, the concurrent increase in TLDC1 mRNA level suggests a role of TUG1 in protection against OS, a key factor in the ASD pathophysiology [83–85].

A neuroprotective function has also been proposed for the Plasmacytoma variant translocation 1 (PVT1) lncRNA [86]. Its expression is altered in some neurological disorders, such as epilepsy [87] and schizophrenia [88,89]. Jiang and colleagues [90] found that PVT1 was downregulated in the serum of children with ASD compared to healthy controls. ROC analysis indicated that PVT1 may be a useful biomarker for early ASD screening, either alone (AUC: 0.848, sensitivity: 85% and specificity: 79.3%) (Table 1) or in association with miR-21-5p, one of its putative targets [91]. Interestingly, combining both PVT1 with miR-21-5p yielded higher diagnostic performance than either marker alone (AUC: 0.954, sensitivity: 83.3%, and specificity: 96.6%) [90], reinforcing the idea that using functionally related markers can improve diagnostic accuracy.

Mitochondrial alterations not only increase the production of Reactive Oxygen Species (ROS) and Reactive Nitrogen Species (RNS) but also impair energy metabolism and calcium homeostasis regulation [92,93]. Several lines of evidence support the crucial role of mitochondrial dysfunction and calcium imbalance in ASD [94–97]. Pourtavakoli and collaborators [98] measured the expression of four calcium signaling lncRNAs, Long intergenic non-protein coding RNA 1231 (LINC01231), Long non-coding RNA solute carrier family 25 member 12 (lnc-SLC25A12), Long non-coding melatonin MT1 receptor (lnc-MTR-1), and Long intergenic non-protein coding RNA 606 (LINC00606), along with their related genes, *Solute Carrier Family 1 Member 1 (SLC1A1), Solute Carrier Family 25 member 12 (SLC25A12), Ryanodine Receptor 2* (*RYR2*), and *ATPase plasma membrane* Ca^{2+} *transporting 2* (*ATP2B2*), in the blood of ASD and healthy subjects. The authors found differences in the expression of two mRNA-lncRNA pairs: SLC1A1-LINC01231 and RYR2-lnc-MTR-1. The underexpression of *SLC1A1*, which encodes for a glutamate transporter, was accompanied by the upregulation of its cognate lncRNA, LINC01231, in ASD cases compared to controls. The authors also found an overexpression of the calcium ion release channel RYR2 in ASD children, while lnc-MTR-1 tended to be downregulated [98]. Notably, copy number variation in the *RYR2* gene has been reported in ASD [99]. Among the lncRNAs studied, LINC01231 showed the highest diagnostic potential (AUC: 0.75) for distinguishing ASD cases from controls [98] (Table 1).

Mitochondrial dysfunctions and OS can trigger the inflammatory response through the activation of intracellular signaling cascades such as nuclear factor kappa B (NF-κB) and phosphoinositide 3-kinase/AKT serine/threonine kinase (PI3K/AKT), which lead to the expression of numerous pro-inflammatory molecules, including interleukin-6 (IL-6), interleukin-1β (IL-1β), interferon-γ (IFN-γ), and tumor necrosis factor α (TNF-α) [100]. On the other hand, immune cells produce large amounts of ROS, highlighting that OS and inflammation sustain each other in a self-reinforcing vicious cycle [101,102]. In the brain, the activation of microglia—the resident immune cells of central nervous system—can lead to chronic neuroinflammation and alterations in neural signaling and cognitive function [103]. Mounting evidence indicates that inflammation plays a key role in ASD pathogenesis [104–113]. In light of these observations, several lncRNAs involved in immune–inflammatory pathways have been investigated as possible biomarkers. Xie and collaborators [114], focusing on the innate immune system abnormalities in ASD, observed elevated TNF-α levels in the blood of children with ASD. This prompted them to investigate the potential association of this cytokine with the expression of TNFα and hnRNPL Related immunoregulatory LincRNA (THRIL), an lncRNA involved in the negative regulation of TNF-α via heterogeneous nuclear ribonucleoprotein L (hnRNPL) [115]. Xie and colleagues [114] found that THRIL expression was significantly lower in children with ASD compared to control children. A previous study by Li and colleagues demonstrated that increased THRIL expression in macrophages could regulate TNF-α expression through epigenetic mechanisms, and conversely, that TNF-α could downregulate THRIL expression via a negative feedback loop [115]. Considering these findings, Xie and collaborators [114] proposed that THRIL might play a role in the dysregulation of the TNF-α signaling observed in ASD; however, the diagnostic performance of this lncRNA has not yet been tested.

The interferon γ–antisense RNA (IFNG-AS1) lncRNA, also known as NeST and TMEVPG1 [116,117], regulates the expression of IFN-γ [118], a cytokine involved in immune-related disorders [119]. IFNG-AS1 is located next to the *IFNG* gene on chromosome 12, where it acts as a regulatory enhancer of *IFNG* expression [120]. Fu and colleagues [121], using a BTBR-ASD mouse model, demonstrated that the administration of IFNG-AS1-containing exosomes, derived from mesenchymal stem cells isolated from adipose tissue, improved neurogenesis, suppressed brain inflammatory microenvironment, and ameliorated ASD-like behavior via the miR-21a-3p/PI3K(p110α)/AKT axis (Figure 2). Fallah and collaborators [122] found a decreased expression of IFNG-AS1 in the peripheral blood of ASD children and an inverse correlation between IFN-γ and IFNG-AS1 expression, highlighting an imbalance in their interactive network (Table 1).

Inappropriate activation of NF-κB has been associated with neuroinflammatory responses [123]. Young and collaborators [124] reported that NF-κB is abnormally expressed in the orbitofrontal cortex of ASD subjects, contributing to inflammation and activation of resident immune cells in brain regions associated with behavioral and clinical symptoms of the disease. Honarmand Tamizkar and colleagues [125] postulated that some genes and lncRNAs associated with NF-κB, previously found dysregulated in schizophrenia [126], might also be aberrantly expressed in the peripheral blood of ASD patients. The authors detected an overexpression of three lncRNAs associated with NF-κB, namely Adipogenic differentiation induced non-coding RNA (ADINR), Antisense non-coding RNA in the INK4 Locus (ANRIL), and NF-κB Interacting lncRNA (NKILA), in ASD subjects compared to controls. None of the identified lncRNAs had previously been linked to ASD, except for ANRIL, for which a tendency for some *ANRIL* haplotypes to associate with ASD risk in the Iranian population was demonstrated [127]. The evaluation of the lncRNA suitability in distinguishing between ASD patients and healthy subjects indicated ANRIL as the lncRNA with the best diagnostic power (AUC: 0.857), followed by NKILA (AUC: 0.757) (Table 1).

Taheri's research group [128] evaluated the expression of Colon Cancer-Associated Transcripts 1 and 2 (CCAT1 and CCAT2) in the whole blood of ASD children and healthy controls, since both lncRNAs are involved in the regulation of immune-related genes and in inflammation [129–132]. The study revealed significantly reduced expression of CCAT1 and elevated levels of CCAT2 in ASD children compared to controls. ROC analysis showed that CCAT2 had stronger diagnostic performance than CCAT1 (AUC: 0.779 and 0.663, respectively) (Table 1) in distinguishing ASD patients from healthy individuals [128].

Table 1. Human dysregulated blood-derived long non-coding RNAs (lncRNAs) in autism spectrum disorders (ASDs).

lncRNAs	Subjects (Gender and Age)	Source	lncRNA Expression	ASD-Related Pathways Suggested by Authors	Molecular Mechanism in ASD	AUC	Sensitivity	Specificity	Ref.
DISC2	30 ASD children (mean age: 6 ± 1.4 years); 41 healthy children (age-, gender-, and ethnicity-matched)	PB	↑	Neuronal differentiation	Unknown	AUC = 0.763 ($p < 0.0001$)	83.33%	73.17%	[53]
PRKAR2A-AS1	30 ASD children (mean age: 6 ± 1.4 years); 41 healthy children (age-, gender-, and ethnicity-matched)	PB	↑	Unknown	Unknown	AUC = 0.794 $p < 0.0001$	86.67%	78.05%	[53]
LOC101928237	30 ASD children (mean age: 6 ± 1.4 years); 41 healthy children (age-, gender-, and ethnicity-matched)	PB	↑	Unknown	Unknown	AUC = 0.9 • ($p < 0.0001$)	90%	82.93%	[53]
LRRC2-AS1	30 ASD children (mean age: 6 ± 1.4 years); 41 healthy children (age-, gender-, and ethnicity-matched)	PB	↓	Unknown	Unknown	AUC = 0.929 • ($p < 0.0001$)	86.67%	100%	[53]
SNHG6	30 ASD children; 41 healthy children (age-, gender-, and ethnicity-matched; age unspecified)	PB	↓	Vitamin D receptor pathway	Interaction with miR-181c	AUC = 0.94 • ($p = 0.002$)	60.00%	73.17%	[57]

Table 1. *Cont.*

lncRNAs	Subjects (Gender and Age)	Source	lncRNA Expression	ASD-Related Pathways Suggested by Authors	Molecular Mechanism in ASD	AUC	Sensitivity	Specificity	Ref.
lincRNA-ROR	30 ASD children (11 females and 19 males, mean age: 6 ± 1.4 years); 41 healthy children (11 females and 30 males, mean age: 6 ± 1.74)	PB	↓	Neuronal differentiation	Unknown	AUC = 0.85 ($p < 0.0001$)	86.67%	65.85%	[60]
LINC-PINT	30 ASD children (11 females and 19 males, mean age: 6 ± 1.4 years); 41 healthy children (11 females and 30 males, mean age: 6 ± 1.74)	PB	↓	Neuronal differentiation	Unknown	AUC = 0.67 ($p = 0.0138$)	/	/	[60]
lincRNAp21	30 ASD children (11 females and 19 males, mean age: 6 ± 1.4 years); 41 healthy children (11 females and 30 males, mean age: 6 ± 1.74)	PB	↓	Neuronal differentiation	Unknown	AUC = 0.64 ($p = 0.0394$)	/	/	[60]

Table 1. *Cont.*

lncRNAs	Subjects (Gender and Age)	Source	lncRNA Expression	ASD-Related Pathways Suggested by Authors	Molecular Mechanism in ASD	AUC	Sensitivity	Specificity	Ref.
PCAT-29	30 ASD children (11 females and 19 males, mean age: 6 ± 1.4 years); 41 healthy children (11 females and 30 males, mean age: 6 ± 1.74)	PB	↓	Neuronal differentiation	Unknown	AUC = 0.74 ($p = 0.0005$)	/	/	[60]
PCAT-1	30 ASD children (11 females and 19 males, mean age: 6 ± 1.4 years); 41 healthy children (11 females and 30 males, mean age: 6 ± 1.74)	PB	↓	Neuronal differentiation	Unknown	AUC = 0.84 ($p < 0.0001$)	80%	70.73%	[60]
MEG3	30 ASD children (mean age: 6.01 ± 1.39 years); 41 healthy children (age-, gender-, and ethnicity-matched)	PB	↑	Neuronal synaptic plasticity; apoptotic pathway	Increases CDH2 expression via EP300, repressing neuronal viability *	AUC = 0.792 ($p < 0.0001$)	83.33%	70.73%	[62]
Shank2-AS	40 ASD children; 40 healthy children (age and gender unspecified)	Lymphocytes from PB	↑	Apoptotic pathway	Decreases Shank2 gene expression *	/	/	/	[69]

Table 1. *Cont.*

lncRNAs	Subjects (Gender and Age)	Source	lncRNA Expression	ASD-Related Pathways Suggested by Authors	Molecular Mechanism in ASD	AUC	Sensitivity	Specificity	Ref.
NEAT1	30 ASD children (mean age: 6.01 ± 1.4 years); 41 healthy children (mean age: 6 ± 1.4 years, gender-matched)	PB	↑	Apoptotic pathway	Interaction with miR-497/BDNF pathway; recruits YY1 to regulate UBE3A expression *	AUC = 0.759 ($p < 0.0001$)	70%	75.61%	[71]
TUG1	30 ASD children (mean age: 6.01 ± 1.4 years); 41 healthy children (mean age: 6 ± 1.4 years, gender-matched)	PB	↑	Apoptotic pathway	Sponge of miR-9	AUC = 0.733 ($p = 0.0001$)	76.67%	65.85%	[71]
	16 ASD children (12 males and 4 females; mean age: 6 ± 1.8 years) 16 healthy controls (11 males and 5 females; mean age: 7.6 ± 2.1 years)	PB mononuclear cells (PBMCs)	↑	Protection against oxidative stress	Regulation of TLDC1 expression	/	/	/	[82]
PVT1	60 ASD children (mean age: 7.07 ± 2.56 years); 58 healthy children (mean age: 7.76 ± 2.67 years)	serum	↓	Neuro-protection	Unknown	AUC = 0.848 (95% CI = 0.78–0.92)	85.0%	79.3%	[90]

Table 1. *Cont.*

lncRNAs	Subjects (Gender and Age)	Source	lncRNA Expression	ASD-Related Pathways Suggested by Authors	Molecular Mechanism in ASD	AUC	Sensitivity	Specificity	Ref.
LINC01231	30 ASD children (11 females mean age: 6 ± 1.73 years and 19 males, mean age: 6 ± 1.33 years); 41 healthy children (11 females, mean age: 5.63 ± 1.28 years and 30 males, mean age: 6.2 ± 1.88 years)	PB	↑	Ca^{2+} signaling	Unknown	AUC = 0.75 ± 0.06 ($p = 0.0003$)	0.77	0.76	[98]
THRIL	10 ASD children; 41 healthy children (age and gender unspecified)	PB	↓	Immune signaling/ inflammation	Regulation of TNF-α expression	/	/	/	[114]
IFNG-AS1	50 ASD children (15 females and 35 males, mean age: 6 ± 1.4 years); 50 healthy controls (14 females and 36 males, mean age: 6 ± 1.74 years)	PB	↓	Immune signaling/ inflammation neurogenesis	Regulation of IFNG; regulation of miR-21a-3p/PI3K (p110α)/ AKT axis *	/	/	/	[122]
ADINR	30 ASD children (11 females and 19 males, mean age: 6 ± 1.39 years); 41 healthy children (11 females and 30 males, age- and ethnicity- matched)	PB	↑	Immune signaling/ inflammation: NF-κB pathway	Unknown	AUC = 0.735	/	/	[125]

Table 1. *Cont.*

lncRNAs	Subjects (Gender and Age)	Source	lncRNA Expression	ASD-Related Pathways Suggested by Authors	Molecular Mechanism in ASD	AUC	Sensitivity	Specificity	Ref.
ANRIL	30 ASD children (11 females and 19 males, mean age: 6 ± 1.39 years); 41 healthy children (11 females and 30 males, age- and ethnicity- matched)	PB	↑	Immune signaling/ inflammation: NF-κB pathway	Unknown	AUC = 0.857	/	/	[125]
NKILA	30 ASD children (11 females and 19 males, mean age: 6 ± 1.39 years); 41 healthy children (11 females and 30 males, age- and ethnicity- matched)	PB	↑	Immune signaling/ inflammation: NF-κB pathway	Unknown	AUC = 0.757	/	/	[125]
CCAT1	30 ASD children (11 females and 19 males mean age: 6.01 ± 1.39 years); 41 healthy controls (11 females and 30 males (age-, sex-, and ethnicity-matched)	PB	↓	Immune signaling/ inflammation	Unknown	AUC = 0.663 (p = 0.016)	54.33%	82.93%	[128]

Table 1. *Cont.*

lncRNAs	Subjects (Gender and Age)	Source	lncRNA Expression	ASD-Related Pathways Suggested by Authors	Molecular Mechanism in ASD	AUC	Sensitivity	Specificity	Ref.
CCAT2	30 ASD children (11 females and 19 males mean age: 6.01 ± 1.39 years); 41 healthy controls (11 females and 30 males (age-, sex-, and ethnicity-matched)	PB	↑	Immune signaling/ inflammation	Unknown	AUC = 0.779 ($p < 0.0001$)	86.67%	73.17%	[128]

"↑" means upregulated lncRNAs in ASD; "↓" means downregulated lncRNAs in ASD; PB, peripheral blood; CDH2, cadherin2; EP300, E1A binding protein p300 transcription factor; BDNF, brain-derived neurotrophic factor; YY1, Yin-Yang 1 transcription factor; UBE3A, ubiquitin protein ligase E3A; TLDC1, TBC/LysM-associated domain containing 1; TNF-α, tumor necrosis factor α; IFNG, interferon gamma; PI3K/AKT, phosphoinositide 3-kinase/AKT serine/threonine kinase; NF-κB, nuclear factor kappa B. Ref. means references; * marks the only validated markers in ASD; • marks highly satisfactory discriminatory accuracy.

2.2. LncRNAs Selected by High-Throughput Analysis

The first attempt to identify a genome-wide differential expression of lncRNAs in ASD blood specimens was carried out by Wang and colleagues [70]. In a Chinese cohort, they compared RNA extracted from peripheral leukocytes of ASD and control children and identified 3929 differentially expressed lncRNAs—2407 upregulated and 1522 downregulated. Functional pathways analysis revealed that deregulated lncRNAs were mainly involved in infection and inflammation pathways, while the upregulated ones were implicated in neurological regulatory pathways, such as long-term depression, long-term potentiation, and synaptic vesicle cycling. In particular, Wang and colleagues [70], reported several dysregulated synapsis-associated lncRNAs in ASD peripheral leucocytes, some of which reside within genes that code for proteins involved in synaptic vesicle transportation and cycling. Notably, these lncRNAs were also analyzed by Fang and colleagues in circulating exosomes isolated from peripheral blood, and they found some to be differentially expressed in ASD subjects compared to controls [133]. In addition, Wang and coworkers [70] identified many dysregulated lncRNAs located at loci where rare high-impact genetic variants are known to increase ASD risk. Since many of these lncRNA originate from antisense, bidirectional, and intragenic traits of homeobox (HOX) genes, the authors proposed that HOX-related lncRNAs could represent a novel group of potential ASD biomarkers. The differential expression of two lncRNAs, named Shank2-AS and BDNF antisense RNA (BDNF-AS), which are the antisense transcripts of the *SHANK2* and *BDNF* genes, respectively, confirms that exploring lncRNAs at ASD-related loci may lead to the identification of new ASD potential markers [70].

Recently, exploiting the huge number of omics datasets, other approaches were employed to find putative biomarkers. Sabaie and colleagues [134] used a bioinformatic approach on microarray dataset to discover lncRNAs-associated ceRNA networks in the peripheral blood of ASD subjects. According to the ceRNA hypothesis, a pool of both coding and non-coding RNAs—including lncRNAs, mRNAs, and circular RNAs– compete for miRNA binding through their miRNA response elements (MREs) [30,58,135], establishing interactive networks [136]. Sabaie and colleagues [134] identified three lncRNAs, named Long intergenic non-protein coding RNA 472 (LINC00472), ANP32A intronic transcript 1 (ANP32A-IT1), and RBM26 antisense RNA 1 (RBM26-AS1), within four different ceRNA networks related to immune response in ASD pathogenesis: LINC00472/hsa-miR-221-3p/PTPN11, ANP32A-IT1/hsa-miR-182-5p/S100A2, LINC00472/hsa-miR-132-3p/S100A2, and RBM26-AS1/hsa-miR-182-5p/S100A2. The three lncRNAs identified were associated with ASD for the first time.

Balasubramanian and colleagues, using brain transcriptomic (RNA-Seq) datasets, predicted the miRNA sponge modules associated with pan-neuropsychiatric disorders, including ASD. Some of these modules were also confirmed in blood transcriptomic dataset from ASD subjects, leading to the identification of two lncRNAs, RP11-448G15.3 and WAC-AS1, as potential ASD biomarkers (AUC > 0.70) [137].

The ceRNA network approach was also used by Jiang and collaborators [138]; by integrating multiple omics data from ASD postmortem brain tissue samples, the authors built a ceRNA network where the MIR600 host gene (MIR600HG) lncRNA emerged as a key player. This lncRNA was consistently downregulated not only in brain tissues but also in peripheral blood, suggesting its putative role as an ASD biomarker. Based on its downstream mRNA targets, the authors suggested that MIR600HG may impair synaptogenesis in ASD, offering novel insights into the molecular mechanisms of the disorder.

3. Discussion

The diagnosis of ASD is currently based only on clinical observation; it usually occurs around three years of age and it may be further delayed due to the large heterogeneity of ASD symptoms [12]. It should be emphasized that early diagnosis can anticipate potential therapeutic interventions during a time window when neuronal plasticity allows for the modification of adverse developmental trajectories [19]. The identification of reliable ASD markers would not only facilitate early diagnosis but also contribute to the discovery of possible therapeutic targets highly demanded in a disorder for which no effective interventions have been found so far. The extensive clinical and biological heterogeneity complicates the search for biomarkers. Although several biomolecules have been proposed, to date, no ASD biomarkers meet the standards required to inform clinical trials [139]. In this review, we focused on dysregulated human blood-derived lncRNAs as potential ASD biomarkers. Owing to their specificity and stability, lncRNAs represent an attractive diagnostic tool for clinical use, especially when isolated from peripheral blood, an accessible tissue requiring cost-effective experimental handling. The identified dysregulated blood-derived lncRNAs are mainly involved in neurodevelopment, synaptic dysregulation, apoptosis, mitochondrial dysfunctions, OS, immune signaling, and inflammation, thus indicating these pathways as candidates for diagnostic interventions in ASD (Figure 3 and Table 1). Although lncRNAs have been suggested to be involved in ASD-related pathways by the authors, limited information has been gathered about the involved molecular mechanisms. Functional studies in cellular or animal ASD models have been carried out only for MEG3, NEAT1, Shank2-AS, and IFNG-AS1 (Figure 2).

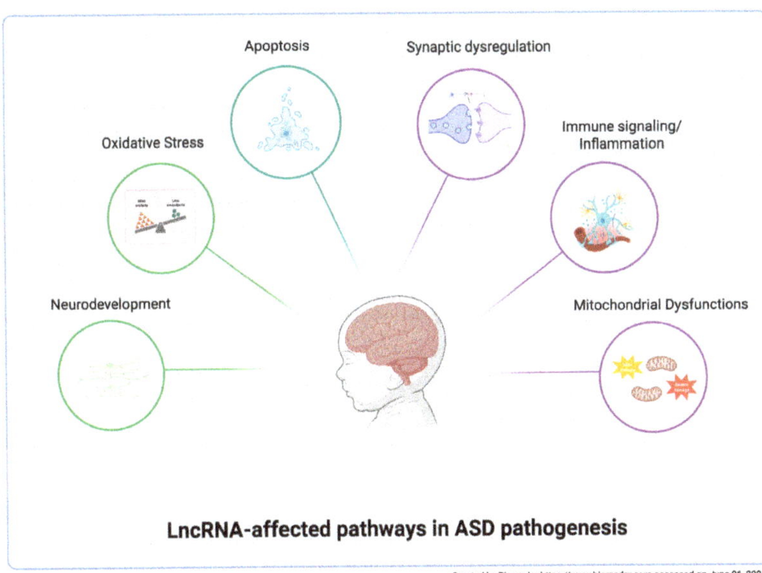

Figure 3. LncRNA-affected pathways in ASD pathogenesis. A schematic representation of the main biological pathways affected by dysregulated lncRNAs in ASD.

In most of the studies reviewed here, transcripts were isolated from total blood, which contains either lncRNAs derived from blood cells or released by other tissues. Some of these lncRNAs are packaged and protected within extracellular vesicles (EVs), nano-sized lipid bilayer particles that reflect the features of the originating tissue. The possibility to isolate brain-derived EVs, which contain molecules related to neuronal functions, paves the way for the detection of reliable markers for neurological diseases by opening a non-invasive molecular window into the brain. Using this strategy, Qin and colleagues [140] identified more than 1700 dysregulated lncRNAs in neuronal-cell-derived EVs in ASD which could be evaluated as candidate biomarkers for diagnostic purposes.

Although aberrant expression is a prerequisite for the selection of putative biomarkers, both specificity and sensitivity are critical requirements for their translation to clinical practice. Many of the studies listed here included ROC analyses for the identification of lncRNAs with sufficient sensitivity and specificity to support ASD diagnosis [141,142]. Notably, the diagnostic power of ROC analysis should not be immediately translated into the clinical context since it only assesses the ability of the tested parameter to discriminate between the two examined populations. Although several lncRNAs showed acceptable AUC values, a satisfactory discriminatory accuracy was achieved only for LOC101928237, LRRC2-AS1, and SNHG6 [53,57]. Diagnostic power could be improved by combining multiple markers to identify a diagnostic signature [143]. This integrated approach has been rarely explored for lncRNAs in ASD research. For instance, a superior diagnostic performance was achieved by combining the expression levels of PVT1 and its target miR-21-5p, with respect to the diagnostic power of each single molecule [90]. Recently, the integration of multiple extensive omics studies has led to the establishment of networks based on lncRNA, mRNA, and miRNA interactions (ceRNAs), which could be employed in clinical practice following a proper validation. Currently, only three studies have investigated the role of ceRNA networks in ASD blood samples [134,137,138]; therefore, this field deserves to be explored.

It is essential to recognize that candidate biomarkers are expected to provide reliable and systematic evidence demonstrating the clinical relevance of the selected lncRNAs for disease diagnosis. Considering this, several important limitations emerge from the reviewed studies. Most of them should be considered as discovery studies, as no subsequent validation was performed to confirm the preliminary results. Every putative biomarker should undergo validation that demonstrates the robustness of results in separate cohorts to prevent sporadic findings and avoid the introduction of tests based on presumed worthless biomarkers. Only a limited number of lncRNAs have been evaluated in independent populations. In particular, the upregulation of Shank2-AS observed in ASD leukocytes was confirmed in peripheral blood lymphocytes of another cohort of ASD subjects [69,70]. In addition, TUG1 was found to be upregulated in both PBMCs and in peripheral blood of ASD subjects from two independent populations [71,82]. Furthermore, the expression of DISC2 was investigated in two independent ASD groups, yielding inconsistent results across studies [53,54]. It should be noted that there is very little information about the demographic and diagnostic criteria, such as symptom severity, co-morbidities, inheritance, and environmental conditions. All these issues are sources of variability, which may account for the lack of reproducibility of the findings. Moreover, sample preparation and RNA processing methods differ across the examined studies. In particular, since lncRNAs are relatively low in abundance in blood samples, bulk detection methods, such as microarrays and NGS, should be preferentially confirmed using more sensitive techniques, like RT-PCR.

Most of the examined studies are based on small sample sizes and single-center cohorts, which inevitably limit the statistical power. Notably, eighteen out of the twenty-three differentially expressed lncRNAs reported in Table 1 were studied by the same research group and were apparently evaluated in the same cohorts of subjects. A few inaccuracies (e.g., inconsistency between AUC values reported in the text with those reported in the figures) make it difficult to assess the relevance of the reported data. The geographical specificity of the recruited subjects further restricts the applicability of these results to broader patient populations with diverse ethnic backgrounds.

In the studies reviewed here, no data are available on the ability of lncRNAs to distinguish ASD subjects from patients affected by diseases which share some of the ASD features. In fact, several dysregulated lncRNAs that emerged as ASD putative biomarkers have also been observed in the peripheral blood of subjects affected by other neurological disorders; for instance, TUG1 has been proposed as a biomarker for the diagnosis of temporal lobe epilepsy [144], while PVT1 is downregulated in schizophrenia [88], and THRIL is underexpressed in epilepsy [145]. Therefore, it would be necessary to estimate the performance of the proposed lncRNA in distinguishing ASD subjects from patients with other neurological disorders to clarify their effective application in clinical practice.

Although these limitations compromise the reliability and the generalization of the conclusions, some of the lncRNAs summarized here are worthy of interest for ASD biomarker development. For instance, MEG3, NEAT1, Shank2-AS, and IFGN-AS1 are relevant due to their experimentally observed implication in ASD pathogenesis, while both TUG1 and Shank2-AS are also important for their expression patterns, which have been consistently reported in more than one study, although a more effective confirmation would be needed. Moreover, LOC101928237, LRRC2-AS1, and SNHG6 seem to be promising according to their AUC values. However, data concerning all lcnRNAs reported here should be considered as preliminary findings deserving deeper exploration to assess their effective translational potential in order to move the field forward.

4. Conclusions and Future Directions

This review highlights the important role of blood-derived lncRNAs in ASD, although this field is still in the early stages of development. Independent studies should be carried out in larger cohorts of subjects with gold-standard enrollment criteria in order to corroborate the reproducibility of the current results. In addition, since many potential biomarkers are altered in several diseases, the obtained results should be validated to test their specificity in subjects affected by similar neurodevelopmental and systemic disorders. Moreover, it would be useful to extend the analysis to younger cohorts to identify whether relevant lncRNAs are expressed in very young children so as to be used as biomarkers for early diagnosis. Finally, the validation of specific biomarkers might also benefit from longitudinal cohort studies where ASD subjects are followed over time across their lifespan. A significant contribution to biomarker discovery could derive from omics approaches, including transcriptomics, proteomics, epigenetics, and metabolomics. The integration of multiple omics would make it possible to set a network of molecular interactions disclosing potential molecular mechanisms which are altered in ASD. It is known that most biomarkers currently applied in clinical practice stem from the deep knowledge of their molecular mechanisms in the pathology, which highlights the need to encourage basic research. To this purpose, new approaches employing cellular and animal ASD models, along with CRISPR-based strategies to selectively modulate the expression of a candidate lncRNA, could improve the assessment of its roles in ASD-related pathways. The CRISPR/Cas system provides innovative tools to investigate the potential roles of lncRNAs through knockout, knockdown, overexpression, and imaging approaches (for a comprehensive overview, see [146–148]). This technology has been successfully applied to study the function several lncRNAs. For example, by using a combination of cellular reprogramming and genome editing, it has been shown that the disruption of PTCHD1-AS impairs excitatory neurotransmission in ASD [149]. By using the CRISPR/Cas system, it is possible to modulate the expression of specific lncRNAs in order to analyze the changes in transcriptome, proteome, and metabolome, thus highlighting the downstream effects of lncRNA regulation [150]. For instance, the deletion of GOMAFU promoter through the CRISPR-Cas9 tool allowed for identifying transcriptomic alterations in numerous genes involved in IFN signaling [151]. Moreover, the integration of lncRNA expression profiles with complementary datasets, including quantitative proteomics and high-resolution imaging, could help decipher cellular functions and the spatial and temporal regulation of lncRNAs [152].

In conclusion, our review suggests that blood-derived lncRNAs deserve further studies aimed at exploiting their potential as candidate ASD biomarkers.

Author Contributions: Conceptualization, C.Z., C.S. and P.M.A.; investigation, C.Z., C.S., P.M.A. and P.D.S.; data curation, C.Z., C.S., P.M.A., P.D.S., M.M. and S.C.; writing—original draft preparation, C.Z., C.S. and P.M.A.; writing—review and editing, C.Z., C.S., P.M.A., P.D.S., M.M. and S.C.; supervision, M.M. All authors have read and agreed to the published version of the manuscript.

Funding: This research received no external funding.

Institutional Review Board Statement: Not applicable.

Informed Consent Statement: Not applicable.

Data Availability Statement: Data sharing is not applicable.

Conflicts of Interest: The authors declare no conflicts of interest.

References

1. Rogge, N.; Janssen, J. The Economic Costs of Autism Spectrum Disorder: A Literature Review. *J. Autism Dev. Disord.* **2019**, *49*, 2873–2900. [CrossRef] [PubMed]
2. Hirota, T.; King, B.H. Autism Spectrum Disorder: A Review. *JAMA* **2023**, *329*, 157–168. [CrossRef]
3. Lugo Marín, J.; Alviani Rodríguez-Franco, M.; Mahtani Chugani, V.; Magán Maganto, M.; Díez Villoria, E.; Canal Bedia, R. Prevalence of Schizophrenia Spectrum Disorders in Average-IQ Adults with Autism Spectrum Disorders: A Meta-analysis. *J. Autism Dev. Disord.* **2018**, *48*, 239–250. [CrossRef] [PubMed]
4. Capal, J.K.; Jeste, S.S. Autism and Epilepsy. *Pediatr. Clin. N. Am.* **2024**, *71*, 241–252. [CrossRef]
5. Santarone, M.E.; Zambrano, S.; Zanotta, N.; Mani, E.; Minghetti, S.; Pozzi, M.; Villa, L.; Molteni, M.; Zucca, C. EEG Features in Autism Spectrum Disorder: A Retrospective Analysis in a Cohort of Preschool Children. *Brain Sci.* **2023**, *13*, 345. [CrossRef] [PubMed]
6. Pizzolorusso, F.; Paparella, M.T.; Pizzolorusso, I.; Masino, F.; Guglielmi, G. Magnetic Resonance Imaging in Autism Spectrum Disorders: Clinical and neuroradiological phenotypes. *Acta Biomed.* **2023**, *94*, e2023027.
7. Maenner, M.J.; Warren, Z.; Williams, A.R.; Amoakohene, E.; Bakian, A.V.; Bilder, D.A.; Durkin, M.S.; Fitzgerald, R.T.; Furnier, S.M.; Hughes, M.M.; et al. Prevalence and Characteristics of Autism Spectrum Disorder Among Children Aged 8 Years—Autism and Developmental Disabilities Monitoring Network, 11 Sites, United States, 2020. *MMWR Surveill. Summ.* **2023**, *72*, 1–14. [CrossRef] [PubMed]
8. Sharma, S.R.; Gonda, X.; Tarazi, F.I. Autism Spectrum Disorder: Classification, diagnosis and therapy. *Pharmacol. Ther.* **2018**, *190*, 91–104. [CrossRef]
9. van 't Hof, M.; Tisseur, C.; van Berckelear-Onnes, I.; van Nieuwenhuyzen, A.; Daniels, A.M.; Deen, M.; Hoek, H.W.; Ester, W.A. Age at autism spectrum disorder diagnosis: A systematic review and meta-analysis from 2012 to 2019. *Autism* **2021**, *25*, 862–873. [CrossRef]
10. Chlebowski, C.; Green, J.A.; Barton, M.L.; Fein, D. Using the childhood autism rating scale to diagnose autism spectrum disorders. *J. Autism Dev. Disord.* **2010**, *40*, 787–799. [CrossRef]
11. Lord, C.; Elsabbagh, M.; Baird, G.; Veenstra-Vanderweele, J. Autism spectrum disorder. *Lancet* **2018**, *392*, 508–520. [CrossRef] [PubMed]
12. Mandy, W.; Midouhas, E.; Hosozawa, M.; Cable, N.; Sacker, A.; Flouri, E. Mental health and social difficulties of late-diagnosed autistic children, across childhood and adolescence. *J. Child Psychol. Psychiatry* **2022**, *63*, 1405–1414. [CrossRef] [PubMed]
13. Frye, R.E.; Vassall, S.; Kaur, G.; Lewis, C.; Karim, M.; Rossignol, D. Emerging biomarkers in autism spectrum disorder: A systematic review. *Ann. Transl. Med.* **2019**, *7*, 792. [CrossRef]
14. Jensen, A.R.; Lane, A.L.; Werner, B.A.; McLees, S.E.; Fletcher, T.S.; Frye, R.E. Modern Biomarkers for Autism Spectrum Disorder: Future Directions. *Mol. Diagn Ther.* **2022**, *26*, 483–495. [CrossRef] [PubMed]
15. Ostrowski, J.; Religioni, U.; Gellert, B.; Sytnik-Czetwertyński, J.; Pinkas, J. Autism Spectrum Disorders: Etiology, Epidemiology, and Challenges for Public Health. *Med. Sci. Monit.* **2024**, *30*, e944161. [CrossRef]
16. Tick, B.; Bolton, P.; Happé, F.; Rutter, M.; Rijsdijk, F. Heritability of autism spectrum disorders: A meta-analysis of twin studies. *J. Child Psychol. Psychiatry* **2016**, *57*, 585–595. [CrossRef]

17. Wiśniowiecka-Kowalnik, B.; Nowakowska, B.A. Genetics and epigenetics of autism spectrum disorder-current evidence in the field. *J. Appl. Genet.* **2019**, *60*, 37–47. [CrossRef]
18. Rylaarsdam, L.; Guemez-Gamboa, A. Genetic Causes and Modifiers of Autism Spectrum Disorder. *Front. Cell. Neurosci.* **2019**, *13*, 385. [CrossRef]
19. Panisi, C.; Guerini, F.R.; Abruzzo, P.M.; Balzola, F.; Biava, P.M.; Bolotta, A.; Brunero, M.; Burgio, E.; Chiara, A.; Clerici, M.; et al. Autism Spectrum Disorder from the Womb to Adulthood: Suggestions for a Paradigm Shift. *J. Pers. Med.* **2021**, *11*, 70. [CrossRef]
20. Hombach, S.; Kretz, M. Non-coding RNAs: Classification, Biology and Functioning. *Adv. Exp. Med. Biol.* **2016**, *937*, 3–17.
21. Fang, Y.; Fullwood, M.J. Roles, Functions, and Mechanisms of Long Non-coding RNAs in Cancer. *Genom. Proteom. Bioinform.* **2016**, *14*, 42–54. [CrossRef]
22. Mattick, J.S.; Amaral, P.P.; Carninci, P.; Carpenter, S.; Chang, H.Y.; Chen, L.L.; Chen, R.; Dean, C.; Dinger, M.E.; Fitzgerald, K.A.; et al. Long non-coding RNAs: Definitions, functions, challenges and recommendations. *Nat. Rev. Mol. Cell Biol.* **2023**, *24*, 430–447. [CrossRef] [PubMed]
23. Bridges, M.C.; Daulagala, A.C.; Kourtidis, A. LNCcation: lncRNA localization and function. *J. Cell Biol.* **2021**, *220*, e202009045. [CrossRef] [PubMed]
24. Huang, H.; Li, L.; Wen, K. Interactions between long non-coding RNAs and RNA-binding proteins in cancer. *Oncol. Rep.* **2021**, *46*, 256. [CrossRef] [PubMed]
25. Ransohoff, J.D.; Wei, Y.; Khavari, P.A. The functions and unique features of long intergenic non-coding RNA. *Nat. Rev. Mol. Cell Biol.* **2018**, *19*, 143–157. [CrossRef]
26. Zhang, X.; Wang, W.; Zhu, W.; Dong, J.; Cheng, Y.; Yin, Z.; Shen, F. Mechanisms and Functions of Long Non-Coding RNAs at Multiple Regulatory Levels. *Int. J. Mol. Sci.* **2019**, *20*, 5573. [CrossRef]
27. Dykes, I.M.; Emanueli, C. Transcriptional and Post-transcriptional Gene Regulation by Long Non-coding RNA. *Genom. Proteom. Bioinform.* **2017**, *15*, 177–186. [CrossRef]
28. Ferrer, J.; Dimitrova, N. Transcription regulation by long non-coding RNAs: Mechanisms and disease relevance. *Nat. Rev. Mol. Cell Biol.* **2024**, *25*, 396–415. [CrossRef]
29. Zhang, J.; Zhu, H.; Li, L.; Gao, Y.; Yu, B.; Ma, G.; Jin, X.; Sun, Y. New mechanism of LncRNA: In addition to act as a ceRNA. *Non-Coding RNA Res.* **2024**, *9*, 1050–1060. [CrossRef]
30. Salmena, L.; Poliseno, L.; Tay, Y.; Kats, L.; Pandolfi, P.P. A ceRNA hypothesis: The Rosetta Stone of a hidden RNA language? *Cell* **2011**, *146*, 353–358. [CrossRef]
31. Yang, C.; Wu, D.; Gao, L.; Liu, X.; Jin, Y.; Wang, D.; Wang, T.; Li, X. Competing endogenous RNA networks in human cancer: Hypothesis, validation, and perspectives. *Oncotarget* **2016**, *7*, 13479–13490. [CrossRef] [PubMed]
32. Tan, Y.T.; Lin, J.F.; Li, T.; Li, J.J.; Xu, R.H.; Ju, H.Q. LncRNA-mediated posttranslational modifications and reprogramming of energy metabolism in cancer. *Cancer Commun.* **2021**, *41*, 109–120. [CrossRef]
33. Karakas, D.; Ozpolat, B. The Role of LncRNAs in Translation. *Noncoding RNA* **2021**, *7*, 16. [CrossRef]
34. Heydarnezhad Asl, M.; Pasban Khelejani, F.; Bahojb Mahdavi, S.Z.; Emrahi, L.; Jebelli, A.; Mokhtarzadeh, A. The various regulatory functions of long noncoding RNAs in apoptosis, cell cycle, and cellular senescence. *J. Cell Biochem.* **2022**, *123*, 995–1024. [CrossRef] [PubMed]
35. Yang, S.; Lim, K.H.; Kim, S.H.; Joo, J.Y. Molecular landscape of long noncoding RNAs in brain disorders. *Mol. Psychiatry* **2021**, *26*, 1060–1074. [CrossRef] [PubMed]
36. Aliperti, V.; Skonieczna, J.; Cerase, A. Long Non-Coding RNA (lncRNA) Roles in Cell Biology, Neurodevelopment and Neurological Disorders. *Non-Coding RNA* **2021**, *7*, 36. [CrossRef]
37. Liaci, C.; Prandi, L.; Pavinato, L.; Brusco, A.; Maldotti, M.; Molineris, I.; Oliviero, S.; Merlo, G.R. The Emerging Roles of Long Non-Coding RNAs in Intellectual Disability and Related Neurodevelopmental Disorders. *Int. J. Mol. Sci.* **2022**, *23*, 6118. [CrossRef]
38. Alammari, F.; Al-Hujaily, E.M.; Alshareeda, A.; Albarakati, N.; Al-Sowayan, B.S. Hidden regulators: The emerging roles of lncRNAs in brain development and disease. *Front. Neurosci.* **2024**, *18*, 1392688. [CrossRef]
39. Anilkumar, A.K.; Vij, P.; Lopez, S.; Leslie, S.M.; Doxtater, K.; Khan, M.M.; Yallapu, M.M.; Chauhan, S.C.; Maestre, G.E.; Tripathi, M.K. Long Non-Coding RNAs: New Insights in Neurodegenerative Diseases. *Int. J. Mol. Sci.* **2024**, *25*, 2268. [CrossRef]
40. Badowski, C.; He, B.; Garmire, L.X. Blood-derived lncRNAs as biomarkers for cancer diagnosis: The Good, the Bad and the Beauty. *NPJ Precis. Oncol.* **2022**, *6*, 40. [CrossRef]
41. Zhang, M.; He, P.; Bian, Z. Long Noncoding RNAs in Neurodegenerative Diseases: Pathogenesis and Potential Implications as Clinical Biomarkers. *Front. Mol. Neurosci.* **2021**, *14*, 685143. [CrossRef] [PubMed]
42. Manna, I.; Quattrone, A.; De Benedittis, S.; Iaccino, E.; Quattrone, A. Roles of Non-Coding RNAs as Novel Diagnostic Biomarkers in Parkinson's Disease. *J. Parkinson's Dis.* **2021**, *11*, 1475–1489. [CrossRef]
43. Yang, Y.; Li, Y.; Yang, H.; Guo, J.; Li, N. Circulating MicroRNAs and Long Non-coding RNAs as Potential Diagnostic Biomarkers for Parkinson's Disease. *Front. Mol. Neurosci.* **2021**, *14*, 631553. [CrossRef]

44. Varesi, A.; Carrara, A.; Pires, V.G.; Floris, V.; Pierella, E.; Savioli, G.; Prasad, S.; Esposito, C.; Ricevuti, G.; Chirumbolo, S.; et al. Blood-Based Biomarkers for Alzheimer's Disease Diagnosis and Progression: An Overview. *Cells* **2022**, *11*, 1367. [CrossRef] [PubMed]
45. Salloum-Asfar, S.; Elsayed, A.K.; Elhag, S.F.; Abdulla, S.A. Circulating Non-Coding RNAs as a Signature of Autism Spectrum Disorder Symptomatology. *Int. J. Mol. Sci.* **2021**, *22*, 6549. [CrossRef] [PubMed]
46. Bourgeron, T. From the genetic architecture to synaptic plasticity in autism spectrum disorder. *Nat. Rev. Neurosci.* **2015**, *16*, 551–563. [CrossRef]
47. Briggs, J.A.; Wolvetang, E.J.; Mattick, J.S.; Rinn, J.L.; Barry, G. Mechanisms of Long Non-coding RNAs in Mammalian Nervous System Development, Plasticity, Disease, and Evolution. *Neuron* **2015**, *88*, 861–877. [CrossRef]
48. Zimmer-Bensch, G. Emerging Roles of Long Non-Coding RNAs as Drivers of Brain Evolution. *Cells* **2019**, *8*, 1399. [CrossRef]
49. Taylor, M.S.; Devon, R.S.; Millar, J.K.; Porteous, D.J. Evolutionary constraints on the Disrupted in Schizophrenia locus. *Genomics* **2003**, *81*, 67–77. [CrossRef]
50. Millar, J.K.; James, R.; Brandon, N.J.; Thomson, P.A. DISC1 and DISC2: Discovering and dissecting molecular mechanisms underlying psychiatric illness. *Ann. Med.* **2004**, *36*, 367–378. [CrossRef]
51. Lin, M.; Pedrosa, E.; Shah, A.; Hrabovsky, A.; Maqbool, S.; Zheng, D.; Lachman, H.M. RNA-Seq of human neurons derived from iPS cells reveals candidate long non-coding RNAs involved in neurogenesis and neuropsychiatric disorders. *PLoS ONE* **2011**, *6*, e23356. [CrossRef]
52. Williams, J.M.; Beck, T.F.; Pearson, D.M.; Proud, M.B.; Cheung, S.W.; Scott, D.A. A 1q42 deletion involving DISC1, DISC2, and TSNAX in an autism spectrum disorder. *Am. J. Med. Genet. A* **2009**, *149A*, 1758–1762. [CrossRef] [PubMed]
53. Tamizkar, K.H.; Ghafouri-Fard, S.; Omrani, M.D.; Pouresmaeili, F.; Arsang-Jang, S.; Taheri, M. Altered expression of lncRNAs in autism spectrum disorder. *Metab. Brain Dis.* **2021**, *36*, 983–990. [CrossRef]
54. Rahmani, Z.; Rahmani, D.; Jazi, M.S.; Ghasemi, M.R.; Sadeghi, H.; Miryounesi, M.; Razjouyan, K.; Fayyazi Bordbar, M.R. Altered expression of Csnk1a1p in Autism Spectrum Disorder in Iranian population: Case-control study. *Sci. Rep.* **2024**, *14*, 28307. [CrossRef] [PubMed]
55. Kholghi Oskooei, V.; Geranpayeh, L.; Omrani, M.D.; Ghafouri-Fard, S. Assessment of functional variants and expression of long noncoding RNAs in vitamin D receptor signaling in breast cancer. *Cancer Manag. Res.* **2018**, *10*, 3451–3462. [CrossRef]
56. Eyles, D.W.; Burne, T.H.; McGrath, J.J. Vitamin D, effects on brain development, adult brain function and the links between low levels of vitamin D and neuropsychiatric disease. *Front. Neuroendocrinol.* **2013**, *34*, 47–64. [CrossRef]
57. Ghafouri-Fard, S.; Eghtedarian, R.; Hussen, B.M.; Motevaseli, E.; Arsang-Jang, S.; Taheri, M. Expression Analysis of VDR-Related LncRNAs in Autism Spectrum Disorder. *J. Mol. Neurosci.* **2021**, *71*, 1403–1409. [CrossRef]
58. Zhang, X.; Liu, Z.; Shu, Q.; Yuan, S.; Xing, Z.; Song, J. LncRNA SNHG6 functions as a ceRNA to regulate neuronal cell apoptosis by modulating miR-181c-5p/BIM signalling in ischaemic stroke. *J. Cell. Mol. Med.* **2019**, *23*, 6120–6130. [CrossRef] [PubMed]
59. Olde Loohuis, N.F.; Kole, K.; Glennon, J.C.; Karel, P.; Van der Borg, G.; Van Gemert, Y.; Van den Bosch, D.; Meinhardt, J.; Kos, A.; Shahabipour, F.; et al. Elevated microRNA-181c and microRNA-30d levels in the enlarged amygdala of the valproic acid rat model of autism. *Neurobiol. Dis.* **2015**, *80*, 42–53. [CrossRef]
60. Sane, S.; Ebrahimi, V.; Shirvani Farsani, Z.; Ghafouri-Fard, S. Assessment of Expression of lncRNAs in Autistic Patients. *J. Mol. Neurosci.* **2024**, *74*, 81. [CrossRef]
61. Tan, M.C.; Widagdo, J.; Chau, Y.Q.; Zhu, T.; Wong, J.J.; Cheung, A.; Anggono, V. The Activity-Induced Long Non-Coding RNA Meg3 Modulates AMPA Receptor Surface Expression in Primary Cortical Neurons. *Front. Cell. Neurosci.* **2017**, *11*, 124. [CrossRef] [PubMed]
62. Taheri, M.; Honarmand Tamizkar, K.; Omrani, S.; Arsang-Jang, S.; Ghafouri-Fard, S.; Omrani, M.D. MEG3 lncRNA is over-expressed in autism spectrum disorder. *Metab. Brain Dis.* **2021**, *36*, 2235–2242. [CrossRef] [PubMed]
63. Liu, X.; Wang, Z.; Zhang, X.; Zhang, D.; Yang, Q.; Hu, P.; Li, F. LncRNA MEG3 activates CDH2 expression by recruitment of EP300 in valproic acid-induced autism spectrum disorder. *Neurosci. Lett.* **2022**, *783*, 136726. [CrossRef] [PubMed]
64. Guang, S.; Pang, N.; Deng, X.; Yang, L.; He, F.; Wu, L.; Chen, C.; Yin, F.; Peng, J. Synaptopathology Involved in Autism Spectrum Disorder. *Front. Cell. Neurosci.* **2018**, *12*, 470. [CrossRef]
65. Wei, H.; Alberts, I.; Li, X. The apoptotic perspective of autism. *Int. J. Dev. Neurosci.* **2014**, *36*, 13–18. [CrossRef]
66. Eftekharian, M.M.; Komaki, A.; Oskooie, V.K.; Namvar, A.; Taheri, M.; Ghafouri-Fard, S. Assessment of Apoptosis Pathway in Peripheral Blood of Autistic Patients. *J. Mol. Neurosci.* **2019**, *69*, 588–596. [CrossRef]
67. Berkel, S.; Marshall, C.R.; Weiss, B.; Howe, J.; Roeth, R.; Moog, U.; Endris, V.; Roberts, W.; Szatmari, P.; Pinto, D.; et al. Mutations in the SHANK2 synaptic scaffolding gene in autism spectrum disorder and mental retardation. *Nat. Genet.* **2010**, *42*, 489–491. [CrossRef]
68. Won, H.; Lee, H.R.; Gee, H.Y.; Mah, W.; Kim, J.I.; Lee, J.; Ha, S.; Chung, C.; Jung, E.S.; Cho, Y.S.; et al. Autistic-like social behaviour in Shank2-mutant mice improved by restoring NMDA receptor function. *Nature* **2012**, *486*, 261–265. [CrossRef]

69. Luo, T.; Liu, P.; Wang, X.Y.; Li, L.Z.; Zhao, L.P.; Huang, J.; Li, Y.M.; Ou, J.L.; Peng, X.Q. Effect of the autism-associated lncRNA Shank2-AS on architecture and growth of neurons. *J. Cell Biochem.* **2019**, *120*, 1754–1762. [CrossRef]
70. Wang, Y.; Zhao, X.; Ju, W.; Flory, M.; Zhong, J.; Jiang, S.; Wang, P.; Dong, X.; Tao, X.; Chen, Q.; et al. Genome-wide differential expression of synaptic long noncoding RNAs in autism spectrum disorder. *Transl. Psychiatry* **2015**, *5*, e660. [CrossRef]
71. Sayad, A.; Omrani, M.D.; Fallah, H.; Taheri, M.; Ghafouri-Fard, S. Aberrant Expression of Long Non-coding RNAs in Peripheral Blood of Autistic Patients. *J. Mol. Neurosci.* **2019**, *67*, 276–281. [CrossRef]
72. Peng, W.; Wang, Z.; Fan, H. LncRNA NEAT1 Impacts Cell Proliferation and Apoptosis of Colorectal Cancer via Regulation of Akt Signaling. *Pathol. Oncol. Res.* **2017**, *23*, 651–656. [CrossRef]
73. Li, K.; Wang, Z. lncRNA NEAT1: Key player in neurodegenerative diseases. *Ageing Res. Rev.* **2023**, *86*, 101878. [CrossRef] [PubMed]
74. He, C.; Zhou, H.; Chen, L.; Liu, Z. NEAT1 Promotes Valproic Acid-Induced Autism Spectrum Disorder by Recruiting YY1 to Regulate UBE3A Transcription. *Mol. Neurobiol.* **2025**, *62*, 846–860. [CrossRef] [PubMed]
75. Li, X.J. Long non-coding RNA nuclear paraspeckle assembly transcript 1 inhibits the apoptosis of retina Müller cells after diabetic retinopathy through regulating miR-497/brain-derived neurotrophic factor axis. *Diab. Vasc. Dis. Res.* **2018**, *15*, 204–213. [CrossRef] [PubMed]
76. Saghazadeh, A.; Rezaei, N. Brain-Derived Neurotrophic Factor Levels in Autism: A Systematic Review and Meta-Analysis. *J. Autism Dev. Disord.* **2017**, *47*, 1018–1029. [CrossRef]
77. Chen, S.; Wang, M.; Yang, H.; Mao, L.; He, Q.; Jin, H.; Ye, Z.M.; Luo, X.Y.; Xia, Y.P.; Hu, B. LncRNA TUG1 sponges microRNA-9 to promote neurons apoptosis by up-regulated Bcl2l11 under ischemia. *Biochem. Biophys Res. Commun.* **2017**, *485*, 167–173. [CrossRef]
78. Xiang, P.; Hu, J.; Wang, H.; Luo, Y.; Gu, C.; Tan, X.; Tu, Y.; Guo, W.; Chen, L.; Gao, L.; et al. miR-204-5p is sponged by TUG1 to aggravate neuron damage induced by focal cerebral ischemia and reperfusion injury through upregulating COX2. *Cell Death Discov.* **2022**, *8*, 89. [CrossRef]
79. Wu, H.; Li, Y.; Wang, X.; Zhang, Z.; Huang, Y. Long non-coding RNA TUG1 knockdown prevents neurons from death to alleviate acute spinal cord injury via the microRNA-338/BIK axis. *Bioengineered* **2021**, *12*, 5566–5582. [CrossRef]
80. He, Z.; Zhao, Y.; Zhu, Y.; Wang, W.; Liu, X.; Lu, F. Interfering TUG1 Attenuates Cerebrovascular Endothelial Apoptosis and Inflammatory injury After Cerebral Ischemia/Reperfusion via TUG1/miR-410/FOXO3 ceRNA Axis. *Neurotox. Res.* **2022**, *40*, 1–13. [CrossRef]
81. Khalil, A.M.; Guttman, M.; Huarte, M.; Garber, M.; Raj, A.; Rivea Morales, D.; Thomas, K.; Presser, A.; Bernstein, B.E.; van Oudenaarden, A.; et al. Many human large intergenic noncoding RNAs associate with chromatin-modifying complexes and affect gene expression. *Proc. Natl. Acad. Sci. USA* **2009**, *106*, 11667–11672. [CrossRef] [PubMed]
82. Zucchini, C.; Serpe, C.; De Sanctis, P.; Ghezzo, A.; Visconti, P.; Posar, A.; Facchin, F.; Marini, M.; Abruzzo, P.M. TLDc Domain-Containing Genes in Autism Spectrum Disorder: New Players in the Oxidative Stress Response. *Int. J. Mol. Sci.* **2023**, *24*, 15802. [CrossRef]
83. Bjørklund, G.; Meguid, N.A.; El-Bana, M.A.; Tinkov, A.A.; Saad, K.; Dadar, M.; Hemimi, M.; Skalny, A.V.; Hosnedlová, B.; Kizek, R.; et al. Oxidative Stress in Autism Spectrum Disorder. *Mol. Neurobiol.* **2020**, *57*, 2314–2332. [CrossRef]
84. Liu, X.; Lin, J.; Zhang, H.; Khan, N.U.; Zhang, J.; Tang, X.; Cao, X.; Shen, L. Oxidative Stress in Autism Spectrum Disorder-Current Progress of Mechanisms and Biomarkers. *Front. Psychiatry* **2022**, *13*, 813304. [CrossRef]
85. Kuźniar-Pałka, A. The Role of Oxidative Stress in Autism Spectrum Disorder Pathophysiology, Diagnosis and Treatment. *Biomedicines* **2025**, *13*, 388. [CrossRef] [PubMed]
86. Chen, L.; Gong, H.Y.; Xu, L. PVT1 protects diabetic peripheral neuropathy via PI3K/AKT pathway. *Eur. Rev. Med. Pharmacol. Sci.* **2018**, *22*, 6905–6911. [PubMed]
87. Li, C.; Li, R.; Bai, X.; Jiang, H. PVT1 regulates hippocampal neuron apoptosis and inflammation in epilepsy by miR-206-3p-dependent regulation of CAMK4. *Gen. Physiol. Biophys.* **2024**, *43*, 423–434. [CrossRef]
88. Safari, M.R.; Komaki, A.; Arsang-Jang, S.; Taheri, M.; Ghafouri-Fard, S. Expression Pattern of Long Non-coding RNAs in Schizophrenic Patients. *Cell. Mol. Neurobiol.* **2019**, *39*, 211–221. [CrossRef] [PubMed]
89. Guo, C.; Li, J.; Guo, M.; Bai, R.; Lei, G.; Sun, H.; Tong, S.; He, K.; He, L. Aberrant expressions of MIAT and PVT1 in serum exosomes of schizophrenia patients. *Schizophr. Res.* **2022**, *240*, 71–72. [CrossRef]
90. Jiang, M.; Chen, G. Investigation of LncRNA PVT1 and MiR-21-5p Expression as Promising Novel Biomarkers for Autism Spectrum Disorder. *J. Mol. Neurosci.* **2023**, *73*, 865–873. [CrossRef]
91. Wu, F.; Huang, W.; Tan, Q.; Guo, Y.; Cao, Y.; Shang, J.; Ping, F.; Wang, W.; Li, Y. ZFP36L2 regulates myocardial ischemia/reperfusion injury and attenuates mitochondrial fusion and fission by LncRNA PVT1. *Cell Death Dis.* **2021**, *12*, 614. [CrossRef] [PubMed]
92. Angelova, P.R.; Abramov, A.Y. Interplay of mitochondrial calcium signalling and reactive oxygen species production in the brain. *Biochem. Soc. Trans.* **2024**, *52*, 1939–1946. [CrossRef]
93. San-Millán, I. The Key Role of Mitochondrial Function in Health and Disease. *Antioxidants* **2023**, *12*, 782. [CrossRef] [PubMed]

94. Citrigno, L.; Muglia, M.; Qualtieri, A.; Spadafora, P.; Cavalcanti, F.; Pioggia, G.; Cerasa, A. The Mitochondrial Dysfunction Hypothesis in Autism Spectrum Disorders: Current Status and Future Perspectives. *Int. J. Mol. Sci.* **2020**, *21*, 5785. [CrossRef]
95. Frye, R.E. Mitochondrial Dysfunction in Autism Spectrum Disorder: Unique Abnormalities and Targeted Treatments. *Semin. Pediatr. Neurol.* **2020**, *35*, 100829. [CrossRef]
96. Khaliulin, I.; Hamoudi, W.; Amal, H. The multifaceted role of mitochondria in autism spectrum disorder. *Mol. Psychiatry* **2025**, *30*, 629–650. [CrossRef]
97. Nguyen, R.L.; Medvedeva, Y.V.; Ayyagari, T.E.; Schmunk, G.; Gargus, J.J. Intracellular calcium dysregulation in autism spectrum disorder: An analysis of converging organelle signaling pathways. *Biochim. Biophys Acta Mol. Cell Res.* **2018**, *1865*, 1718–1732. [CrossRef]
98. Pourtavakoli, A.; Ghafouri-Fard, S.; Eslami, S.; Brand, S.; Taheri, M. Expression assay of calcium signaling related lncRNAs in autism. *Mol. Biol. Rep.* **2024**, *51*, 185. [CrossRef] [PubMed]
99. Soueid, J.; Kourtian, S.; Makhoul, N.J.; Makoukji, J.; Haddad, S.; Ghanem, S.S.; Kobeissy, F.; Boustany, R.M. RYR2, PTDSS1 and AREG genes are implicated in a Lebanese population-based study of copy number variation in autism. *Sci. Rep.* **2016**, *6*, 19088. [CrossRef]
100. Morris, G.; Gevezova, M.; Sarafian, V.; Maes, M. Redox regulation of the immune response. *Cell. Mol. Immunol.* **2022**, *19*, 1079–1101. [CrossRef]
101. Biswas, S.K. Does the Interdependence between Oxidative Stress and Inflammation Explain the Antioxidant Paradox? *Oxidative Med. Cell. Longev.* **2016**, *2016*, 5698931. [CrossRef]
102. Hussain, T.; Tan, B.; Yin, Y.; Blachier, F.; Tossou, M.C.; Rahu, N. Oxidative Stress and Inflammation: What Polyphenols Can Do for Us? *Oxidative Med. Cell. Longev.* **2016**, *2016*, 7432797. [CrossRef] [PubMed]
103. Xiong, Y.; Chen, J.; Li, Y. Microglia and astrocytes underlie neuroinflammation and synaptic susceptibility in autism spectrum disorder. *Front. Neurosci.* **2023**, *17*, 1125428. [CrossRef]
104. Matta, S.M.; Hill-Yardin, E.L.; Crack, P.J. The influence of neuroinflammation in Autism Spectrum Disorder. *Brain Behav. Immun.* **2019**, *79*, 75–90. [CrossRef] [PubMed]
105. Saghazadeh, A.; Ataeinia, B.; Keynejad, K.; Abdolalizadeh, A.; Hirbod-Mobarakeh, A.; Rezaei, N. A meta-analysis of pro-inflammatory cytokines in autism spectrum disorders: Effects of age, gender, and latitude. *J. Psychiatr. Res.* **2019**, *115*, 90–102. [CrossRef]
106. Eissa, N.; Sadeq, A.; Sasse, A.; Sadek, B. Role of Neuroinflammation in Autism Spectrum Disorder and the Emergence of Brain Histaminergic System. Lessons Also for BPSD? *Front. Pharmacol.* **2020**, *11*, 886. [CrossRef] [PubMed]
107. Zhao, H.; Zhang, H.; Liu, S.; Luo, W.; Jiang, Y.; Gao, J. Association of Peripheral Blood Levels of Cytokines With Autism Spectrum Disorder: A Meta-Analysis. *Front. Psychiatry* **2021**, *12*, 670200. [CrossRef]
108. Pangrazzi, L.; Cerilli, E.; Balasco, L.; Khurshid, C.; Tobia, C.; Dall'O', G.M.; Chelini, G.; Perini, S.; Filosi, M.; Barbieri, A.; et al. The interplay between oxidative stress and inflammation supports autistic-related behaviors in Cntnap2 knockout mice. *Brain Behav. Immun.* **2025**, *127*, 5–71. [CrossRef]
109. Than, U.T.T.; Nguyen, L.T.; Nguyen, P.H.; Nguyen, X.H.; Trinh, D.P.; Hoang, D.H.; Nguyen, P.A.T.; Dang, V.D. Inflammatory mediators drive neuroinflammation in autism spectrum disorder and cerebral palsy. *Sci. Rep.* **2023**, *13*, 22587. [CrossRef]
110. Kordulewska, N.K.; Kostyra, E.; Piskorz-Ogórek, K.; Moszyńska, M.; Cieślińska, A.; Fiedorowicz, E.; Jarmołowska, B. Serum cytokine levels in children with spectrum autism disorder: Differences in pro- and anti-inflammatory balance. *J. Neuroimmunol.* **2019**, *337*, 577066. [CrossRef]
111. Lochman, I.; Švachová, V.; Mílková Pavlíková, K.; Medřická, H.; Novák, V.; Trilecová, L.; Pavliska, L.; Procházka, V. Serum Cytokine and Growth Factor Levels in Children with Autism Spectrum Disorder. *Med. Sci. Monit.* **2018**, *24*, 2639–2646. [CrossRef] [PubMed]
112. Abruzzo, P.M.; Matté, A.; Bolotta, A.; Federti, E.; Ghezzo, A.; Guarnieri, T.; Marini, M.; Posar, A.; Siciliano, A.; De Franceschi, L.; et al. Plasma peroxiredoxin changes and inflammatory cytokines support the involvement of neuro-inflammation and oxidative stress in Autism Spectrum Disorder. *J. Transl. Med.* **2019**, *17*, 332. [CrossRef]
113. Petrelli, F.; Pucci, L.; Bezzi, P. Astrocytes and Microglia and Their Potential Link with Autism Spectrum Disorders. *Front. Cell. Neurosci.* **2016**, *10*, 21. [CrossRef]
114. Xie, J.; Huang, L.; Li, X.; Li, H.; Zhou, Y.; Zhu, H.; Pan, T.; Kendrick, K.M.; Xu, W. Immunological cytokine profiling identifies TNF-α as a key molecule dysregulated in autistic children. *Oncotarget* **2017**, *8*, 82390–82398. [CrossRef]
115. Li, Z.; Chao, T.C.; Chang, K.Y.; Lin, N.; Patil, V.S.; Shimizu, C.; Head, S.R.; Burns, J.C.; Rana, T.M. The long noncoding RNA THRIL regulates TNFα expression through its interaction with hnRNPL. *Proc. Natl. Acad. Sci. USA* **2014**, *111*, 1002–1007. [CrossRef] [PubMed]
116. Gomez, J.A.; Wapinski, O.L.; Yang, Y.W.; Bureau, J.F.; Gopinath, S.; Monack, D.M.; Chang, H.Y.; Brahic, M.; Kirkegaard, K. The NeST long ncRNA controls microbial susceptibility and epigenetic activation of the interferon-γ locus. *Cell* **2013**, *152*, 743–754. [CrossRef] [PubMed]

117. Wang, J.; Peng, H.; Tian, J.; Ma, J.; Tang, X.; Rui, K.; Tian, X.; Wang, Y.; Chen, J.; Lu, L.; et al. Upregulation of long noncoding RNA TMEVPG1 enhances T helper type 1 cell response in patients with Sjögren syndrome. *Immunol. Res.* **2016**, *64*, 489–496. [CrossRef]
118. Spurlock, C.F.; Shaginurova, G.; Tossberg, J.T.; Hester, J.D.; Chapman, N.; Guo, Y.; Crooke, P.S.; Aune, T.M. Profiles of Long Noncoding RNAs in Human Naive and Memory T Cells. *J. Immunol.* **2017**, *199*, 547–558. [CrossRef]
119. De Giacomo, A.; Gargano, C.D.; Simone, M.; Petruzzelli, M.G.; Pedaci, C.; Giambersio, D.; Margari, L.; Ruggieri, M. B and T Immunoregulation: A New Insight of B Regulatory Lymphocytes in Autism Spectrum Disorder. *Front. Neurosci.* **2021**, *15*, 732611. [CrossRef]
120. Li, H.; Hao, Y.; Zhang, D.; Fu, R.; Liu, W.; Zhang, X.; Xue, F.; Yang, R. Aberrant expression of long noncoding RNA TMEVPG1 in patients with primary immune thrombocytopenia. *Autoimmunity* **2016**, *49*, 496–502. [CrossRef]
121. Fu, Y.; Zhang, Y.L.; Liu, R.Q.; Xu, M.M.; Xie, J.L.; Zhang, X.L.; Xie, G.M.; Han, Y.T.; Zhang, X.M.; Zhang, W.T.; et al. Exosome lncRNA IFNG-AS1 derived from mesenchymal stem cells of human adipose ameliorates neurogenesis and ASD-like behavior in BTBR mice. *J. Nanobiotechnol.* **2024**, *22*, 66. [CrossRef]
122. Fallah, H.; Sayad, A.; Ranjbaran, F.; Talebian, F.; Ghafouri-Fard, S.; Taheri, M. IFNG/IFNG-AS1 expression level balance: Implications for autism spectrum disorder. *Metab. Brain Dis.* **2020**, *35*, 327–333. [CrossRef]
123. Shih, R.H.; Wang, C.Y.; Yang, C.M. NF-kappaB Signaling Pathways in Neurological Inflammation: A Mini Review. *Front. Mol. Neurosci.* **2015**, *8*, 77. [CrossRef] [PubMed]
124. Young, A.M.; Campbell, E.; Lynch, S.; Suckling, J.; Powis, S.J. Aberrant NF-kappaB expression in autism spectrum condition: A mechanism for neuroinflammation. *Front. Psychiatry* **2011**, *2*, 27. [CrossRef] [PubMed]
125. Honarmand Tamizkar, K.; Badrlou, E.; Aslani, T.; Brand, S.; Arsang-Jang, S.; Ghafouri-Fard, S.; Taheri, M. Dysregulation of NF-κB-Associated LncRNAs in Autism Spectrum Disorder. *Front. Mol. Neurosci.* **2021**, *14*, 747785. [CrossRef] [PubMed]
126. Safa, A.; Badrlou, E.; Arsang-Jang, S.; Sayad, A.; Taheri, M.; Ghafouri-Fard, S. Expression of NF-κB associated lncRNAs in schizophrenia. *Sci. Rep.* **2020**, *10*, 18105. [CrossRef]
127. Safa, A.; Noroozi, R.; Taheri, M.; Ghafouri-Fard, S. Association Analysis of ANRIL Polymorphisms and Haplotypes with Autism Spectrum Disorders. *J. Mol. Neurosci.* **2021**, *71*, 187–192. [CrossRef]
128. Taheri, M.; Younesi, Z.; Moradi, S.; Tamizkar, K.H.; Razjouyan, K.; Arsang-Jang, S.; Ghafouri-Fard, S.; Neishabouri, S.M. Altered expression of CCAT1 and CCAT2 lncRNAs in autism spectrum disorder. *Gene Rep.* **2021**, *23*, 101172. [CrossRef]
129. Kaur, A.; Lee, L.H.; Chow, S.C.; Fang, C.M. IRF5-mediated immune responses and its implications in immunological disorders. *Int. Rev. Immunol.* **2018**, *37*, 229–248. [CrossRef]
130. Zhao, C.; Qiao, C.; Zong, L.; Chen, Y. Long non-coding RNA-CCAT2 promotes the occurrence of non-small cell lung cancer by regulating the Wnt/β-catenin signaling pathway. *Oncol. Lett.* **2018**, *16*, 4600–4606. [CrossRef]
131. Liu, Z.; Ma, C.; Tang, X.; Tang, Q.; Lou, L.; Yu, Y.; Zheng, F.; Wu, J.; Yang, X.B.; Wang, W.; et al. The Reciprocal Interaction Between LncRNA CCAT1 and miR-375-3p Contribute to the Downregulation of IRF5 Gene Expression by Solasonine in HepG2 Human Hepatocellular Carcinoma Cells. *Front. Oncol.* **2019**, *9*, 1081. [CrossRef] [PubMed]
132. Li, X.; Xiang, Y.; Li, F.; Yin, C.; Li, B.; Ke, X. WNT/β-Catenin Signaling Pathway Regulating T Cell-Inflammation in the Tumor Microenvironment. *Front. Immunol.* **2019**, *10*, 2293. [CrossRef]
133. Fang, Y.; Wan, C.; Wen, Y.; Wu, Z.; Pan, J.; Zhong, M.; Zhong, N. Autism-associated synaptic vesicle transcripts are differentially expressed in maternal plasma exosomes of physiopathologic pregnancies. *J. Transl. Med.* **2021**, *19*, 154. [CrossRef] [PubMed]
134. Sabaie, H.; Dehghani, H.; Shiva, S.; Asadi, M.R.; Rezaei, O.; Taheri, M.; Rezazadeh, M. Mechanistic Insight Into the Regulation of Immune-Related Genes Expression in Autism Spectrum Disorder. *Front. Mol. Biosci.* **2021**, *8*, 754296. [CrossRef] [PubMed]
135. Tay, Y.; Rinn, J.; Pandolfi, P.P. The multilayered complexity of ceRNA crosstalk and competition. *Nature* **2014**, *505*, 344–352. [CrossRef]
136. Sun, J.J.; Chen, B.; Yu, T. Construction of an immune-related ceRNA network to screen for potential diagnostic markers for autism spectrum disorder. *Front. Genet.* **2022**, *13*, 1025813. [CrossRef]
137. Balasubramanian, R.; Vinod, P.K. Inferring miRNA sponge modules across major neuropsychiatric disorders. *Front. Mol. Neurosci.* **2022**, *15*, 1009662. [CrossRef]
138. Jiang, M.; Wang, Z.; Lu, T.; Li, X.; Yang, K.; Zhao, L.; Zhang, D.; Li, J.; Wang, L. Integrative analysis of long noncoding RNAs dysregulation and synapse-associated ceRNA regulatory axes in autism. *Transl. Psychiatry* **2023**, *13*, 375. [CrossRef]
139. Parellada, M.; Andreu-Bernabeu, Á.; Burdeus, M.; San José Cáceres, A.; Urbiola, E.; Carpenter, L.L.; Kraguljac, N.V.; McDonald, W.M.; Nemeroff, C.B.; Rodriguez, C.I.; et al. In Search of Biomarkers to Guide Interventions in Autism Spectrum Disorder: A Systematic Review. *Am. J. Psychiatry* **2023**, *180*, 23–40. [CrossRef]
140. Qin, Y.; Cao, L.; Zhang, J.; Zhang, H.; Cai, S.; Guo, B.; Wu, F.; Zhao, L.; Li, W.; Ni, L.; et al. Whole-Transcriptome Analysis of Serum L1CAM-Captured Extracellular Vesicles Reveals Neural and Glycosylation Changes in Autism Spectrum Disorder. *J. Mol. Neurosci.* **2022**, *72*, 1274–1292. [CrossRef]
141. Metz, C.E. Basic principles of ROC analysis. *Semin. Nucl. Med.* **1978**, *8*, 283–298. [CrossRef] [PubMed]
142. Perlis, R.H. Translating biomarkers to clinical practice. *Mol. Psychiatry* **2011**, *16*, 1076–1087. [CrossRef] [PubMed]

143. Abruzzo, P.M.; Ghezzo, A.; Bolotta, A.; Ferreri, C.; Minguzzi, R.; Vignini, A.; Visconti, P.; Marini, M. Perspective Biological Markers for Autism Spectrum Disorders: Advantages of the Use of Receiver Operating Characteristic Curves in Evaluating Marker Sensitivity and Specificity. *Dis. Markers* **2015**, *2015*, 329607. [CrossRef] [PubMed]
144. Li, C.; Zheng, X.; Liu, P.; Li, M. Clinical value of lncRNA TUG1 in temporal lobe epilepsy and its role in the proliferation of hippocampus neuron via sponging miR-199a-3p. *Bioengineered* **2021**, *12*, 10666–10673. [CrossRef]
145. Mirzajani, S.; Ghafouri-Fard, S.; Habibabadi, J.M.; Arsang-Jang, S.; Omrani, M.D.; Fesharaki, S.S.H.; Sayad, A.; Taheri, M. Expression Analysis of lncRNAs in Refractory and Non-Refractory Epileptic Patients. *J. Mol. Neurosci.* **2020**, *70*, 689–698. [CrossRef]
146. Awwad, D.A. Beyond classic editing: Innovative CRISPR approaches for functional studies of long non-coding RNA. *Biol. Methods Protoc.* **2019**, *4*, bpz017. [CrossRef]
147. Hazan, J.; Bester, A.C. CRISPR-Based Approaches for the High-Throughput Characterization of Long Non-Coding RNAs. *Non-Coding RNA* **2021**, *7*, 79. [CrossRef]
148. Zibitt, M.S.; Hartford, C.C.R.; Lal, A. Interrogating lncRNA functions via CRISPR/Cas systems. *RNA Biol* **2021**, *18*, 2097–2106. [CrossRef]
149. Ross, P.J.; Zhang, W.B.; Mok, R.S.F.; Zaslavsky, K.; Deneault, E.; D'Abate, L.; Rodrigues, D.C.; Yuen, R.K.C.; Faheem, M.; Mufteev, M. Synaptic Dysfunction in Human Neurons With Autism-Associated Deletions in PTCHD1-AS. *Biol. Psychiatry* **2020**, *87*, 139–149. [CrossRef]
150. Zhang, T.N.; Wen, R.; Yang, Y.H.; Yang, N.; Liu, C.F. Integration of transcriptomic, proteomic, and metabolomic data to identify lncRNA rPvt1 associations in lipopolysaccharide-treated H9C2 cardiomyocytes. *Front. Genet.* **2023**, *14*, 1278830. [CrossRef]
151. Teng, P.; Li, Y.; Ku, L.; Wang, F.; Goldsmith, D.R.; Wen, Z.; Yao, B.; Feng, Y. The human lncRNA GOMAFU suppresses neuronal interferon response pathways affected in neuropsychiatric diseases. *Brain Behav. Immun.* **2023**, *112*, 175–187. [CrossRef] [PubMed]
152. Chen, B.; Deng, S.; Ge, T.; Ye, M.; Yu, J.; Lin, S.; Ma, W.; Songyang, Z. Live cell imaging and proteomic profiling of endogenous NEAT1 lncRNA by CRISPR/Cas9-mediated knock-in. *Protein Cell* **2020**, *11*, 641–660. [CrossRef] [PubMed]

biomolecules

MDPI

Review

Unveiling Metabolic Signatures as Potential Biomarkers in Common Cancers: Insights from Lung, Breast, Colorectal, Liver, and Gastric Tumours

Kha Wai Hon and Rakesh Naidu *

Jeffrey Cheah School of Medicine and Health Sciences, Monash University Malaysia, Jalan Lagoon Selatan, Bandar Sunway 47500, Selangor, Malaysia; dominichon93@gmail.com
* Correspondence: rakesh.naidu@monash.edu; Tel.: +603-5514 6345

Abstract: Reprogramming is a hallmark of cancer, enabling tumour cells to sustain rapid proliferation, resist cell death, and adapt to hostile microenvironments. This review explores the expression profiles of key metabolic enzymes and transporters involved in glucose, amino acid, and lipid metabolism across the five most deadly cancers worldwide: lung, breast, colorectal, liver, and gastric cancers. Through a comparative analysis, we identify consistent upregulation of glycolytic enzymes such as LDHA, PKM2, and HK2, as well as nutrient transporters like GLUT1, ASCT2, and LAT1, which contribute to cancer progression, metastasis, and therapy resistance. The role of enzymes involved in glutaminolysis (e.g., GLS1, GDH), one-carbon metabolism (e.g., SHMT2, PHGDH), and fatty acid synthesis (e.g., FASN, ACLY) is also examined, with emphasis on their emerging relevance as diagnostic, prognostic, and predictive biomarkers. While several metabolic proteins show strong potential for clinical translation, only a few, such as tumour M2-pyruvate kinase (TuM2-PK) and serum LDH measurement, have progressed into clinical use or trials. This review addresses some of the challenges in biomarker development. Ultimately, our findings underscore the importance of metabolic proteins not only as functional drivers of malignancy but also as promising candidates for biomarker discovery. Advancing their clinical implementation could significantly enhance early detection, treatment stratification, and personalized oncology.

Keywords: biomarker; breast cancer; colorectal cancer; gastric cancer; lung cancer; liver cancer; metabolic proteins

1. Introduction

Cancer is a leading global health challenge, accounting for a significant proportion of morbidity and mortality worldwide. According to the latest Global Cancer Statistics 2022, an estimated 20 million new cancer cases were diagnosed globally, highlighting its pervasive impact, with lung, breast, colorectal (CRC), prostate, stomach, liver and thyroid cancers ranking as the most common malignancies [1]. Cancer also remains a significant cause of mortality, contributing to approximately 9.7 million deaths globally in 2022 [1]. Lung cancer is the top contributor to cancer-related deaths (1.82 million deaths, 18.7% of total), followed by CRC (90,000 deaths, 9.3%), liver cancer (760,000 deaths, 7.8%), breast cancer (670,000 deaths, 6.9%), and stomach cancer (660,000 deaths, 6.8%) [1]. Projections indicate that the global cancer burden will continue to rise, with annual new cases expected to reach 35.3 million by 2050, which is almost an 80% increase from 2020 [2]. Additionally, 18.5 million cancer deaths are projected by 2050, which is an 89.7% increase from the

2022 estimate of 9.7 million [2]. These statistics and projections underscore the urgent need for improved diagnostic and therapeutic strategies, mainly by identifying reliable cancer biomarkers.

Cancer biomarkers are molecules that indicate the presence or progression of cancer, and their identification and application have significantly advanced cancer diagnosis, prognosis, and treatment. These biomarkers include proteins, DNA, RNA, metabolites, and cellular changes, which can be detected in body fluids, tissues, or cells [3–5]. The advent of high-throughput technologies such as next-generation sequencing (NGS), proteomics, and metabolomics has accelerated the discovery of novel biomarkers [6–9]. For example, NGS enables comprehensive analysis of genomic alterations, facilitating the identification of cancer-specific mutations like BRCA1/2 in breast and ovarian cancers, TP53 in multiple cancer types, KRAS in CRC and lung cancers, BRAF in melanoma and thyroid cancer, PIK3CA in breast cancer, and APC in CRC [10,11]. More importantly, cancer biomarkers based on metabolic enzymes, drug transporters, and metabolic proteins are gaining prominence due to their critical roles in tumour growth, metabolism, drug resistance, and therapy response [12–14]. These biomarkers are primarily associated with cancer-specific alterations in cellular metabolism (such as the Warburg effect, glutamine and lipid metabolism) and drug metabolism [12–14].

Metabolic reprogramming is a key hallmark of cancers, in which cancer cells adapt their metabolism to sustain rapid proliferation, resist stress, evade immune responses, and adapt to the tumour microenvironment [15]. One of the most well-known metabolic alterations in cancer cells is the Warburg effect, which refers to the preference of cancer cells for aerobic glycolysis, even in the presence of sufficient oxygen [16]. This shift from oxidative phosphorylation to glycolysis can provide cancer cells with rapid ATP production and rapid synthesis of intermediates, such as nucleotides, amino acids, and lipids, which are crucial for proliferation [16]. Additionally, mitochondrial metabolism still plays an essential role in the metabolic reprogramming of cancer cells by contributing to biosynthesis, redox balance, and suppression of apoptosis [17,18]. Cancer cells also demonstrate a pronounced preference for glutamine metabolism, which is an amino acid critical for replenishing tricarboxylic acid (TCA) cycle intermediates, providing nitrogen for biosynthetic pathways, and producing glutathione to manage reactive oxygen species (ROS) [19]. Cancer cells also tend to reprogram lipid metabolism by enhancing fatty acid synthesis and uptake to support membrane biosynthesis, energy storage and regulation of signalling pathways [20].

Redox homeostasis is often altered in the metabolic reprogramming of cancer cells, in which elevated metabolic activity and ROS production necessitate robust antioxidant systems, such as increased glutathione synthesis and upregulation of the pentose phosphate pathway (PPP), to maintain cellular survival under oxidative stress [21–24]. Similarly, the tumour microenvironment, which is usually hypoxic, further drives metabolic shifts through the activation of hypoxia-inducible factors (HIFs), which enhance glycolysis and lactate production while suppressing mitochondrial oxidative phosphorylation [25,26]. Key enzymes such as hexokinase 2 (HK2), pyruvate kinase M2 (PKM2), and glutaminase (GLS) regulate glycolysis, glutaminolysis, and biosynthetic pathways, while transporters like GLUT1 and ASCT2 mediate increased glucose and amino acid uptake, respectively, highlighting their roles in sustaining cancer cell metabolism [27–32]. Additionally, enzymes like fatty acid synthase (FASN) and glucose-6-phosphate dehydrogenase (G6PD) support lipid biosynthesis and antioxidant defence, respectively, to fulfil the biochemical demands of rapidly proliferating tumours [33–35].

Membrane transporters such as monocarboxylate transporters (MCTs) facilitate lactate export, contributing to an acidic microenvironment that promotes invasion and immune evasion, while drug transporters, including ABCC1, BCRP and P-gp, modulate the uptake and efflux of chemotherapeutic agents, impacting drug efficacy and resistance [12,36,37]. These molecules are frequently dysregulated in cancers, making them ideal biomarkers for early detection, prognosis, and therapeutic response monitoring. These metabolic components can be leveraged to identify cancer-specific signatures by targeting their specific expression patterns or functional activity, improving diagnostic precision and developing novel, metabolism-focused therapeutic strategies. Therefore, this review aims to provide insights into the discovery of metabolic enzymes, proteins, and transporters as potential biomarkers for the top five most deadly cancers, including lung, CRC, breast, liver and stomach.

2. The Warburg Effect and Glucose Metabolism

The Warburg effect is defined by the preference of cancer cells for aerobic glycolysis over oxidative phosphorylation. Several key glycolytic enzymes that regulate glucose metabolism and associated pathways have been widely reported in cancers, including ALDOA, ENO1, HK2, LDH, PFK1, PGAM1, PGK1, PKM2 and TPI1 [38–41]. Lactate dehydrogenase (LDH) is the most widely studied in cancer diagnosis, while different isoforms could be associated with distinct responses to cancer treatment [42]. Lactate dehydrogenase A (LDHA) converts pyruvate to lactate, regenerating NAD+ and supporting the production of lactate, which acidifies the tumour microenvironment to promote invasion and immune evasion [43]. Lactate dehydrogenase B (LDHB) could be dysregulated in specific cancer types depending on the glycolytic requirements, while most of the time, LDHB is silenced by promoter methylation [44–46].

Aldolase A (ALDOA) is a key glycolytic enzyme responsible for the cleavage of fructose-1,6-bisphosphate into glyceraldehyde-3-phosphate and dihydroxyacetone phosphate [47]. Beyond its metabolic role, ALDOA has been increasingly recognised for its involvement in cancer progression by promoting EMT, metastasis and tumour progression [48–50]. ALDOA is often upregulated in cancers, particularly those aggressive phenotypes with high glycolytic activity and associated with poor prognosis [48,50–53]. Alpha enolase (ENO1) catalyses the interconversion of 2-phosphoglycerate to phosphoenolpyruvate in both gluconeogenesis and glycolytic metabolic pathways [54]. In cancers, ENO1 has a non-metabolic function as a plasminogen receptor in the tumour cell surface, enhancing proliferation, migration, invasion, and metastasis [55,56]. Overexpression of ENO1 is mainly associated with chemoresistance, poor prognosis, reduced survival and stemness in cancers [55,57–61].

Hexokinase 2 (HK2) catalyses the initial phosphorylation of glucose to sustain glycolytic flux into cancer cells, while its overexpression enhances glycolytic flux, supporting drug resistance and tumour growth besides associated with poor prognosis [14,54,62–66]. Phosphofructokinase-1 (PFK1) controls the rate-limiting step, converting fructose-6-phosphate to fructose-1,6-bisphosphate [38]. Elevated PFK1 expression, particularly the PFKP isoform, is often linked to enhanced glycolysis, tumour growth, and resistance to therapies, making it a potential target for cancer treatment strategies [67,68]. Moreover, phosphoglycerate kinase 1 (PGK1) catalyses the reversible transfer of a phosphate group from 1,3-bisphosphoglycerate (1,3-BPG) to ADP, producing 3-phosphoglycerate (3-PG) and ATP in the first rate-limiting step of glycolysis [54]. It has non-glycolytic roles in tumour progression by promoting angiogenesis and metastasis, while its overexpression is associated with poor prognosis and drug resistance [69–71]. Phosphoglycerate mutase 1 (PGAM1) is a glycolytic enzyme that catalyses the conversion of 3-phosphoglycerate (3-PG) to 2-phosphoglycerate (2-PG).

Pyruvate kinase M2 (PKM2) is a cancer-specific isoform that facilitates the conversion of phosphoenolpyruvate (PEP) to pyruvate, with its less efficient activity allowing glycolytic intermediates to accumulate for biosynthesis [72]. Overexpression of PKM2 isoform in cancers is mainly associated with treatment response, metastasis, poor prognosis and reduced survival, making it a promising biomarker for diagnosis and treatment monitoring [73–75]. Triose phosphate isomerase (TPI) catalyses the interconversion between dihydroxyacetone phosphate and glyceraldehyde-3-phosphate in glycolysis and gluconeogenesis [76]. Elevated TPI levels in patient samples correlate with poor prognosis, tumour proliferation, migration, and invasion [77–80]. In addition, glucose transporters (GLUTs) play a crucial role in the Warburg effect by increasing glucose uptake for glycolysis and anabolic processes [54]. The most relevant GLUTs in cancer include GLUT1 (SLC2A1), GLUT3 (SLC2A3) and GLUT4 (SLC2A4), which are highly expressed in breast, CRC and lung cancers, with a high affinity for glucose and commonly associated with poor prognosis and therapy resistance [32,81,82]. The upregulation of glycolytic enzymes and glucose transporters collectively enables the metabolic reprogramming necessary for rapid energy production and biosynthesis to support cancer cell proliferation, stress adaptation, and tumour progression. However, the expression levels of key glycolytic enzymes vary across different cancer types, reflecting the metabolic adaptations of various tumour types to support their growth and survival.

2.1. Breast Cancer and Glucose Metabolism

In breast cancer, elevated LDHA levels have been detected in cancer tissue and patient serum, suggesting that LDHA could be released from breast cancer cells into the body circulation and remain stable for clinical detection [83–85]. In metastatic breast cancer, LDHA overexpression in patient serum was correlated with poor prognosis and unfavourable survival after first-line chemotherapy treatment, which could be an independent prognostic biomarker [83]. Patients with basal-like/triple negative BC (TNBC) expressed increased levels of LDHA and LDHB in serum measurement and tissue microarray analysis, which were significantly associated with shorter survival, poor prognosis, pathological complete response (pCR) to neoadjuvant chemotherapy and cancer relapse, suggesting the potential of LDHA and LDHB as biomarkers for treatment monitoring [86,87]. Dong et al. also reported that LDHA overexpression in cancer tissue via immunohistochemistry and serum LDH status was positively correlated with a higher incidence of brain metastasis among TNBC patients, suggesting the potential of LDHA as a prognostic factor in TNBC [88]. Collectively, elevated levels of LDHA and LDHB in breast cancer tissue could be developed into tissue-based assays. The measurement of LDHA in patient serum may facilitate less invasive testing methods for routine monitoring.

Both PKM2 and HK2 were significantly upregulated in breast cancer tissues compared to adjacent normal tissues [64,89,90]. Elevated expression levels of these glycolytic enzymes have been associated with advanced clinical stages, positive lymph node involvement, and distant metastasis [64,89,90]. Based on microarray data and immunohistochemical (IHC) analysis, HK2 protein expression was significantly correlated with HIF-1α immunoreactivity, Ki67 status and cancer recurrence in breast cancer patients [64]. Similarly, PKM2 overexpression was observed in tissue samples of various breast cancer subtypes, including TNBC, luminal, and HER2-positive cancers, significantly associated with poor prognosis, EMT and increased chemosensitivity to epirubicin and 5-fluorouracil [89,90]. These findings suggest that PKM2 and HK2 play crucial roles in breast cancer progression and may serve as potential biomarkers or therapeutic targets. Additionally, PFK1 levels were higher in breast cancer tissues than in adjacent non-cancerous tissues, with a prominent shift in isoform expression [91]. Breast cancer tissues predominantly express the platelet-type isoform (PFKP), whereas adjacent tissues mainly express the liver-type isoform (PFKL) [91]. This switch was associated with increased glycolytic activity in tumour cells, highlighting the potential role of cancer-specific PFK1 isoform, predominantly PFKP, in clinical diagnosis. Coelho et al. also demonstrated that the intracellular distribution and enzymatic activities of HK2 and PFK1 were positively correlated with the aggressiveness and invasiveness of different breast carcinoma subtypes, highlighting the great potential of these enzymes as prognostic biomarkers in breast cancer [92].

IHC analysis revealed that PGAM1 was significantly overexpressed in tissue samples of invasive ductal breast carcinomas and associated with poor overall survival [93]. PGAM1 exhibited metabolic and non-metabolic roles in breast cancer cells, in which it promoted cell proliferation by upregulating glycolysis besides interacting with cytoskeletal protein ACTA2 to induce cell migration and metastasis [93]. Furthermore, the expression of glucose transporters, especially GLUT1, has been extensively investigated in breast cancer samples due to their role in facilitating increased glucose uptake in metabolic reprogramming. Previous studies have shown that GLUT1 was overexpressed in invasive ductal carcinoma and TNBC based on IHC staining [94–96]. This overexpression was significantly associated with higher nuclear grade and negative status for EGFR, HER2, oestrogen and progesterone receptors [94,96]. Patients with GLUT1-positive tumours had shorter disease-free and overall survival periods, suggesting that GLUT1 expression correlates with more aggressive tumour behaviour and poorer prognosis [94].

Another IHC study by Sun et al. reported that GLUT1 was significantly overexpressed in the tissue of invasive ductal breast carcinoma (IDBC) with type 2 diabetes mellitus (T2DM) compared to those without T2DM, suggesting enhanced glucose uptake and metabolic reprogramming in IDC [97]. GLUT1 overexpression was closely associated with the upregulation of matrix metalloproteinases MMP2 and MMP9 in IDBC tissue, which is crucial for extracellular matrix degradation and tumour cell invasion [97]. This study establishes a mechanistic link between glucose metabolism and tumour invasiveness, where GLUT1 overexpression promotes a more aggressive cancer phenotype in breast cancer [97]. These findings highlight the importance of GLUT1 as a potential target for future therapeutic development to reduce tumour invasiveness and metastasis in breast cancer patients with metabolic disorders such as diabetes.

Pinheiro et al. investigated the expression profiles of GLUT1 and carbonic anhydrase IX (CAIX) in breast cancer tissue via IHC and their correlation with adverse prognostic factors, particularly monocarboxylate transporter 1 (MCT1) overexpression [98]. GLUT1 and CAIX were significantly upregulated in aggressive breast cancer subtypes and strongly associated with poor prognostic factors, including high tumour grade, hypoxic tumour microenvironment, and increased metabolic reprogramming [98]. This study highlights a positive correlation between GLUT1, CAIX, and MCT1 overexpression, suggesting a cooperative role in tumour progression by facilitating glycolytic metabolism and acid-base balance in hypoxic conditions. These findings underscore the potential of GLUT1, CAIX, and MCT1 as prognostic biomarkers and possible therapeutic targets in breast cancer. Their co-expression may help refine patient stratification, guide treatment decisions, and support the development of novel metabolic inhibitors to improve outcomes in aggressive breast cancer cases.

HER2-positive ($HER2^+$) breast cancer is another highly aggressive subtype and is commonly associated with poor prognosis [99]. Previously, Wang et al. investigated the role of sodium-dependent glucose transporter 1 (SGLT1) in $HER2^+$ breast cancer, focusing on its stabilisation by HER2 and its involvement in oncogenic signalling [100]. SGLT1 and GLUT1 are involved in glucose transport but differ in their mechanisms, functions, and expression patterns. SGLT1 is an active transporter that uses sodium gradients to import glucose into cells, especially under low-glucose conditions actively [101]. In contrast, GLUT1 is a passive transporter that allows glucose to move down its concentration gradient, supporting rapid glucose uptake in highly glycolytic tumours [54]. GLUT1 is more commonly overexpressed in cancer cells, whereas SGLT1 expression in cancers is little known. SGLT1 was significantly overexpressed in HER2+ breast cancer tissues as confirmed via IHC staining, which positively correlated with HER2 status, lymph node metastasis, shorter disease-free survival and poor overall survival [100]. Further in-depth analysis showed that HER2 stabilisation prevented SGLT1 degradation, maintaining intracellular glucose levels critical for cancer cell survival [100]. Inhibition of HER2 using targeted inhibitors reduced SGLT1 expression in HER2+ breast cancer cells, while suppressing SGLT1 function using phlorizin or HER2 inhibitors led to a significant decrease in glycolytic activity with downregulation of PI3K/Akt/mTOR signalling, reducing breast cancer cell proliferation [100]. This study identified SGLT1 as a key metabolic and prognostic marker in HER2+ breast cancer, linking glucose transport with oncogenic signalling.

The expression of other glucose transporters has been explored in breast cancer studies. Krzeslak et al. examined GLUT1 and GLUT3 protein expression in breast cancer tissue using Western blot, which reported the presence of GLUT1 in 48.7% and GLUT3 in 22.8% of tumour samples [102]. Though not statistically significant, GLUT1 expression was more prevalent in poorly differentiated tumours (grades 2 and 3) than in well-differentiated ones [102]. Compared to normal breast tissue, GLUT1 was detected in 22.2% and GLUT3 in 16.7% of samples, suggesting increased expression in cancerous tissues. These findings highlight GLUT1 as a more dominant glucose transporter in breast cancer, while GLUT3 may play a more selective role, potentially linked to aggressive tumour phenotypes, warranting further investigation.

Monocarboxylate transporters (MCTs) are integral membrane proteins that facilitate the transmembrane transport of monocarboxylates, such as lactate and pyruvate. MCT1, MCT2, and MCT4 expression patterns have been investigated in breast cancer patient samples to understand their roles in tumour progression and potential as therapeutic targets. Based on an IHC study using high-density tissue microarray, MCT1 (SLC16A1) was overexpressed in glycolytic breast tumours, which was associated with increased glycolytic activity and poor prognosis in breast cancer patients [103]. Inhibition of MCT1 impairs the proliferation of glycolytic breast cancer cells by disrupting pyruvate export, leading to enhanced oxidative metabolism and reduced tumour growth [103]. Similarly, Johnson et al. investigated the MCT1 expression in different subtypes of IDC using tissue microarrays [104]. High MCT1 expression was observed in TNBC cases, compared to ER+ and/or PR+ and HER2+ tumours [104]. MCT1 overexpression was positively correlated with large tumour size, high-grade tumours, shorter progression-free survival, and cancer recurrence, irrespective of breast cancer subtypes [104]. MCT1 could be a prognostic biomarker in IDC.

IHC analyses have demonstrated MCT2 expression in breast carcinoma samples, with some evidence suggesting a correlation between high MCT2 expression and poor patient prognosis. Elevated MCT2 (SLC16A7) levels were associated with increased expression of interleukin-1β (IL-1β) and lipocalin-2 (LCN2), both of which are linked to tumour progression and adverse outcomes [105]. Combined high expression of MCT2 with IL-1β or LCN2 has been proposed as a potential biomarker for poor prognosis in breast cancer patients [105]. Meanwhile, MCT4 (SLC16A3) was upregulated in breast cancer tissue in IHC analysis, particularly TNBC subtypes, and associated with high tumour grade and poor prognosis [82]. Functional studies have shown that the PI3K regulates the MCT4–Akt signalling pathway to support breast cancer cell survival under hypoxic conditions, contributing to tumour growth and disease progression [82]. The authors also proposed that tumoral MCT4 expression could serve as an independent prognostic survival factor for TNBCs and node-negative breast cancer [82]. The differential expression patterns and functional roles of MCT1, MCT2 and MCT4 highlight their potential as prognostic biomarkers and therapeutic targets in breast cancer. In summary, evidence suggests that GLUT1, HK2, LDHA, MCT1, PKM2 and PFK1 could become potential biomarkers, either diagnostic or prognostic, in breast cancer, depending on the molecular subtypes and treatment involved.

2.2. Lung Cancer and Glucose Metabolism

LDHA and LDHB subunits play pivotal roles in the metabolic reprogramming of lung cancer cells. IHC analyses revealed elevated LDHA protein levels in both lung adenocarcinoma (LUAD) and lung squamous carcinoma (LUSC) compared to adjacent normal tissues [106]. This overexpression correlates with poor prognosis, radiotherapy resistance, enhanced glycolysis and hypoxia within the tumour microenvironment, as evidenced by positive associations between LDHA expression and markers such as HIF1A and GLUT-1. The overexpression of LDHA in non-small cell lung cancer (NSCLC) is primarily regulated by hypoxia-inducible factors (HIFs), which activate LDHA transcription under hypoxic conditions prevalent in tumours [106]. This regulatory mechanism underscores the enzyme's contribution to the metabolic adaptations of cancer cells. Meanwhile, elevated LDHB levels in lung adenocarcinoma tissues are significantly associated with poor prognosis and reduced survival rates in cancer patients with KRAS-wild-type lung cancer [107]. Silencing LDHB in lung cancer cells has been shown to reduce invasion and migration capabilities, likely due to decreased intracellular glutathione levels and altered mitochondrial metabolism [107]. This highlights LDHB's role in maintaining lung cancer cells' redox balance and metastatic potential. The differential expression patterns of LDHA and LDHB in lung cancer suggest their potential as prognostic biomarkers.

Elevated expression of ALDOA has been observed in lung cancer patient samples of different subtypes, with studies indicating its association with tumour progression and poor prognosis. In an IHC analysis of NSCLC tumour samples, higher ALDOA levels correlate with increased metastatic potential and reduced survival rates [108]. Mechanistically, ALDOA promotes lung cancer metastasis by activating the HIF-1α/MMP9 axis, enhancing cell invasion and migration [108]. In lung adenocarcinoma, overexpression of ALDOA is linked to advanced tumour stages and unfavourable outcomes, suggesting its role as an independent prognostic factor [53]. In lung squamous cell carcinoma, ALDOA is significantly upregulated in metastatic tissues compared to non-metastatic and adjacent normal tissues, implicating its involvement in tumourigenesis and metastasis [50]. These findings highlight ALDOA's potential as a biomarker for lung cancer progression and as a target for therapeutic intervention. Meanwhile, ENO1 has been implicated in the progression and metastasis of lung cancer. IHC and Western blot analyses have shown that ENO1 protein is overexpressed in lung cancer tissues, and associated with increased tumour growth, migration, and invasion [109–111]. Mechanistically, ENO1 interacts with hepatocyte growth factor receptor (HGFR) and activates HGFR and Wnt signaling pathways, leading to EMT and metastasis [110]. In NSCLC, ENO1 promotes cell proliferation through the focal adhesion kinase (FAK)-mediated PI3K/AKT pathway, further contributing to tumour progression [111]. These findings suggest that ENO1 plays a significant role in lung cancer development and may serve as a potential therapeutic target.

PFKP, PGAM1 and PGK1 are key metabolic enzymes in lung cancer, contributing to tumour progression and patient prognosis. PFKP was significantly overexpressed in lung cancer tissues as confirmed via IHC staining, which correlated with larger tumour size and poorer patient prognosis [112]. Functional studies confirmed that reducing PFKP expression in lung cancer cells decreased glucose uptake, lactate production, and cell proliferation, highlighting its role in tumour metabolism and growth. Furthermore, PGK1 was significantly upregulated in lung cancer tissues and associated with increased migratory and invasive capabilities of cancer cells [69]. Elevated PGK1 expression in IHC analysis was significantly correlated with poor outcomes in lung cancer patients [69]. Mechanistically, PGK1 promotes metastasis by interacting with HIV Tat-specific factor 1 (HTATSF1) involved in RNA metabolism, affecting multiple genes in tumour progression [69].

Several IHC studies on NSCLC tissues also reported that PGAM1 was significantly overexpressed compared to adjacent normal lung tissues, and associated with poor prognosis, increased tumour growth, and enhanced metastatic potential [113,114]. Li et al. demonstrated that PGAM1 functions as an oncogene by activating the transforming growth factor-β (TGF-β) signalling pathway in NSCLC [114]. PGAM1 enhances cancer cell proliferation and invasion by upregulating TGF-β signalling, contributing to tumour progression and EMT [114]. Another study revealed that PGAM1 is a critical regulator of mTOR-mediated tumour growth and is associated with aggressive NSCLC phenotypes [113]. PGAM1 promotes glycolytic metabolism, facilitating ATP production and biosynthetic processes essential for rapid tumour cell proliferation [113]. IHC analyses of NSCLC patient samples confirmed a significant correlation between high PGAM1 expression and poor survival [113]. The overexpression of PGAM1 in NSCLC patient samples and its roles in TGF-β and mTOR signalling pathways underscore its potential as a prognostic biomarker and therapeutic target.

In NSCLC, the protein expression levels of GLUT1, MCT1, and MCT4 have been extensively studied. An IHC study analysing 445 NSCLC patients demonstrated that those with positive expression of both GLUT1 and PKM2 (G+/P+) had significantly poorer overall and disease-free survival rates following curative R0 resection [115]. Elevated GLUT1 expression correlated with advanced tumour stage, lymphatic invasion, and pleural invasion [115]. These findings suggest that assessing GLUT1 and PKM2 expression could provide valuable prognostic information for NSCLC patients post-R0 resection. In addition, MCT1 protein was expressed in cancer cells and stromal cells within NSCLC tissues, as validated with tissue microarray and IHC staining [116]. Interestingly, MCT1 overexpression in cancer cells was associated with improved disease-specific survival, suggesting a less aggressive tumour phenotype [116]. Conversely, elevated MCT1 expression in stromal cells correlates with poorer prognosis, potentially due to its role in facilitating lactate uptake and promoting a supportive microenvironment for tumour growth [116]. Furthermore, MCT4 was upregulated in NSCLC tissue as shown by IHC staining, particularly in tumours exhibiting high glycolytic activity [117]. Elevated MCT4 protein levels have been linked to increased tumour aggressiveness and worse overall survival [117]. High MCT4 expression could be a negative prognostic marker in NSCLC patients [117].

2.3. *CRC and Glucose Metabolism*

The overexpression of glycolytic enzymes, glucose transporters, and monocarboxylate transporters in colorectal cancer (CRC) has been extensively documented. HK2 is frequently overexpressed in CRC tissues compared to normal colonic epithelium based on IHC studies [63,118,119]. This upregulation is associated with enhanced glycolytic activity, promoting tumour growth and survival under hypoxic conditions [118]. Elevated HK2 expression correlates with larger tumour size, deeper invasion, liver metastasis, and advanced TNM stages in CRC patients [63,118,119]. Moreover, high HK2 level is often associated with increased recurrence rates and overall mortality, establishing HK2 as an independent prognostic factor for both disease-free and overall survival in CRC [63,118,119]. Conversely, He et al. conducted IHC analysis of two independent cohorts using CRC tissue [120]. HK1 was overexpressed in 30% of CRC cases, and significantly correlated with advanced tumour stages and poorer overall survival [120]. Multivariate analysis further identified HK1 overexpression as an independent prognostic factor for CRC, suggesting that HK1 could serve as a valuable biomarker for predicting patient prognosis [120].

ENO1 was significantly upregulated in CRC tissues as reported in IHC staining, with higher expression levels correlating with advanced TNM stages, lymph node involvement, vascular invasion, and distant metastasis [58,121,122]. Elevated ENO1 expression was associated with poor overall survival in CRC patients, serving as an independent prognostic factor [58,121,122]. ENO1 enhances CRC cell proliferation, drug resistance, EMT, migration, and invasion by regulating oncogenic pathways, such as AMPK/mTOR signaling [58,121,122]. These findings suggest that ENO1 may serve as a valuable biomarker for CRC prognosis. Additionally, LDHB was downregulated in CRC tissue and associated with good prognosis, progression-free survival in CRC patients [44,81]. LDHB was suppressed by ribosomal protein RPS7 and Krüppel-like transcription factor 14 (KLF14) to inhibit glycolysis in CRC tissue [44,81].

PKM2 was significantly upregulated in CRC tissues as IHC staining and Western blot analyses reveal its predominant localization in the cytoplasm and nucleus, where it influences metabolic reprogramming and oncogenic signaling pathways, such as STAT3, β-catenin, and HIF-1α [123–129]. High PKM2 expression correlates with tumour grade, invasion depth, lymph node metastasis, and poor prognosis [123–129]. The dimeric form (TuM2-PK) of PKM2 is detectable in the serum or plasma of CRC patients and serves as a non-invasive biomarker [130–132]. Elevated serum/plasma PKM2 levels have been reported in CRC patients compared to healthy controls, showing potential for early detection and disease monitoring [130–132]. However, its diagnostic accuracy varies, requiring combination with other clinical biomarkers, such as CEA and CA19–9 to enhance sensitivity and specificity [130–132]. On the other hand, an IHC study observed that PGK1 expression was significantly higher in metastatic colon cancer tissues compared to non-metastatic ones [71]. PGK1 undergoes arginine methylation at residue R206 by protein arginine methyltransferase 1 (PRMT1), leading to increased phosphorylation at serine 203 [133]. These modifications enhance glycolytic activity and promote tumour growth in CRC cells. Elevated levels of PRMT1 and methylated PGK1 correlate with advanced tumour stages and poorer survival rates in CRC patients [133].

In CRC, several glycolytic transporters, namely, GLUT1, MCT1, and MCT4, exhibit altered expression in patients' tissue and serum. Elevated GLUT1 in CRC tissues significantly correlated with adverse features, including poor differentiation, lymph node metastasis, and advanced tumour stages [134]. High GLUT1 expression in CRC tissue was associated with poorer overall and disease-free survival rates, underscoring its potential as a prognostic marker. Martins et al. presented a comprehensive IHC study highlighting the significance of MCT expression in various types of tissue samples among CRC patients, including primary tissue, lymph node and hepatic metastasis [135]. Their findings demonstrated that MCT1 and MCT4 were overexpressed in primary CRC tissues compared to adjacent normal tissues [135]. Specifically, MCT1 was overexpressed in primary tumours but a significant decrease in expression in lymph node and hepatic metastases [135]. Conversely, MCT4 was consistently overexpressed in primary tumours, lymph node metastases, and hepatic metastases, suggesting its role in facilitating metastatic progression [135]. The differential expression of these transporters correlated with various clinicopathological features. MCT1 expression was associated with older patient age, while MCT4 positivity correlated with tumour location in the colon and deeper tumour invasion [135]. Positive MCT1 expression in stage IV CRC tissue was linked to improved cumulative survival, whereas MCT4 expression was associated with poor prognosis [135]. These findings suggest that MCT1 and MCT4 are potential biomarkers for CRC progression and prognosis.

2.4. Liver Cancer and Glucose Metabolism

In hepatocellular carcinoma (HCC), several glycolytic enzymes and glucose transporters have been identified as potential biomarkers. Overexpression of GLUT1 and GLUT3 has been validated with IHC staining in HCC tissue samples, correlating with increased glucose uptake and glycolysis, thereby supporting rapid tumour growth [32,136,137]. Additionally, elevated GLUT3 expression was associated with poor prognosis in HCC patients [136]. A study involving IHC analysis of 157 HCC tumour samples reported that elevated HK2 levels were significantly associated with higher tumour grade and advanced cancer stage [138]. Increased HK2 expression correlated with reduced overall survival, suggesting its potential as a prognostic marker in HCC [138]. Another interesting investigation by Guzman et al. analysed HK2 expression in various types of liver tissues using IHC, including normal liver, non-dysplastic cirrhosis (NDC), liver cell dysplasia (LCD), and HCC [139]. Their study reported a progressive increase in HK2 protein expression from normal liver to NDC, LCD, and HCC. Elevated HK2 expression was also associated with more aggressive histological features in HCC, suggesting that HK2 upregulation may be an early event in hepatocarcinogenesis [66]. Thus, HK2 could serve as a diagnostic and prognostic biomarker in HCC.

LDH-C4 was significantly upregulated in HCC tissues compared to adjacent non-cancerous tissues in IHC staining, correlating positively with advanced clinical stages and larger tumour sizes [140]. HCC patients exhibiting high LDH-C4 expression demonstrated poorer prognosis, with reduced overall survival compared to those with low expression levels [140]. Meanwhile, an elevated preoperative serum LDH level was associated with poor overall survival and disease-free survival in HCC patients undergoing partial hepatectomy [141]. Multivariate analyses indicated that serum LDH could be an independent prognostic factor for both overall survival and disease-free survival in HCC [141]. On the other hand, among HCC patients treated with sorafenib, a high pre-treatment serum LDH level was associated with reduced progression-free survival and overall survival [142]. An increase in LDH levels after three months of treatment predicted poorer survival outcomes, suggesting that serum LDH could serve as a potential biomarker for treatment efficacy and disease progression [142]. These findings underscore the potential of LDH, particularly tissue-specific expression of LDH-C4 and elevated serum LDH level, for diagnosis, prognosis, and monitoring therapeutic responses in HCC.

Several IHC studies demonstrated that PKM2 was overexpressed in HCC tissues and associated with unfavourable clinicopathological features and poor patient prognosis [143–146]. HCC patients with high PKM2 expression exhibited significantly shorter survival times and a higher risk of recurrence compared to those with low PKM2 expression, highlighting its potential as a prognostic biomarker [143–146]. Meanwhile, MCTs also exhibit distinct expression patterns in HCC. Jeon et al. reported that MCT1 expression in HCC tissues varied depending on the tumour's glucose uptake characteristics [147]. Specifically, MCT1 expression was higher in tumours with low [^{18}F] FDG uptake but decreased in tumours with high [^{18}F] fluorodeoxyglucose (FDG) uptake [147]. This suggests a potential inverse relationship between MCT1 expression and glycolytic activity in HCC tumours. Additionally, researchers investigated the expression of MCTs and GLUT1 in HCC samples [148]. MCT4 and GLUT1 expression increased progressively from non-neoplastic liver tissues to primary HCC and was highest in metastatic lesions [148]. This pattern suggests that elevated GLUT1 and MCT4 expression may play a role in tumour progression and metastasis in HCC. In contrast, MCT2 expression decreased progressively from non-neoplastic liver tissues to primary HCC and further in metastatic lesions [148]. This trend indicated that reduced MCT2 expression may be associated with HCC progression. The progressive decrease in MCT2 and increase in MCT4 expression from non-neoplastic tissues to metastatic lesions suggest that these transporters may serve as biomarkers for HCC progression and potential therapeutic targets.

2.5. Gastric Cancer and Glucose Metabolism

In gastric cancer (GC), several key enzymes and transporters, namely ALDOA, GLUTs, LDHA, and PKM2, have been widely investigated as potential biomarkers. ALDOA is frequently upregulated in GC tissues compared to adjacent normal tissues in IHC studies, associated with enhanced glycolytic activity, contributing to tumour progression [48,149,150]. Clinically, elevated ALDOA expression correlates with deeper tumour invasion, lymph node metastasis, and advanced clinical stages, while proteomic analysis suggests that ALDOA may be associated with gastric cancer stem cells (CSCs) [48,149–151]. ALDOA activates the epidermal growth factor receptor (EGFR) pathway, leading to downstream signaling that promotes GC cell proliferation and drug resistance [149]. Thus, ALDOA could serve as a valuable biomarker for prognosis and therapeutic monitoring in GC.

Several IHC studies demonstrated the overexpression of GLUT1 in GC tissues compared to normal gastric mucosa, which significantly correlated with tumour aggressiveness, invasion depth, lymph node metastasis, and advanced TNM stage [152–154]. Elevated GLUT1 expression was also significantly associated with short overall survival and disease progression in GC, suggesting the potential of GLUT1 as a poor prognostic biomarker [152–154]. In addition, another IHC study investigated the expression of GLUTs (1, 3, 6, and 10) in GC tissues and their association with clinical outcomes [155]. The findings revealed that all four GLUTs are expressed in GC tissue, with GLUT3 showing a significant correlation to both the UICC (Union for International Cancer Control) stage and overall survival in patients [155]. Overexpression of GLUT3 was associated with advanced tumour stages and poorer survival rates, suggesting its potential as a prognostic biomarker [155]. This study underscores the role of GLUTs in the metabolic reprogramming of gastric cancer cells, highlighting GLUT3 as a key player in tumour progression.

Yang et al. also reported the overexpression of GLUT3 in primary GC tumour and metastasis tumour tissues using IHC staining [156]. In depth analysis demonstrated that GLUT3 could promote histone modifications by regulating LDHA in GC, suggesting the possible interactions between glycolytic enzymes to facilitate non-metabolic activities in tumour progression [156]. Similarly, previous IHC studies have demonstrated the elevated LDHA expression in GC tissue compared to normal gastrointestinal mucosa, often correlated with advanced TNM staging, lymph node metastasis, poor prognosis, and low survival rates in GC patients [157–160]. Overexpression of LDHA in GC cells enhances glycolytic flux, contributing to tumour progression [157–159]. Elevated LDHA expression has been linked to poorer overall survival and disease-free survival in GC patients, suggesting its potential as a prognostic marker and therapeutic target. Additionally, PKM2 expression was significantly elevated in GC tissues and was mostly associated with poor prognosis and shorter survival, as shown in multiple IHC studies [74,161,162]. Notably, well-differentiated GC adenocarcinomas exhibited higher PKM2 expression compared to undifferentiated signet-ring cell GC carcinomas [162]. Elevated PKM2 expression in signet-ring cell gastric cancers correlated with shorter overall survival, suggesting its potential as an independent prognostic marker in this subtype [162]. Table 1 summarizes the key enzymes and proteins in the Warburg effect and glucose metabolism of the top five cancers.

Table 1. Key enzymes and proteins in the Warburg effect and glucose metabolism.

Protein/Enzyme	Function in Glycolysis/Metabolism	Cancer Types	Biomarker Role
ALDOA	Cleaves fructose-1,6-bisphosphate into triose phosphates	Breast, Lung, CRC, Gastric	Prognostic, EMT, invasion
ENO1	Catalyzes conversion of 2-PG to PEP	Breast, Lung, CRC, Gastric	Prognostic, chemoresistance, stemness
GLUT1/GLUT3/GLUT4	Facilitates glucose uptake	Breast, CRC, Lung, Gastric	Diagnostic, prognostic, associated with aggressiveness
HK2/HK1	Initiates glycolysis via phosphorylation of glucose	Breast, CRC, Liver, Lung	Prognostic, linked to HIF-1α, recurrence
LDHA/LDHB	Converts pyruvate to lactate; supports NAD+ regeneration	Breast, Lung, CRC, Gastric, Liver	Prognostic, treatment monitoring, serum biomarker (LDHA)
MCT1/MCT4	Lactate and pyruvate export	Breast, CRC, Lung	Prognostic, therapy resistance, hypoxia response
PFKP	Rate-limiting conversion of F6P to F1,6BP	Breast, Lung	Prognostic, therapy resistance
PGAM1	Converts 3-PG to 2-PG	Breast, Lung	Prognostic, migration, EMT
PGK1	Generates ATP via 1,3-BPG to 3-PG conversion	CRC, Lung	Prognostic, promotes angiogenesis, metastasis
PKM2	Catalyzes PEP to pyruvate; allows accumulation of glycolytic intermediates	Breast, Lung, CRC, Gastric, Liver	Prognostic, metastasis, therapy response, serum marker
SGLT1	Sodium-dependent glucose transporter	$HER2^+$ Breast	Prognostic, linked with HER2 stabilization
TPI1	Isomerizes DHAP and G3P	Breast, CRC	Prognostic, proliferation, invasion

3. Amino Acid Metabolism in Cancers

Amino acid metabolism is another key component of cancer metabolic reprogramming, supporting tumour growth, immune evasion, redox balance and drug resistance [163]. Cancer cells rewire amino acid pathways to meet increased bioenergetic and biosynthetic demands [163]. For instance, glutamine is the most abundant amino acid in plasma, in which different types of tumours rely on glutamine metabolism to facilitate the TCA cycle, nucleotide synthesis, and redox homeostasis [164]. Serine and glycine metabolism contribute to one-carbon metabolism which is essential for DNA synthesis and epigenetic modifications [165]. Branched-chain amino acids (BCAAs) activate mTORC1 signaling, promoting tumour proliferation [166]. Proline and arginine metabolism regulate oxidative stress and immune suppression, enhancing tumour progression [167,168]. Identifying biomarkers associated with dysregulated amino acid metabolism, such as amino acid transporters and critical enzymes, could improve cancer diagnosis, prognosis, and treatment strategies. Among these, ASCT2, GLS1, LAT1, SLC7A11, SLC3A2, and SHMT2 are key regulators of amino acid metabolism and have been implicated in tumour progression, therapy resistance, and immune evasion.

ASCT2 (SLC1A5) is a sodium-dependent neutral amino acid transporter that facilitates glutamine uptake, fuelling the TCA cycle and mTORC1 activation, making it essential for glutamine-addicted tumours [169]. Once inside the cell, glutaminase (GLS) converts glutamine to glutamate, which feeds into the TCA cycle and supports nucleotide and lipid biosynthesis [170]. GLS also maintains redox homeostasis by producing glutamate for glutathione synthesis, protecting cancer cells from oxidative stress [171]. Glutamate dehydrogenase (GDH) catalyses the reversible conversion of glutamate to α-ketoglutarate (α-KG), which is further utilised in the TCA cycle, contributing to energy production and biosynthesis in cancer cells [172]. SLC7A5 (LAT1) transports essential amino acids to activate mTORC1 signaling, promoting protein synthesis and proliferation [173]. SLC7A11 (xCT) enables cystine uptake for glutathione synthesis, protecting cancer cells from oxidative stress and ferroptosis [174]. SLC3A2, also known as CD98 heavy chain, is integral to amino acid transport and integrin signaling [175]. SLC3A2 acts as a chaperone for LAT1 and xCT, enhancing amino acid transport while supporting integrin-mediated tumour invasion [176]. SHMT2 catalyses serine-to-glycine conversion, sustaining nucleotide biosynthesis and redox homeostasis, especially in hypoxic tumours [177]. Overexpression of these regulators is linked to aggressive tumour behaviour, therapy resistance, and poor prognosis. Targeting these metabolic vulnerabilities through inhibitors or dietary interventions holds promise for cancer therapy by disrupting essential nutrient uptake and biosynthetic pathways.

3.1. Breast Cancer and Amino Acid Metabolism

Several studies have investigated the proteomic profiling of breast cancer tissues, revealing the subtype-specific patterns of amino acid transporters and metabolic enzymes, highlighting their potential role as prognostic biomarkers. A comprehensive proteomic analysis employing reverse-phase protein array (RPPA) technology on 289 breast cancer samples demonstrated higher ASCT2 and SHMT2 protein expression in aggressive subtypes, including HER2-positive and TNBC, compared to luminal subtypes [178]. Elevated ASCT2 and SHMT2 levels in these subtypes were associated with shorter recurrence-free survival, highlighting their potential as independent prognostic factors [178]. Kim et al. conducted another tissue microarray profiling of 702 breast cancer patients using IHC staining for glutamine metabolism-related proteins, including GLS1, GDH, and ASCT2, in stromal and tumoral sites [179]. $HER2^+$ breast cancer had the highest expression of stromal GLS1, tumoral GDH, stromal GDH, and tumoral ASCT2, compared to TNBC and luminal types [179]. These elevated expressions suggest increased glutamine metabolic activity in $HER2^+$ tumours. In addition, ASCT2 expression was inversely correlated with oestrogen receptor (ER) and progesterone receptor (PR) positivity [179]. Tumours negative for ER and PR showed higher ASCT2 expression, indicating increased glutamine uptake in these subtypes [179]. Collectively, these findings highlight the metabolic heterogeneity among breast cancer subtypes, with specific transporters and enzymes upregulated in more aggressive forms like $HER2^+$ and TNBC.

TNBC is regarded as an aggressive subtype of breast cancer with limited choices of target therapies, while tumour-infiltrating lymphocytes (TILs) are often correlated with better outcomes in TNBC [180]. However, metabolic adaptations, particularly glutamine metabolism, may counteract the benefits of TILs [181]. The prognostic significance of GLS expression has been evaluated in node-positive TNBC patients with high TILs, which demonstrated that GLS overexpression correlated with worse disease-free survival (DFS) and overall survival (OS) [182]. Despite the presence of abundant TILs, patients with elevated GLS expression exhibited significantly poorer outcomes, suggesting that metabolic reprogramming through glutaminolysis undermines the protective role of immune infiltration [182]. GLS expression could serve as a negative prognostic biomarker in node-positive TNBC, even in cases with high TILs. In addition, Zhou et al. performed tissue microarray and discovered that the prognostic value of GLS1 in breast cancer patients varied based on H3K27me3 (trimethylation of histone H3 at lysine 27) expression and menopausal status [183]. High GLS1 expression was associated with worse prognosis, particularly in tumours with low H3K27me3 levels, suggesting epigenetic regulation of glutamine metabolism may influence tumour progression [183]. Postmenopausal patients with high GLS1 expression exhibited poorer survival, whereas premenopausal patients showed no significant correlation, indicating metabolic dependencies shift with hormonal changes [183]. GLS1 is a potential prognostic biomarker in breast cancer, but its impact is modulated by H3K27me3 expression and menopausal status. These findings underscore the importance of integrating metabolic and epigenetic profiling in breast cancer prognosis.

Oestrogen receptor-positive (ER+) breast cancer constitutes most breast cancer cases, with endocrine therapy being the standard treatment [184]. However, resistance to endocrine therapy remains a significant clinical challenge. Alfarsi et al. investigated the co-expression of SLC7A5 and SLC3A2 as potential predictors of response to endocrine therapy in ER+ breast cancer patients via IHC staining [185]. High co-expression of SLC7A5 and SLC3A2 in breast cancer tissue was significantly associated with aggressive tumour phenotypes, such as increased proliferation, invasion, and metastasis [185]. More importantly, among patients receiving endocrine therapy alone, those with SLC7A5+/SLC3A2+ tumours experienced poorer prognosis, higher recurrence and mortality rates [185]. Knockdown of SLC7A5 and SLC3A2 resulted in reduced proliferation of ER+ breast cancer cells and enhanced the sensitivity of ER+ breast cancer cells to tamoxifen [185]. The same group of researchers also investigated the co-expression of ASCT2 and glycolytic enzyme TALDO1 in ER+ luminal breast cancer [184]. Similarly, high expression of SLC1A5 and TALDO1 correlated with aggressive tumour features, including higher grade and increased proliferation [184]. Patients with high SLC1A5 expression experienced poorer outcomes following endocrine therapy, while the co-expression of SLC1A5 and TALDO1 was significantly associated with a higher risk of recurrence and mortality in breast cancer patients receiving endocrine therapy [184]. Previous IHC studies by Ansari et al. also demonstrated that the high expression of SLC3A2 and SLC7A5 in TNBC, $HER2^+$ and luminal B subtypes was significantly associated with poor patient outcome and shorter survival especially in patients with ER+ breast tumours [186,187]. Thus, the co-expression of ASCT2, SLC7A5 and SLC3A2 may serve as potential biomarkers for identifying ER+ highly proliferative subclass and $HER2^+$ subclass of breast cancer patients at higher risk of poor response to endocrine therapy.

Serine hydroxymethyltransferases (SHMTs) are pivotal enzymes in one-carbon metabolism, facilitating the reversible conversion of serine to glycine and contributing to nucleotide biosynthesis [188]. Two primary isoforms exist: SHMT1, located in the cytoplasm, and SHMT2, found in mitochondria [188]. In breast cancer, previous IHC studies demonstrated that SHMT2 protein was significantly overexpressed in breast cancer tissues compared to noncancerous tissues, which correlated with increased tumour aggressiveness, including advanced TNM staging and higher Elston grade [189,190]. Higher SHMT2 expression was observed in grade III tumours compared to grades I–II, suggesting its role in tumour progression [189]. Elevated SHMT2 expression has been linked to poorer prognosis in breast cancer patients [178,189,190]. Notably, SHMT2's prognostic value appears more significant in ER- breast cancer cases, indicating its potential as a biomarker for this subgroup [190]. SHMT2 promotes breast cancer cell proliferation through the activation of vascular endothelial growth factor (VEGF) and mitogen-activated protein kinase (MAPK) signaling pathways [191]. By regulating serine/glycine metabolism, SHMT2 supports nucleotide biosynthesis and redox balance, which are essential for rapidly proliferating cancer cells [191]. The association between SHMT2 overexpression and poor prognosis underscores its potential as a prognostic biomarker in breast cancer management.

3.2. Lung Cancer and Amino Acid Metabolism

Argininosuccinate synthase 1 (ASS1) is a key enzyme in the urea cycle, facilitating the synthesis of arginine from citrulline and aspartate to support cellular metabolism [168]. Its expression patterns in lung cancer tissues have been the subject of various studies, revealing potential implications for tumour behaviour and patient prognosis. ASS1 was significantly overexpressed in NSCLC tissue samples based on IHC staining [192]. The absence of ASS1 expression was associated with increased angiogenesis, suggesting that ASS1 deficiency may contribute to tumour progression [192]. Moreover, the presence of ASS1 in tumour cells correlated with a better prognosis in NSCLC patients, especially when cancer-associated fibroblasts (CAFs) did not express arginase-2 (ARG2) [192]. This association underscores the potential of ASS1 as a prognostic biomarker in lung cancer. Additionally, several amino acid transporters, namely ASCT2, LAT1 and SLC3A2, are dysregulated in lung cancer, contributing to tumour growth, survival, and metabolic reprogramming.

In lung cancer, ASCT2 exhibits distinct expression patterns across various histological subtypes, primarily overexpression in NSCLC subtypes. Hassanein et al. demonstrated that SLC1A5 was overexpressed in both adenocarcinoma (ADC) and squamous cell carcinoma (SCC) subtypes of NSCLC when compared to normal lung tissues in IHC analysis [193]. Further analysis revealed that ASCT2 was predominantly localized to the cytoplasmic membrane of cancer cells [193]. ASCT2 expression was significantly higher in SCC compared to ADC, while male patients also expressed higher SLC1A5 compared to female patients [193]. Most of the glutamine transport in lung cancer cells is sodium-dependent, with ASCT2 accounting for about half of this activity, highlighting the prognostic value of ASCT2 in supporting cancer cell metabolism and proliferation [193,194]. Another study by Shimizu et al. also supported that ASCT2 was significantly overexpressed in NSCLC, which correlated with advanced disease stage, lymphatic permeation, vascular invasion, increased cell proliferation, angiogenesis, and mTOR phosphorylation [195]. These associations were particularly pronounced in adenocarcinoma patients [195]. Furthermore, validation with an independent cohort confirmed that ASCT2 serves as an independent marker for poor outcomes in lung adenocarcinoma patients [195].

Yazawa et al. investigated the co-expression of ASCT2 and LAT1 in lung adenocarcinoma [196]. Their findings indicated that high expression levels of both transporters were significantly associated with tumour aggressiveness and poor survival outcomes [196]. This co-expression suggests a coordinated role between these two proteins in facilitating tumour growth and progression. In addition, IHC analysis of 295 NSCLC cases revealed that LAT1 protein was detected in the cytoplasm and/or on the plasma membrane of tumour cells [197]. Plasma membrane expression of LAT1 was associated with tumour histology, differentiation grade, pathologic stage, T classification, pleural invasion, lymph-vessel invasion, and overall survival rate [197]. Imai et al. also investigated different subtypes of stage I NSCLC tissues, which demonstrated that LAT1 expression was more prominent in squamous cell carcinoma and large cell carcinoma, compared to adenocarcinoma [198]. A significant correlation was observed between LAT1 expression and the Ki67 labelling index, indicating a link to proliferative activity [198]. The 5-year survival rate for LAT1-positive patients was significantly lower compared to LAT1-negative patients [198]. Multivariate analysis confirmed that positive LAT1 expression was an independent factor for predicting poor prognosis in these patients [198]. These findings suggest that LAT1 expression could serve as a valuable prognostic marker in NSCLC cases.

Elevated serum levels of SLC3A2 have been detected in patients with lung adenocarcinoma (LUAD) and lung squamous cell carcinoma (LUSC), which also corresponded to the increased mRNA level of SLC3A2 in cancer tissues [199]. This overexpression was associated with enhanced tumourigenesis via the MEK/ERK signaling pathway, underscoring the importance of SLC3A2 in cancer cell proliferation and survival [199]. High SLC3A2 expression also correlated with poorer overall survival in LUAD patients, suggesting its potential as a non-invasive prognostic biomarker [199]. SLC3A2 influences the tumour microenvironment by promoting the polarization of tumour-associated macrophages (TAMs) toward the M2 phenotype, which supports tumour growth [200]. This effect is mediated through metabolic reprogramming involving arachidonic acid pathways [200]. Taken together, SLC3A2 is significantly overexpressed in lung cancer tissues and patient serum, highlighting its potential as a diagnostic and prognostic biomarker.

Serine metabolism plays a critical role in cellular function by contributing to biosynthetic processes, redox balance, and signaling pathways [201]. Serine is a non-essential amino acid involved in nucleotide biosynthesis, lipid metabolism, and methylation reactions [165]. Many cancer cells rely on de novo synthesis through the serine synthesis pathway (SSP) to support rapid proliferation and survival [165]. Phosphoglycerate dehydrogenase (PHGDH) is the rate-limiting enzyme involved in the SSP, catalysing the conversion of 3-phosphoglycerate to 3-phosphohydroxypyruvate [201]. A study involving 319 NSCLC patients demonstrated significantly increased PHGDH expression at both mRNA and protein levels in tumour tissues compared to adjacent non-tumour tissues [202]. IHC analysis revealed that high PHGDH expression was significantly associated with lymph node metastasis and advanced TNM stage [202]. Moreover, elevated PHGDH levels correlated with poorer 5-year overall survival rates, suggesting its potential as a prognostic marker in NSCLC [202]. Additionally, another study involving bioinformatics analysis of 720 LUAD patients and tissue microarray analysis of 75 LUAD samples indicated that high PHGDH expression defines a metabolic subtype characterized by enhanced glycolysis and increased serine production from glucose [203]. This metabolic reprogramming supports rapid cell proliferation and migration by facilitating nucleotide biosynthesis [203]. PHGDH overexpression has been linked to resistance against the tyrosine kinase inhibitor erlotinib in LUAD cell lines, with inhibition of PHGDH re-sensitizing resistant cells to treatment [203]. While PHGDH is significantly overexpressed in NSCLC and LUAD subtypes, its regulatory mechanisms and functional roles underscore its potential as a prognostic marker in lung cancer management.

Serine metabolism may play an important role in lung cancer progression. Previous studies demonstrated that SHMT1 was significantly overexpressed in lung cancer tissues, which was associated with increased cell proliferation and tumour progression [204]. Knockdown experiments of SHMT1 in lung cancer cell lines led to cell cycle arrest and p53-dependent apoptosis [204]. This suggests that SHMT1 plays a crucial role in maintaining the proliferative capacity of lung cancer cells. Moreover, SHMT2 was significantly upregulated in lung adenocarcinoma tissues, and associated with advanced stages of lung cancer, suggesting its involvement in tumour aggressiveness and potential as a prognostic marker [205]. While both SHMT1 and SHMT2 are overexpressed in lung cancer tissues, their utility as independent prognostic markers remains uncertain, which warrants further validation in a larger cohort.

3.3. CRC and Amino Acid Metabolism

Numerous IHC and tissue microarray studies have reported elevated GLS expression in CRC tissues compared to adjacent normal tissues, indicating its role in the metabolic adaptation of CRC cells to support rapid growth and survival [171,206–209]. The overexpression of GLS is particularly evident in highly proliferative regions of the tumour, such as invasive margins and areas with tumour budding [207]. This suggests that GLS may support the invasive properties of CRC by providing energy and metabolites needed for cell motility and invasion. High GLS expression correlated with advanced tumour stages, lymph node metastasis, and poor prognosis in CRC patients [171,206–209]. The upregulation of GLS in CRC tissues reflects its important role in maintaining the metabolic flexibility of cancer cells, supporting their growth and survival. This makes GLS a promising biomarker in CRC. On the other hand, GDH was significantly overexpressed in CRC tissues and metastatic lesions, which correlated with advanced tumour stages and poor differentiation in IHC staining [210]. High GDH expression was also significantly associated with poorer overall survival and disease-free survival rates in CRC patients, while multivariate analyses identified GDH expression as an independent prognostic factor [210].

Functionally, GDH promoted CRC cell proliferation, migration, and invasion via STAT3-mediated EMT, while knockdown of GDH inhibited these processes [210]. The findings suggest that GDH could be a prognostic biomarker for identifying cancer metastasis in CRC patients.

Protein expression of a few amino acid transporters has been investigated in CRC clinical samples. SLC3A2 was significantly overexpressed in CRC tissues based on IHC staining and associated with enhanced tumour growth and metastasis [211]. Functionally, macrophage migration inhibitory factor (MIF) bound directly to SLC3A2, enhancing its expression and subsequently promoting CRC progression [211]. The MIF/SLC3A2 axis activated the AKT/GSK-3β signaling pathway, leading to increased tumour cell proliferation and metastasis [211]. The overexpression of SLC3A2 in CRC tissues and its involvement in critical signaling pathways highlight its potential as a biomarker and therapeutic target in CRC. Meanwhile, the upregulation of SLC1A5 in CRC tissues was significantly associated with enhanced tumour growth and survival [212,213]. SLC1A5-mediated glutamine uptake supports the metabolic demands of rapidly proliferating CRC cells, contributing to tumour growth and survival [213]. More importantly, Toda et al. found a significant correlation between KRAS mutational status and SLC1A5 expression based on tissue microarray, when elevated SLC1A5 levels were notably associated with tumour depth and vascular invasion in KRAS-mutant CRC cases [212]. Knockdown of SLC1A5 suppressed the migration of KRAS-mutant CRC cells, indicating that SLC1A5 could be a novel therapeutic target against KRAS-mutant CRC [212]. Interestingly, clinical samples from CRC patients resistant to cetuximab exhibited significantly higher SLC1A5 expression in IHC staining compared to those responsive to the treatment, suggesting the potential of SLC1A5 as a prognostic biomarker to predict chemo-response in CRC [214]. These findings highlight the critical role of SLC1A5 as a potential biomarker in CRC cases, particularly with KRAS mutations or resistance to cetuximab.

SLC7A5 was significantly overexpressed in early-stage (stages I and II) CRC cases, although it was less prevalent in tumours with mucinous histology and those exhibiting lymphovascular invasion in IHC staining [215]. Elevated SLC7A5 expression correlated with improved OS and DFS in early-stage CRC patients, especially stage II patients, where high expression was an independent indicator of better clinical outcomes [215]. SLC7A5 was also expressed in activated lymphocytes, including NK, T, and B cells [215]. Its overexpression in early-stage CRC may enhance anti-tumour immune responses, contributing to improved patient outcomes [215]. SLC7A5 overexpression could serve as a favourable prognostic marker in early-stage CRC. Similarly, another study by Najumudeen et al. reported that KRAS-mutant CRCs exhibited increased SLC7A5 expression, suggesting a link between oncogenic KRAS signaling and amino acid transporter regulation [216]. SLC7A5 facilitated glutamine consumption in CRC cells to sustain the metabolic demands of rapid proliferation [216]. By maintaining intracellular amino acid levels, SLC7A5 promotes bulk protein synthesis, underpinning the enhanced proliferation of KRAS-mutant CRC cells [216]. SLC7A5 plays a critical role in the metabolic reprogramming and progression of CRC, particularly in the context of KRAS mutations.

In contrast, SLC25A21 expression was significantly reduced in KRAS-mutant CRC tissues based on IHC staining, while patients with lower SLC25A21 expression exhibited poorer survival outcomes, particularly in KRAS-mutant CRC cases [217]. SLC25A21 is a mitochondrial carrier involved in transporting 2-oxoadipate and α-ketoglutarate across the inner mitochondrial membrane [218]. Reduced SLC25A21 expression led to decreased efflux of α-ketoglutarate from mitochondria, enhancing glutamine replenishment to support the metabolic demands of rapidly proliferating KRAS-mutant CRC cells [217]. The increased glutamine metabolism resulting from SLC25A21 downregulation provides substrates for GTP synthesis, maintaining persistent KRAS activation and promoting tumour progression [217]. Downregulation of SLC25A21 was correlated with increased invasion, migration, metastasis and drug resistance of KRAS-mutant CRC cells [217]. Restoring SLC25A21 expression reduced KRAS signaling and sensitized CRC tumours to cetuximab, suggesting that targeting SLC25A21 could enhance therapeutic efficacy [217]. Thus, SLC25A21 could serve as an independent prognostic marker for metabolic intervention strategies in CRC.

Since KRAS mutations are found in 40 to 50% of CRC cases, identifying the expression levels of SLC1A5, SLC7A5 and SLC25A21 could help stratify patient risk [219,220]. SLC1A5 (ASCT2) is mostly upregulated, facilitating glutamine uptake to support KRAS-driven tumour growth, making it a prognostic and therapeutic biomarker [212]. SLC7A5 (LAT1) is also overexpressed in CRC tissue, promoting mTOR activation and tumour aggressiveness, with high expression correlating with poor survival and resistance to EGFR inhibitors [216]. In contrast, SLC25A21 is downregulated in CRC tissue, enhancing glutamine metabolism and sustaining KRAS activity, with low expression linked to poor prognosis and cetuximab resistance [217]. Collectively, these transporters offer valuable insights into CRC pathophysiology, and their combined evaluation could improve risk stratification, therapy selection, and personalized treatment strategies in KRAS-mutant CRC. Further clinical validation is required to establish their diagnostic, prognostic, and predictive utility in routine practice.

3.4. *Liver Cancer and Amino Acid Metabolism*

Kidney-type GLS1 and liver-type GLS2 play distinct roles in HCC. GLS1 is frequently upregulated in HCC tissues, while in contrast, GLS2 expression is often reduced in HCC. A study involving 1140 HCC patients demonstrated that GLS1 expression had a sensitivity of 74.6% and specificity of 84.2% for diagnosing HCC [221]. When combined with glypican 3 (GPC3) and alpha-fetoprotein (AFP), the diagnostic performance improved, achieving a sensitivity of 84.6% and specificity of 88.6% [221]. Additionally, high GLS1 expression correlated with advanced tumour stages and poorer disease-free survival, underscoring its prognostic value [221]. Another IHC investigation reported that GLS1 was highly expressed in HCC tissues but minimally present in non-cancerous liver tissues [222]. This differential expression suggests that GLS1 could serve as a reliable marker for distinguishing HCC from non-cancerous liver conditions [222]. In contrast to GLS1, GLS2 expression was typically diminished in HCC, which could be attributed to the CpG methylation of GLS2 promoter [223]. Further research is warranted to explore the utility of these proteins for early detection and therapeutic strategies in HCC.

GDH1 expression was significantly upregulated in HCC tissues compared to adjacent non-cancerous liver tissues in IHC studies [224]. This upregulation supports the notion that GDH1 contributes to tumourigenesis by promoting glutaminolysis, which sustains cellular survival under conditions of glucose deprivation [224]. In addition, glutamine synthetase (GS) is an ATP-dependent enzyme that catalyses the conversion of glutamate and ammonia to glutamine [225]. Elevated GS expression was observed in HCC tissues and patient serum, which was associated with tumour invasion and metastasis [226]. In vitro and in vivo studies demonstrated that GS facilitated HCC cell migration and invasion through EMT modulation [226]. Notably, GS demonstrates high diagnostic accuracy for alpha-fetoprotein (AFP)-negative HCC cases, with an area under the receiver operating characteristic curve of 0.913 [226]. Combining GS with AFP enhances diagnostic sensitivity and specificity, suggesting its value in early detection of HCC. Further research is warranted to fully validate the potential of GDH and GS as early biomarkers in HCC.

Asparagine synthetase (ASNS) is an enzyme that catalyses the ATP-dependent conversion of aspartate and glutamine into asparagine and glutamate [208]. This reaction is crucial for protein synthesis, nucleotide biosynthesis, and cellular response to amino acid deprivation [208]. IHC studies have demonstrated that ASNS expression is significantly elevated in HCC tissues and associated with various clinicopathological features, including serum AFP levels, tumour size, microscopic vascular invasion, tumour encapsulation, TNM stage, and Barcelona Clinic Liver Cancer (BCLC) stage [227]. Notably, patients exhibiting low ASNS expression levels have been observed to have poorer overall survival, suggesting that ASNS serves as an independent prognostic factor for OS in HCC patients [227]. On the other hand, glutamic-oxaloacetic transaminase 2 (GOT2) is a mitochondrial enzyme that plays a pivotal role in amino acid metabolism and the malate-aspartate shuttle [228]. This shuttle is essential for transferring reducing equivalents across the mitochondrial membrane, thereby linking cytosolic and mitochondrial metabolic pathways [228]. GOT2 expression is often downregulated in HCC tumour tissues in IHC studies and associated with advanced disease progression and poorer patient prognosis [229]. In vivo studies further demonstrate that the tumour-suppressive role of GOT2 in HCC is linked to its influence on glutamine metabolism [229]. Loss of GOT2 leads to enhanced glutaminolysis, nucleotide synthesis, and glutathione production [229]. This reprogramming supports the cellular antioxidant system, thereby activating the PI3K/AKT/mTOR signaling pathway, which contributes to HCC progression [229]. Therefore, assessing GOT2 expression levels could provide valuable prognostic information in HCC.

The expression levels of ASCT2 and LAT1 were significantly upregulated in HCC tissues and correlated with larger tumour size, poorer histological differentiation, and advanced tumour staging in IHC analysis [32,230]. Elevated levels of ASCT2 and LAT1 are also associated with reduced overall survival, suggesting their potential as independent prognostic factors in HCC [32,230]. Meanwhile, xCT (SLC7A11) is frequently overexpressed in HCC tissues, contributing to tumour development and metastasis [231–233]. The elevated expression of xCT is also significantly associated with the overexpression of HIF1α, programmed death ligand 1 (PD-L1), colony-stimulating factor 1 (CSF1), tumour-associated macrophage (TAM) and myeloid-derived suppressor cell (MDSC) infiltration [231,232]. Functionally, xCT promotes PD-L1 and CSF1 expression through αKG-HIF1α cascade, resulting in intratumoral TA and MDSC infiltration and regulation of ferroptosis and the tumour microenvironment [231,232]. High expression of xCT also correlates with advanced tumour stage and unfavourable survival in HCC patients, especially those with strong PD-L1, suggesting that xCT may serve as an independent prognostic marker in HCC [231,233].

3.5. Gastric Cancer and Amino Acid Metabolism

In GC, GLS2 expression was significantly downregulated in cancer tissues compared to adjacent non-cancerous tissues as confirmed in IHC staining [234]. This reduction in GLS2 correlated with enhanced tumour cell proliferation and migration, alongside decreased apoptosis [234]. Overexpression of GLS2 suppressed these malignant behaviours, implying a tumour-suppressive function for GLS2 in GC [234]. Further research is warranted to elucidate whether GLS2 could serve as a potential biomarker for diagnostic or prognostic purposes in GC [234]. Moreover, GDH was significantly overexpressed in GC tissue compared to the normal gastric mucosa tissues based on IHC analysis, while it was primarily expressed in the cytoplasmic membrane of GC cells [235]. The expression of GDH in GC tissue was closely related to the rate of CD34 positive, cancer cell differentiation degree, infringement of capsular, vascular tumour emboli, lymph node metastasis, and nerve infiltration [235]. More importantly, elevated expression of GDH was correlated with better prognosis and longer survival in GC patients, suggesting its potential as an independent prognostic biomarker [235].

ASCT2 was significantly overexpressed in GC tissues compared to adjacent non-cancerous gastric mucosa, as shown by IHC and Western blotting [236,237]. ASCT2 was predominantly localized to the cell membrane of gastric carcinoma cells, facilitating efficient glutamine uptake essential for tumour metabolism [236]. Silencing ASCT2 expression significantly reduced the proliferation of gastric cancer cell lines, underscoring its role in tumour growth [237]. The overexpression of ASCT2 in GC tissues and its involvement in key metabolic pathways highlight its potential as a biomarker. In addition, ASS1 expression is significantly elevated in GC tissues compared to normal gastric mucosa in IHC studies [238,239]. ASS1 influences autophagic processes in GC cells [239]. High ASS1 protein expression in GC tissue is positively correlated with aggressive tumour phenotype and unfavourable prognosis, supporting the utility of ASS1 as a useful prognostic marker for GC [238,239].

Meanwhile, IHC studies have demonstrated that LAT1 is significantly overexpressed in GC cells, which is predominantly localized to the plasma membranes of carcinoma cells [240,241]. High LAT1 expression in GC tissues has been significantly associated with adverse clinicopathologic features, including larger tumour size, lymph node metastasis, deeper local invasion, and advanced TNM stage [241]. Downregulation of LAT1 expression in GC cell lines has been shown to inhibit cell proliferation and induce cell cycle arrest in the G0/G1 phase [240]. Patients with high LAT1-expressing non-scirrhous GC have been reported to have a significantly poorer prognosis compared to those with low LAT1 expression [241]. Multivariate analysis has identified LAT1 expression as an independent prognostic factor in this subgroup [241]. Furthermore, IHC studies have shown that xCT is significantly overexpressed in GC tissues compared to adjacent normal tissues [242]. This overexpression is associated with enhanced tumour growth and poor prognosis [242]. Functionally, xCT activates the PI3K/AKT signaling pathway to inhibit ferroptosis and support GC progression by enhancing cell proliferation, migration, and invasion [242]. Table 2 summarizes the key enzymes and proteins in the amino acid metabolism of the top five cancers.

Table 2. Key enzymes and proteins in the amino acid metabolism.

Protein/Enzyme	Function in Amino Acid Metabolism	Cancer Types	Biomarker Role
ASCT2 (SLC1A5)	Glutamine transporter, supports mTOR activation	Breast, CRC, Lung, Gastric, Liver	Prognostic, therapy resistance (e.g., anti-EGFR)
GDH (GLUD1)	Converts glutamate to α-KG	CRC, Gastric, Liver	Prognostic, metastasis, links to EMT
GLS1/GLS2	Converts glutamine to glutamate	Breast, CRC, Liver	Prognostic, therapeutic target (GLS inhibitors)
GOT2	Links amino acid metabolism to TCA cycle	Liver	Tumour suppressor, poor prognosis when lost
LAT1 (SLC7A5)/CD98 (SLC3A2)	Transports essential amino acids (e.g., leucine), mTOR activation	Breast, CRC, Gastric, Liver, Lung	Prognostic, survival marker, endocrine resistance
xCT (SLC7A11)	Cystine/glutamate antiporter, regulates redox and ferroptosis	Gastric, Liver	Prognostic, immune escape, therapeutic target
SHMT1/SHMT2	Converts serine to glycine (one-carbon metabolism)	Breast, Lung	Prognostic, linked to hypoxia and cell proliferation
PHGDH	Catalyzes serine biosynthesis from 3-PG	Lung, Breast	Prognostic, therapy resistance, metabolic reprogramming

4. Lipid Metabolism

Lipid metabolism plays a pivotal role in cancer metabolic reprogramming by fuelling energy production, promoting membrane biosynthesis, and enhancing survival under stress [20]. Cancer cells often alter several key enzymes and transporters involved in lipid uptake, synthesis, and oxidation to meet their energetic and biosynthetic needs. Unlike normal cells, which primarily obtain fatty acids (FAs) from dietary sources, cancer cells enhance de novo fatty acid synthesis by modulating ACLY, ACC, FASN and SCD [20]. ATP citrate lyase (ACLY) converts citrate to acetyl-CoA as precursor for lipid synthesis, while acetyl-CoA carboxylase (ACC) catalyses the carboxylation of acetyl-CoA to malonyl-CoA [243]. Fatty acid synthase (FASN) is a rate-limiting enzyme in FA synthesis, while stearoyl-CoA desaturase-1 (SCD1) converts saturated fatty acids to monounsaturated fatty acids (MUFAs), promoting membrane fluidity and tumour survival [33,244]. Cancer cells upregulate lipid transporters, namely CD36 (FAT) and fatty acid-binding proteins (FABPs) to increase fatty acid uptake [245,246]. Fatty acid oxidation (FAO) provides ATP, NADH, and $FADH_2$ to support oxidative phosphorylation in mitochondria, which key enzymes involved are namely carnitine palmitoyltransferase 1 (CPT1A) and acyl-CoA dehydrogenases (ACADs) [243]. CPT1A facilitates FA transport into mitochondria, while ACADs catalyse FA oxidation in the mitochondria [243].

4.1. *Breast Cancer and Lipid Metabolism*

Several studies have investigated the expression of lipid metabolism-related proteins across different breast cancer subtypes and metastatic sites, providing insights into their potential roles in tumour behaviour and patient prognosis [247,248]. Kim et al. analysed the expression of lipid metabolism-related proteins in 476 invasive breast cancer cases of these four different subtypes using tissue microarray and IHC staining [247]. The proteins examined included perilipin 1 (PLIN1), CPT1A, FASN, FABP4, and acyl-CoA oxidase 1 (ACOX1). Notably, HER2-enriched tumours exhibited the highest expression of these proteins, highlighting that $HER2^+$ breast tumours demonstrate higher metabolic plasticity with elevated lipid metabolism compared to other subtypes [247]. ACOX-1 positivity was associated with significantly shorter overall survival in all the subtypes, while particularly FABP4 positivity was significantly associated with shorter disease-free survival and overall survival in TNBC. These observations suggest that ACOX-1 and FABP4 are poor prognostic biomarkers in metastatic breast cancer, which can be specific to certain subtypes.

Another IHC study by Cha et al. compared the expression profiles of lipid metabolism-related proteins between invasive lobular carcinoma (ILC) and invasive ductal carcinoma (IDC) [249]. ILC and IDC are the two most common histological subtypes of breast cancer. IDC is the most common type of breast cancer (70–80% of cases) originating from the epithelial cells of the milk ducts, whereas ILC arises from the epithelial cells of the lobules (milk-producing glands), accounting for 10–15% of breast cancers [78,250]. ILC is more often ER+ and PR+ but $HER2^-$, whereas IDC has more heterogeneous receptor expression, with some cases being $HER2^+$ or TNBC [78,250]. Their findings indicated that hormone-sensitive lipase (HSL) and FABP4 were more frequently expressed in ILC compared to IDC, particularly in Luminal A-type ILC [249]. Conversely, PLIN1 was more commonly expressed in IDC than in ILC [249]. These differences suggest distinct lipid metabolic pathways between these histological subtypes, which may influence tumour behaviour and response to therapy. Hormone therapy is more effective in ILC than in IDC, but chemotherapy is less effective in ILC [250]. IDC can be more aggressive, especially $HER2^+$ and triple-negative subtypes, requiring more aggressive treatment [250].

In addition, Jung et al. investigated 149 cases of metastatic breast cancer to determine the expression of lipid metabolism-related proteins across various metastatic sites, including bone, brain, liver, and lung, using IHC staining [248]. Their analysis revealed that FASN expression varied significantly according to molecular subtype, with higher expression in $HER2^+$ types and lower expression in Luminal A and TNBC subtypes [248]. Additionally, PLIN1 positivity was identified as an independent poor prognostic factor, associated with shorter overall survival [248]. These results highlight the potential of lipid metabolism-related proteins as biomarkers and therapeutic targets in metastatic breast cancer. Furthermore, the same team of researchers also investigated the expression patterns of lipid metabolism-related proteins in phyllodes tumours of the breast and their associations with tumour grade and patient prognosis [251]. Phyllodes tumours (PTs) are rare fibroepithelial breast tumours that originate from the periductal stromal cells rather than from the epithelial components of the breast [252]. These tumours account for less than 1% of all breast tumours and can exhibit a wide spectrum of biological behaviour, ranging from benign to highly aggressive malignant forms [252]. Higher-grade PTs exhibited increased stromal expression of HSL, perilipin 2, FABP4, CPT-1, and FASN [251]. This suggests that lipid metabolism alterations are associated with tumour aggressiveness. Perilipin 2, FABP4, CPT-1, and FASN were associated with shorter overall survival in PT patients [252]. Notably, stromal perilipin 2 positivity emerged as an independent prognostic factor for reduced disease-free survival. These findings highlight the potential role of lipid metabolism-related proteins as prognostic biomarkers in PT progression. Collectively, these studies underscore the heterogeneity of lipid metabolism-related protein expression across breast cancer subtypes and their potential implications for prognosis and targeted therapy. High expression of FASN also significantly correlated with $HER2^+$ status and brain metastasis, suggesting FASN could serve as a potential biomarker for distinguishing HER2-enriched breast cancer and predicting metastasis at later stage of malignancy. High expression of PLIN also positively correlated with poor prognosis, particularly in $HER2^+$ tumours and TNBC, which warrants further research to validate its clinical significance in these breast cancer subtypes.

4.2. Lung Cancer and Lipid Metabolism

Lipid metabolism alterations are increasingly recognized as significant contributors to lung cancer pathogenesis, with distinct expression patterns observed across different histological subtypes. Lung adenocarcinoma (LUAD) is a predominant subtype of NSCLC that exhibits distinct alterations in lipid metabolism, with specific proteins playing pivotal roles in tumour development and progression [253]. A proteomic analysis of primary LUAD tumours using global mass spectrometry identified the upregulation of enzymes involved in fatty acid synthesis and oxidation, underscoring their potential role as early biomarkers [254]. Key lipid metabolism-related proteins, such as FASN and ACC, showed elevated expression in LUAD tissues compared to normal lung tissue [254]. These enzymes are crucial for de novo lipogenesis in cancer cells to meet increased lipid demands for membrane synthesis and energy production. Additionally, IHC studies have demonstrated increased nuclear expression of SREBP1 in lung adenocarcinoma (LUAD) tissues compared to normal lung tissues [253]. This upregulation correlates with enhanced lipogenesis and tumour growth, indicating its role in lung cancer metabolism [253]. Proteomic analyses using LC-MS have identified higher expression levels of ACSL1, ACSL4, and ACSL5 significantly correlated with LUAD [255]. ACSL5 expression was inversely associated with Ki67 in low-grade tumours, while ACSL3 was positively associated with Ki67 in high-grade tumours [255]. These findings from patient-derived samples emphasize the critical role of lipid metabolism-related proteins in LUAD pathogenesis. The consistent upregulation of enzymes like FASN and ACC suggests their potential as biomarkers for diagnosis and as targets for therapeutic intervention in LUAD. A recent LC-MS/MS proteomic analysis on small cell lung cancer (SCLC) tissue samples identified two distinct categories of primary tumours based on their protein expression profiles, revealing the importance of lipid metabolism, particularly phospholipid metabolism, and immune response pathways in SCLC metastasis [256]. Both high-density and low-density lipoprotein cholesterol levels were found to be independent prognostic factors for disease-free survival in SCLC patients [256]. High stromal expression of phospholipase A2 group IIA (sPLA2-IIA) was associated with delayed disease recurrence in patients with limited-stage SCLC [256]. The expression patterns of lipid metabolism-related proteins vary notably between NSCLC and SCLC, offering insights into tumour biology and potential clinical applications. Given the growing evidence that phospholipid remodelling plays a critical role in tumour biology and cancer metabolism, integrating phospholipidomic profiling with enzyme-focused analyses offers a promising avenue for future research. Such multi-omics approaches could provide a more comprehensive understanding of the metabolic reprogramming in lung cancer, potentially uncovering novel biomarkers and therapeutic targets.

4.3. CRC and Lipid Metabolism

IHC analysis of CRC specimens revealed higher ACLY staining scores in cancerous tissues, with even greater expression observed in metastatic cases [257]. A positive correlation between ACLY and β-catenin (CTNNB1) protein levels has been identified, suggesting a collaborative role in CRC progression [257]. ACLY stabilizes β-catenin to promote its nuclear translocation and enhance its transcriptional activity, thereby facilitating CRC cell migration and invasion [257]. High ACLY expression correlates with increased risk of lymph node and distant metastasis, poorer tumour differentiation, and advanced clinical stages [258]. Patients with elevated ACLY levels exhibit higher recurrence rates and reduced overall survival [258]. These findings underscore the potential of ACLY as a biomarker for CRC aggressiveness and as a target for therapeutic intervention.

Acetyl-CoA synthetase short-chain family member 2 (ACSS2) is an enzyme that catalyses the conversion of acetate into acetyl-CoA, playing a crucial role in lipid biosynthesis and energy metabolism [243]. IHC analyses have demonstrated that ACSS2 protein expression is significantly higher in CRC tissues compared to normal colorectal tissues [259]. ACSS2 exhibits distinct subcellular localization patterns in benign versus malignant tissues. In benign tissues, ACSS2 immunoreactivity is predominantly diffuse within the cytosol [259]. In contrast, CRC samples often display localized cytosolic staining, with a lower cytosolic/nuclear ratio, suggesting a preferential nuclear presence of ACSS2 in cancer cells [259]. Current research on tissue-based analyses highlights the need for further investigation into developing ACSS2 as a potential biomarker for CRC diagnosis and prognosis.

On the other hand, FASN protein levels are markedly increased in CRC tissues compared to adjacent normal tissues in IHC staining [260–264]. This overexpression correlates with advanced tumour stages, lymph node metastasis, and poorer patient prognosis [262,264]. Mechanistic studies suggest that FASN may modulate the AMPK/mTOR and Wnt signaling pathways, contributing to tumour progression, cell proliferation, invasion, and metastasis [260,261]. Additionally, CRC patients have significantly higher FASN concentrations in the serum compared to healthy individuals [265]. Elevated serum FASN levels are associated with tumour extent, lymph node involvement, distant metastasis, and advanced clinical stages [265]. High serum FASN levels have been identified as an independent predictor of reduced overall and disease-free survival rates in CRC patients [265]. In summary, FASN protein is overexpressed in CRC tissues and elevated in patient serum, correlating with disease progression and poorer outcomes. These findings underscore the potential of FASN as both a diagnostic and prognostic biomarker in CRC.

In CRC, differential expression of specific ACSL isoforms has been observed, influencing tumour behaviour and patient outcomes. Overexpression of ACSL1 and ACSL4 has been linked to enhanced tumour aggressiveness [266]. In a tissue microarray study of 77 stage II CRC patients, simultaneous overexpression of ACSL1, ACSL4, and stearoyl-CoA desaturase (SCD) correlated with poorer disease-free survival, suggesting a cooperative role in promoting EMT and metastasis [266]. In contrast, low ACSL5 expression in CRC tissues is associated with early tumour recurrence [267]. In an IHC study involving

72 patients, those with reduced ACSL5 levels experienced significantly increased tumour recurrence within one year, independent of TNM and UICC stages [267]. Collectively, dysregulated expression of ACSL1, ACSL4 and ACSL5 isoforms in CRC tissues is associated with tumour progression and patient outcomes, warranting additional research for biomarker development.

The expression of various FABPs has been studied in CRC tissues and patient serum, revealing significant associations with disease progression and potential diagnostic implications. Zhang et al. have demonstrated that FABP4 is highly expressed in CRC tissues and patient serum, which correlates with advanced tumour stages and lymph node metastasis [268]. Notably, FABP4 levels in tumour tissues are inversely related to E-cadherin expression and positively associated with Snail protein levels as shown in IHC studies, suggesting a role in promoting EMT and tumour progression [268]. Similarly, FABP6 expression is significantly upregulated in CRC tissues compared to adjacent normal tissues [269]. The increased presence of FABP6 in tumour cells indicates its potential involvement in CRC carcinogenesis [269]. Increased serum levels of FABP6 have also been observed in CRC patients [269]. Notably, serum concentrations of both FABP4 and FABP6 significantly decreased in two weeks post-surgery, suggesting their potential utility in monitoring treatment response [269]. When combined with CEA, the sensitivity and specificity of FABP4 and FABP6 have been improved, suggesting that a panel of these biomarkers could enhance CRC detection [269]. The overexpression of FABP4 and FABP6 in CRC tissues and their elevated levels in patient serum are significantly associated with disease progression and metastasis [268,269]. These findings underscore the potential of FABP4 and FABP6 as biomarkers for CRC diagnosis, prognosis, and monitoring therapeutic responses. Further research is warranted to validate these biomarkers in larger CRC patient cohorts.

4.4. Liver Cancer and Lipid Metabolism

In HCC, key enzymes and transporters involved in de novo lipogenesis, fatty acid uptake, and cholesterol synthesis, such as FASN, SCD1, SREBPs, and fatty acid transport proteins, are frequently dysregulated in tumour tissues, contributing to tumour growth, immune evasion and therapy resistance. FASN protein expression is notably elevated in HCC tissue compared to adjacent non-cancerous liver tissues [270]. This upregulation is associated with enhanced tumour cell proliferation and progression [270]. ACLY is also significantly upregulated in HCC tissues and associated with poor prognosis and advanced tumour stages [271]. Functionally, ACLY contributes to HCC progression by activating the Wnt/β-catenin signaling pathway, thereby promoting cancer stemness and metastasis [271].

On the other hand, FABP4 and FABP5 have been implicated in HCC tumour progression and patient prognosis as reported by several IHC and tissue microarray studies. FABP4 expression is significantly reduced in HCC tissues compared to adjacent normal tissues [272]. Lower FABP4 levels are associated with larger tumour size, presence of portal vein tumour thrombus (PVTT), and shorter OS and DFS [272]. In contrast, overexpression of FABP5 in HCC tissues is associated with poorer tumour differentiation, vascular invasion, advanced TNM stages, and reduced OS and DFS, suggesting its importance as a prognostic marker in HCC [273,274]. Interestingly, elevated serum FABP5 levels in HCC patients are significantly associated with postoperative survival, independent of its tissue expression levels [273]. Although FABP4 and FABP5 demonstrate opposite expression levels in HCC tissues and serum, these proteins are promising biomarkers and therapeutic targets in HCC, as their dysregulation reflects the metabolic dependency of liver cancer on fatty acid synthesis and utilization.

ACSL3 and ACSL4 play significant roles in HCC development and progression. Their expression patterns in tumour tissues have prognostic implications and may serve as potential therapeutic targets. Ndiaye et al. assessed the expression and localization of ACSL3 and ACSL4 in HCC, cholangiocarcinoma (CCA), and hepatic metastases by performing IHC analyses on tissue microarrays and subcellular fractionation of HepG2 cells [275]. ACSL3 and ACSL4 exhibited increased expression in HCC tissues compared to normal liver. ACSL4 expression was significantly higher in HCC than in all other tumour types, while ACSL3 expression was similar in HCC and hepatic metastases. Increased ACSL3 expression distinguished HCC from CCA, while combined ACSL3 and ACSL4 staining scores differentiated HCC from hepatic metastases. ACSL4 localized to the plasma membrane, lipid droplets, and endoplasmic reticulum (ER), while ACSL3 was associated with lipid droplets and the ER. Taken together, the differential expression and localization of ACSL3 and ACSL4 can aid in distinguishing HCC from other hepatic tumours, suggesting their potential as diagnostic biomarkers.

Several studies also highlight the clinical significance of ACSL4 in HCC. High ACSL4 expression is associated with poor survival in HCC patients, particularly those treated with postoperative adjuvant transarterial chemoembolization (TACE) [276–279]. High ACSL4 expression promotes HCC progression by stabilizing the oncoprotein c-Myc through the ERK/FBW7/c-Myc axis [276,279]. Additionally, patients with high ACSL4 expression in pre-treatment tumour tissues also exhibited complete or partial responses to sorafenib, which is a first-line treatment for advanced HCC [280]. These findings underscore the potential of ACSL4 as a biomarker to predict sorafenib responsiveness in HCC patients [280]. Clinically, ACSL4 was positively correlated with the expression of SREBP1 in HCC patients [276]. Another IHC study by Li et al. revealed that SREBP1 was significantly upregulated in HCC tissues compared to adjacent non-cancerous liver tissues, which was associated with larger tumour size, higher histological grade, and advanced TNM stage [281]. Patients with elevated SREBP1 levels have poorer overall and disease-free survival rates, suggesting the prognostic potential of this protein in HCC [281]. Functionally, SREBP1 enhances tumour cell proliferation, migration, and invasion, suggesting its role in promoting HCC progression [281].

SCD1 is significantly overexpressed in HCC tissues, which correlates with enhanced tumour growth, metastasis, immune infiltration and poor clinical outcomes [282]. Elevated serum levels of SCD1 have been observed in HCC patients, suggesting its potential as a non-invasive biomarker for HCC diagnosis and prognosis [282]. Chen et al. have further demonstrated the diagnostic accuracy of serum SCD1 levels in distinguishing HCC patients from healthy individuals, validating its utility in clinical settings [282]. SCD1 overexpression in HCC tissues and elevated serum levels underscore its significance as a biomarker for disease progression and a potential therapeutic target. Meanwhile, SLC27A5 (FATP5) was significantly downregulated in the IHC analysis of HCC tissues [283,284]. This downregulation is linked to DNA hypermethylation of the SLC27A5 promoter region [283,284]. Lower SLC27A5 levels correlate with poorer patient prognosis, including larger tumour size and higher recurrence rates [283,284]. Functionally, SLC27A5 deficiency leads to increased lipid peroxidation, activating the NRF2/TXNRD1 pathway, which contributes to tumour progression and resistance to therapies like sorafenib [283,284]. In summary, the dysregulated expression of ACLY, ACSL3, ACSL4, FABP4, FABP5, SCD1, SLC27A5, and SREBP1 in HCC underscores their critical roles in lipid metabolism alterations contributing to tumour development and progression. These proteins hold potential as biomarkers for prognosis and as targets for therapeutic intervention in HCC.

4.5. Gastric Cancer and Lipid Metabolism

In GC, FASN overexpression has been consistently reported in cancer tissues compared to adjacent normal mucosa [285–288]. This upregulation correlates with advanced TNM stage, lymph node metastasis, poor differentiation, and increased tumour invasiveness [285–287]. Clinically, high FASN protein expression is an unfavourable prognostic indicator, as it is associated with reduced OS and DFS in GC patients [285–288]. Similarly, elevated serum levels of FASN have been observed in GC patients compared to healthy controls [289]. Notably, increased serum FASN levels were detected even in early-stage gastric cancer patients, suggesting its utility as a non-invasive biomarker for early detection [289]. ACLY is also significantly overexpressed in GC tissues compared to adjacent non-cancerous tissues in IHC studies [290–292]. High ACLY expression correlates with advanced tumour stages and lymph node metastasis, indicating its role in tumour aggressiveness [290–292]. Elevated ACLY expression has been associated with poorer overall survival in GC patients, suggesting its potential as a prognostic biomarker [290–292]. On the other hand, CD36 protein expression in GC tissue has been linked to tumour progression and metastasis. Studies using IHC staining have shown that higher CD36 expression correlates with poorer prognosis in GC patients [293]. Additionally, hypoxia-induced CD36 expression has been observed in peritoneal metastases, suggesting its role in fatty acid uptake and tumour spread [294]. CD36 expression is associated with increased migration and invasion of GC cells, particularly in adipocyte-rich environments [294]. This highlights its potential as a biomarker for aggressive disease and a therapeutic target for intervention.

Studies have demonstrated that SCD1 is significantly overexpressed in GC tissues compared to adjacent non-cancerous tissues [295]. This upregulation is associated with enhanced tumour growth, migration, and resistance to ferroptosis [295]. High SCD1 expression levels in GC tissues also correlate with poor patient prognosis, suggesting its potential as a prognostic biomarker [295]. In addition, CPT1A is significantly upregulated in GC tissues, which correlates with enhanced tumour proliferation, invasion, and EMT [296]. Elevated CPT1A expression is significantly associated with advanced TNM stages, lymph node metastasis, and poorer overall survival in GC patients [296]. IHC analyses have demonstrated higher CPT1A levels in tumours with distant metastasis and *Helicobacter pylori* positivity [296]. Moreover, studies have reported increased CPT1A levels in the serum of GC patients compared to healthy controls [297]. This elevation is associated with poor prognosis and lymph node metastasis, suggesting its potential as a non-invasive biomarker for GC diagnosis and prognosis [297]. The overexpression of CPT1A in GC tissues and its elevated levels in patient serum underscore its pivotal role in tumour metabolism and progression. Its association with adverse clinical outcomes positions CPT1A as a promising biomarker and potential therapeutic target in gastric cancer management.

The overexpression of ACC1 has been reported in GC tissue using IHC staining, underscoring its importance in tumour progression [298]. ACC1 upregulation in GC tissue is associated with enhanced fatty acid synthesis, promoting tumour growth and progression [298]. More importantly, increased expression of phosphorylated ACC1 (p-ACC1) has been identified as an independent prognostic factor in gastric cancer patients without lymph node metastasis, indicating its potential role in monitoring tumour aggressiveness [298]. Meanwhile, lipoprotein lipase (LPL) is a key enzyme in lipid metabolism, hydrolysing triglycerides in circulating lipoproteins to release free fatty acids for cellular uptake [243]. LPL is significantly overexpressed in GC tissues, especially in cases with lymph node metastasis [299]. This overexpression facilitates increased lipid uptake, supporting tumour growth and metastatic potential [299]. Mechanistically, leptin, an adipokine often elevated in obesity, suppresses the cleavage of LPL, leading to enhanced lipid uptake and promoting lymph node metastasis in GC [299]. This highlights a link between metabolic disorders and cancer progression. More importantly, alterations in serum lipid profiles, such as decreased high-density lipoprotein cholesterol (HDL-C) and increased low-density lipoprotein cholesterol (LDL-C), have been observed in GC patients, which may indirectly reflect changes in LPL activity [300]. Further research is needed to elucidate the potential of LPL as a non-invasive biomarker in GC through serum level assessments.

Peroxisome proliferator-activated receptors (PPARs) are nuclear hormone receptors that regulate lipid metabolism, inflammation, and cellular differentiation. The three isoforms, including PPARα, PPARδ/PPARβ, and PPARγ, exhibit distinct expression patterns and functional roles in GC. PPARδ is overexpressed in approximately 90% of GC tissues, including both intestinal and diffuse subtypes, while its expression is low or absent in normal gastric epithelium [301]. Elevated PPARδ expression correlates with increased tumour cell proliferation, invasion, and metastasis [301]. It interacts with the Hippo pathway coactivator YAP1 to promote SOX9 expression, contributing to tumourigenesis [301]. Similarly, Ma et al. revealed significantly higher PPARγ expression in GC tissues compared to adjacent non-cancerous and normal gastric mucosa [302]. Activation of PPARγ has been shown to inhibit gastric cancer cell proliferation, migration, and invasion [302]. It downregulates β-catenin signaling and may induce apoptosis, suggesting a tumour-suppressive role [302]. The expression patterns and functional implications of PPARα in GC remain less defined. Further research is needed to elucidate the roles of PPARs in GC fully and to assess their potential as non-invasive biomarkers. Table 3 summarizes the key enzymes and proteins in the amino acid metabolism of top five cancers.

Table 3. Key enzymes and proteins in lipid metabolism.

Protein/Enzyme	Function in Lipid Metabolism	Cancer Types	Biomarker Role
FASN	De novo fatty acid synthesis from acetyl-CoA	Breast, Lung, CRC, Prostate	Prognostic, therapeutic target (FASN inhibitors)
ACLY	Converts citrate to acetyl-CoA, fuels lipogenesis	Breast, Lung	Prognostic, linked to growth and metastasis
ACC	Converts acetyl-CoA to malonyl-CoA	Breast, Liver	Prognostic, therapeutic target, FA synthesis regulation
SCD1	Converts saturated FA to monounsaturated FA	Breast, Lung, CRC	Prognostic, linked to membrane fluidity and survival
CPT1A	Controls fatty acid oxidation (FAO)	Breast, Prostate	Prognostic, metabolic plasticity, therapy resistance
FABP5/FABP4	Fatty acid-binding proteins aiding lipid uptake	Prostate, Breast	Prognostic, associated with aggressiveness
LPL	Hydrolyzes triglycerides for fatty acid uptake	Prostate, Breast	Prognostic, lipid-driven tumour progression

5. Metabolic Biomarkers in Clinical Trials and Applications

Validation in prospective clinical trials is essential for biomarkers to reach clinical implementation. Companion diagnostic development must meet regulatory requirements set by agencies like the United States Food and Drug Administration (FDA) and the European Medicines Agency (EMA), focusing on assay standardization, analytical reproducibility, and clinical utility [303]. Despite the vast array of metabolic proteins and enzymes identified as being dysregulated in cancers, only a limited number have advanced to clinical application. For metabolic biomarkers to reach this stage, these molecules must demonstrate not only mechanistic relevance and diagnostic accuracy but also reproducibility and clinical utility in prospective studies [304]. Several metabolic biomarkers, mainly those involved in glycolysis, have been approved or are currently undergoing evaluation in clinical trials for cancer detection, prognosis, or treatment monitoring.

5.1. *Tumour M2-Pyruvate Kinase (TuM2-PK)*

One of the most extensively studied metabolic biomarkers in clinical settings is TuM2-PK, an isoform of pyruvate kinase M2 (PKM2) secreted by tumour cells. TuM2-PK is elevated in the plasma or stool of patients with various cancers, including breast, CRC, gastric, lung, and pancreatic cancers [130,305–308]. Commercial assays, such as the ScheBo® Tumour M2-PK Test, (Schebo Biotech AC, Giessen, Germany), have been approved in Europe and used clinically to aid CRC detection via stool or plasma samples [309]. Studies report increased sensitivity when TuM2-PK is combined with carcinoembryonic antigen (CEA) or faecal occult blood testing (FOBT), making it valuable in early detection and post-surgical monitoring [130,309,310].

5.2. *Lactate Dehydrogenase (LDH)*

Serum LDH level is considered a nonspecific marker of tissue damage and is elevated in many conditions, including cancers, haemolysis, myocardial infarction, and liver disease. In oncology, elevated serum LDH may be used as a prognostic indicator or for monitoring disease progression or response to therapy, especially in lymphoma and melanoma [311–313]. Although nonspecific, its elevation often correlates with high tumour burden and aggressive disease phenotypes. In solid tumours such as CRC, HCC and NSCLC, serum LDH has prognostic utility, especially in the context of hypoxic or highly glycolytic tumours [141,314,315]. Moreover, pre-treatment LDH levels have been used to stratify cancer patients for targeted therapies like immunotherapy and tyrosine kinase inhibitors [316,317].

5.3. *Isocitrate Dehydrogenase 1 (IDH1) and 2-Hydroxyglutarate (2-HG)*

IDH1 mutations are among the most studied metabolic alterations in cancers, particularly in gliomas, acute myeloid leukaemia, cholangiocarcinoma, and chondrosarcoma [318–321]. These mutations confer a neo-morphic enzymatic activity that converts alpha-ketoglutarate to the oncometabolite 2-HG, which accumulates at high levels in tumour cells [322]. Elevated 2-HG induces widespread epigenetic changes, contributing to tumourigenesis through aberrant DNA and histone methylation [320]. Clinically, 2-HG serves as a valuable biomarker for IDH1-mutant tumours and can be detected in tissues and, to a lesser extent, in blood or cerebrospinal fluid using advanced techniques such as magnetic resonance spectroscopy and mass spectrometry [319,320,322]. Importantly, inhibitors targeting mutant IDH1 that reduce 2-HG levels show promise in preclinical and clinical settings, highlighting the therapeutic and diagnostic significance of this metabolic axis in cancer [323,324].

5.4. *Liquid Biopsy Approaches and Analytical Challenges in Clinical Settings*

Liquid biopsy has emerged as a transformative tool in cancer diagnostics, offering minimally invasive access to circulating tumour components such as cell-free DNA (cfDNA), circulating tumour cells (CTCs), and extracellular vesicles. This approach enables real-time monitoring of tumour dynamics, early detection, and therapeutic response assessment, making it highly valuable for clinical chemists and cancer researchers. Despite promising advances, several analytical challenges remain critical for successful clinical translation. Sample handling is a paramount concern, as pre-analytical variables such as blood collection tubes, processing times, storage conditions, and cfDNA isolation methods significantly influence biomarker integrity and assay reproducibility [325–327]. Standardization of protocols is essential to minimize variability and ensure reliable downstream analyses. Assay standardization is another challenge due to the diverse technologies used, including PCR-based methods, digital droplet PCR, and next-generation sequencing. Each platform varies in sensitivity, specificity, and quantification accuracy. Establishing consensus on quality control measures, reference materials, and reporting standards is vital to facilitate inter-laboratory comparability and regulatory acceptance. Thus, overcoming these challenges through rigorous method validation and harmonization efforts will be crucial for integrating liquid biopsy into routine clinical practice, thereby enhancing personalized oncology care.

6. Future Directions

Emerging technologies in spatial metabolomics and single-cell metabolomics are revolutionizing our understanding of metabolic heterogeneity within tumours, which is essential in tumour progression, tumour microenvironment, and therapeutic response. Spatial metabolomics enables the precise localization of metabolites directly within tissue sections, preserving the native architectural context of the tumour and its microenvironment [328,329]. This approach uncovers metabolic gradients, niches, and spatial interactions among cancer cells, stromal cells, and immune infiltrates not seen in bulk tissue analyses [329]. Single-cell metabolomics complements this by providing detailed metabolic profiles at the resolution of individual cells [328]. This high-resolution approach reveals previously unrecognized cellular heterogeneity and dynamic metabolic states within tumours, clarifying the metabolic plasticity that underlies phenotypic diversity, metastasis, and drug resistance [330]. The integration of spatial and single-cell metabolomics holds great promise for advancing biomarker discovery, enabling the identification of cell-type specific metabolic vulnerabilities and spatially defined metabolic signatures that can inform patient stratification and guide precision oncology interventions [330–333]. As these technologies continue to evolve, they are poised to transform cancer research and clinical diagnostics by linking metabolism with cellular identity and spatial context.

7. Conclusions

The intricate relationship between cancer metabolism and disease progression has become increasingly apparent, with metabolic reprogramming now recognized as a hallmark of cancer. This review comprehensively highlights how key metabolic enzymes, transporters, and proteins involved in glucose, amino acid, and lipid metabolism are differentially expressed across the top five cancers worldwide, namely lung, breast, CRC, liver, and gastric cancers. These molecules not only support tumour proliferation, survival, and therapy resistance but also show considerable promise as diagnostic, prognostic, and predictive biomarkers. Glycolytic enzymes such as LDHA, PKM2, HK2, and ENO1, along with nutrient transporters like GLUT1, ASCT2, and LAT1, are consistently upregulated across multiple cancer types and have been linked to poor prognosis and therapy resistance. Similarly, enzymes involved in glutamine metabolism (e.g., GLS1, GDH) and lipid biosynthesis (e.g., FASN, ACLY) further underscore the metabolic dependencies that distinguish cancer cells from normal counterparts. These findings not only support the role of metabolic proteins as potential tissue or serum-based biomarkers but also highlight their value as therapeutic targets in the era of precision oncology. Despite significant progress, few metabolic biomarkers have reached clinical implementation. Only a handful, such as TuM2-PK and serum LDH, are currently utilized in diagnostic or therapeutic settings. This gap between discovery and clinical application underscores the urgent need for rigorous validation strategies, including large-scale multi-omics integration, prospective biomarker trials, and standardization of assay platforms. In conclusion, advancing our understanding of cancer-specific metabolic rewiring and translating this knowledge into clinically actionable biomarkers will be essential for enhancing early detection, personalizing treatment strategies, and improving outcomes across high-burden malignancies. Future efforts must focus on refining biomarker validation pipelines, leveraging artificial intelligence to mine complex metabolic datasets, and developing combination strategies that co-target metabolic vulnerabilities alongside conventional therapies.

Author Contributions: Conceptualization, K.W.H. and R.N.; writing—original draft preparation, K.W.H.; writing—review and editing, R.N. All authors have read and agreed to the published version of the manuscript.

Funding: This research received no external funding.

Institutional Review Board Statement: Not applicable.

Informed Consent Statement: Not applicable.

Data Availability Statement: No new data were created or analyzed in this study.

Acknowledgments: The authors would like to thank the Jeffrey Cheah School of Medicine and Health Sciences, Monash University Malaysia, for providing the research facilities and support to conduct this study.

Conflicts of Interest: The authors declare no conflicts of interest.

Abbreviations

The following abbreviations are used in this manuscript:

2-HG	2-hydroxyglutarate
ACC	Acetyl-CoA carboxylase
ACAD	Acyl-CoA dehydrogenase
ACOX	Acyl-coenzyme A oxidase 1
ACLY	ATP citrate lyase
ACSL	Acyl-CoA synthetase long-chain
ACSS	Acyl-CoA synthetase short-chain
ADP	Adenosine diphosphate
AFP	Alpha-fetoprotein
AKT	Protein kinase B
ALDOA	Aldolase A
AMPK	AMP-activated protein kinase
ARG2	Arginase-2
ASCT2	Alanine serine cysteine transporter 2 (SLC1A5)
ASNS	Asparagine synthetase
ASS1	Argininosuccinate synthetase 1
ATP	Adenosine triphosphate
BCAT1	Branched-chain amino acid transaminase 1
BRCA	Breast cancer gene
CAF	Cancer-associated fibroblast
CD36	Cluster of differentiation 36
CEA	Carcinoembryonic antigen
CPT1A	Carnitine palmitoyltransferase 1A
CRC	Colorectal cancer
CSF	Cerebrospinal fluid
CSF1	Colony-stimulating factor 1
CTNNB1	Catenin beta 1
DBC	Ductal breast carcinoma
EMA	European Medicines Agency
EMT	Epithelial–mesenchymal transition
ENO1	Alpha-enolase 1
EGCG	Epigallocatechin gallate

EGFR	Epidermal growth factor receptor
ER	Endoplasmic reticulum
FA	Fatty acid
FABP	Fatty acid-binding protein
FAO	Fatty acid oxidation
FASN	Fatty acid synthase
FAT	Fatty acid translocase
FATP	Fatty acid transporter
FBP1	Fructose-1,6-bisphosphatase 1
FDA	Food and Drug Administration
FDG	Fluorodeoxyglucose
FOBT	Faecal occult blood test
FOXM1	Forkhead box protein M1
G6PD	Glucose-6-phosphate dehydrogenase
GC	Gastric cancer
GDH (GLUD1)	Glutamate dehydrogenase
GLS1/2	Glutaminase 1/2
GLUT1	Glucose transporter 1
GOT2	Glutamic-oxaloacetic transaminase 2
GS	Glutamine synthetase
GSH	Glutathione
HCC	Hepatocellular carcinoma
HGFR	Hepatocyte growth factor receptor
HIF-1α	Hypoxia-inducible factor 1-alpha
HK2	Hexokinase 2
HTATSF1	HIV Tat-specific factor 1
IDC	Invasive ductal carcinoma
IHC	Immunohistochemistry
ILC	Invasive lobular carcinoma
IL-1β	Interleukin-1 beta
LAT1	L-type amino acid transporter 1 (SLC7A5)
LDH/LDHA/LDHB	Lactate dehydrogenase A/B
LPL	Lipoprotein lipase
MAPK	Mitogen-activated protein kinase
MCT1	Monocarboxylate transporter 1
MUFA	Monounsaturated fatty acid
mTOR	Mechanistic target of rapamycin
MYC	MYC proto-oncogene
NAD+/NADH	Nicotinamide adenine dinucleotide (oxidized/reduced)
NADPH	Nicotinamide adenine dinucleotide phosphate (reduced)
NGS	Next-generation sequencing
NSCLC	Non-small cell lung cancer
OXPHOS	Oxidative phosphorylation
PD-L1	Programmed death-ligand 1
PDH	Pyruvate dehydrogenase
PDK1	Pyruvate dehydrogenase kinase 1
PFK1	Phosphofructokinase-1
PFKP	Phosphofructokinase, platelet,
PGAM1	Phosphoglycerate mutase 1
PGK1	Phosphoglycerate kinase 1
PHGDH	Phosphoglycerate dehydrogenase
PI3K	Phosphoinositide 3-kinase
PKM2	Pyruvate kinase M2
PLIN1	Perilipin1

PPARs	Peroxisome proliferator-activated receptors
PPP	Pentose phosphate pathway
PTM	Post-translational modification
PTs	Phyllodes tumours
PVTT	Portal vein tumour thrombus
ROS	Reactive oxygen species
SCD1	Stearoyl-CoA desaturase 1
SGLT1	Solute carrier family 5 member 1
SHMT1/2	Serine hydroxymethyltransferase 1/2
SLC	Solute carrier
SREBPs	Sterol regulatory element-binding proteins
TALDO	Transaldolase
TAM	Tumour-associated macrophage
TCA	Tricarboxylic acid (cycle)
TME	Tumour microenvironment
TNM	Tumour, Node, Metastasis (staging)
TP53	Tumour protein p53
TPI1	Triosephosphate isomerase
TuM2-PK	Tumour M2-pyruvate kinase
VEGF	Vascular endothelial growth factor
UICC	Union for International Cancer Control
xCT	Cystine/glutamate antiporter (SLC7A11)
YAP	Yes-associated protein

References

1. Bray, F.; Laversanne, M.; Sung, H.; Ferlay, J.; Siegel, R.L.; Soerjomataram, I.; Jemal, A. Global cancer statistics 2022: GLOBOCAN estimates of incidence and mortality worldwide for 36 cancers in 185 countries. *CA Cancer J. Clin.* **2024**, *74*, 229–263. [CrossRef] [PubMed]
2. Bizuayehu, H.M.; Ahmed, K.Y.; Kibret, G.D.; Dadi, A.F.; Belachew, S.A.; Bagade, T.; Tegegne, T.K.; Venchiarutti, R.L.; Kibret, K.T.; Hailegebireal, A.H.; et al. Global Disparities of Cancer and Its Projected Burden in 2050. *JAMA Netw. Open* **2024**, *7*, e2443198. [CrossRef]
3. Gao, W.; Chen, Y.; Yang, J.; Zhuo, C.; Huang, S.; Zhang, H.; Shi, Y. Clinical Perspectives on Liquid Biopsy in Metastatic Colorectal Cancer. *Front. Genet.* **2021**, *12*, 634642. [CrossRef]
4. Menzel, M.; Ossowski, S.; Kral, S.; Metzger, P.; Horak, P.; Marienfeld, R.; Boerries, M.; Wolter, S.; Ball, M.; Neumann, O.; et al. Multicentric pilot study to standardize clinical whole exome sequencing (WES) for cancer patients. *npj Precis. Oncol.* **2023**, *7*, 106. [CrossRef]
5. Loktionov, A. Biomarkers for detecting colorectal cancer non-invasively: DNA, RNA or proteins? *World J. Gastrointest. Oncol.* **2020**, *12*, 124–148. [CrossRef] [PubMed]
6. Hristova, V.A.; Chan, D.W. Cancer biomarker discovery and translation: Proteomics and beyond. *Expert Rev. Proteom.* **2019**, *16*, 93–103. [CrossRef] [PubMed]
7. Bhardwaj, M.; Terzer, T.; Schrotz-King, P.; Brenner, H. Comparison of Proteomic Technologies for Blood-Based Detection of Colorectal Cancer. *Int. J. Mol. Sci.* **2021**, *22*, 1189. [CrossRef]
8. Lin, T.-T.; Zhang, T.; Kitata, R.B.; Liu, T.; Smith, R.D.; Qian, W.-J.; Shi, T. Mass spectrometry-based targeted proteomics for analysis of protein mutations. *Mass Spectrom. Rev.* **2023**, *42*, e21741. [CrossRef]
9. Erben, V.; Bhardwaj, M.; Schrotz-King, P.; Brenner, H. Metabolomics Biomarkers for Detection of Colorectal Neoplasms: A Systematic Review. *Cancers* **2018**, *10*, 246. [CrossRef]
10. Nicolussi, A.; Belardinilli, F.; Mahdavian, Y.; Colicchia, V.; D'Inzeo, S.; Petroni, M.; Zani, M.; Ferraro, S.; Valentini, V.; Ottini, L.; et al. Next-generation sequencing of BRCA1 and BRCA2 genes for rapid detection of germline mutations in hereditary breast/ovarian cancer. *PeerJ* **2019**, *7*, e6661. [CrossRef]
11. Kinnersley, B.; Sud, A.; Everall, A.; Cornish, A.J.; Chubb, D.; Culliford, R.; Gruber, A.J.; Lärkeryd, A.; Mitsopoulos, C.; Wedge, D.; et al. Analysis of 10,478 cancer genomes identifies candidate driver genes and opportunities for precision oncology. *Nat. Genet.* **2024**, *56*, 1868–1877. [CrossRef] [PubMed]
12. Pasello, M.; Giudice, A.M.; Scotlandi, K. The ABC subfamily A transporters: Multifaceted players with incipient potentialities in cancer. *Semin. Cancer Biol.* **2020**, *60*, 57–71. [CrossRef]

13. Guo, D.; Meng, Y.; Jiang, X.; Lu, Z. Hexokinases in cancer and other pathologies. *Cell Insight* **2023**, *2*, 100077. [CrossRef]
14. Lu, Z.; Hunter, T. Metabolic Kinases Moonlighting as Protein Kinases. *Trends Biochem. Sci.* **2018**, *43*, 301–310. [CrossRef]
15. Sun, L.; Zhang, H.; Gao, P. Metabolic reprogramming and epigenetic modifications on the path to cancer. *Protein Cell* **2021**, *13*, 877–919. [CrossRef] [PubMed]
16. Pascale, R.M.; Calvisi, D.F.; Simile, M.M.; Feo, C.F.; Feo, F. The Warburg Effect 97 Years after Its Discovery. *Cancers* **2020**, *12*, 2819. [CrossRef]
17. Tait, S.W.; Green, D.R. Mitochondrial regulation of cell death. *Cold Spring Harb. Perspect. Biol.* **2013**, *5*, a008706. [CrossRef]
18. Sharma, A.; Virmani, T.; Kumar, G.; Sharma, A.; Virmani, R.; Gugulothu, D.; Singh, K.; Misra, S.K.; Pathak, K.; Chitranshi, N.; et al. Mitochondrial signaling pathways and their role in cancer drug resistance. *Cell. Signal.* **2024**, *122*, 111329. [CrossRef]
19. Nan, D.; Yao, W.; Huang, L.; Liu, R.; Chen, X.; Xia, W.; Sheng, H.; Zhang, H.; Liang, X.; Lu, Y. Glutamine and cancer: Metabolism, immune microenvironment, and therapeutic targets. *Cell Commun. Signal.* **2025**, *23*, 45. [CrossRef] [PubMed]
20. Zheng, M.; Wang, W.; Liu, J.; Zhang, X.; Zhang, R. Lipid Metabolism in Cancer Cells. In *Lipid Metabolism in Tumor Immunity*; Li, Y., Ed.; Springer: Singapore, 2021; pp. 49–69.
21. Sies, H.; Mailloux, R.J.; Jakob, U. Fundamentals of redox regulation in biology. *Nat. Rev. Mol. Cell Biol.* **2024**, *25*, 701–719. [CrossRef]
22. Panieri, E.; Santoro, M.M. ROS homeostasis and metabolism: A dangerous liason in cancer cells. *Cell Death Dis.* **2016**, *7*, e2253. [CrossRef]
23. Moloney, J.N.; Cotter, T.G. ROS signalling in the biology of cancer. *Semin. Cell Dev. Biol.* **2018**, *80*, 50–64. [CrossRef] [PubMed]
24. Tretter, V.; Hochreiter, B.; Zach, M.L.; Krenn, K.; Klein, K.U. Understanding Cellular Redox Homeostasis: A Challenge for Precision Medicine. *Int. J. Mol. Sci.* **2021**, *23*, 106. [CrossRef]
25. Li, J.; Eu, J.Q.; Kong, L.R.; Wang, L.; Lim, Y.C.; Goh, B.C.; Wong, A.L.A. Targeting Metabolism in Cancer Cells and the Tumour Microenvironment for Cancer Therapy. *Molecules* **2020**, *25*, 4831. [CrossRef]
26. Dmello, R.S.; To, S.Q.; Chand, A.L. Therapeutic Targeting of the Tumour Microenvironment in Metastatic Colorectal Cancer. *Int. J. Mol. Sci.* **2021**, *22*, 2067. [CrossRef]
27. Garcia, S.N.; Guedes, R.C.; Marques, M.M. Unlocking the Potential of HK2 in Cancer Metabolism and Therapeutics. *Curr. Med. Chem.* **2019**, *26*, 7285–7322. [CrossRef]
28. Li, T.; Han, J.; Jia, L.; Hu, X.; Chen, L.; Wang, Y. PKM2 coordinates glycolysis with mitochondrial fusion and oxidative phosphorylation. *Protein Cell* **2019**, *10*, 583–594. [CrossRef] [PubMed]
29. Saha, S.K.; Islam, S.M.R.; Abdullah-Al-Wadud, M.; Islam, S.; Ali, F.; Park, K.S. Multiomics Analysis Reveals that GLS and GLS2 Differentially Modulate the Clinical Outcomes of Cancer. *J. Clin. Med.* **2019**, *8*, 355. [CrossRef]
30. Li, L.; Peng, G.; Liu, X.; Zhang, Y.; Han, H.; Liu, Z.R. Pyruvate Kinase M2 Coordinates Metabolism Switch between Glycolysis and Glutaminolysis in Cancer Cells. *iScience* **2020**, *23*, 101684. [CrossRef]
31. Kodama, M.; Oshikawa, K.; Shimizu, H.; Yoshioka, S.; Takahashi, M.; Izumi, Y.; Bamba, T.; Tateishi, C.; Tomonaga, T.; Matsumoto, M.; et al. A shift in glutamine nitrogen metabolism contributes to the malignant progression of cancer. *Nat. Commun.* **2020**, *11*, 1320. [CrossRef] [PubMed]
32. Sun, H.-W.; Yu, X.-J.; Wu, W.-C.; Chen, J.; Shi, M.; Zheng, L.; Xu, J. GLUT1 and ASCT2 as Predictors for Prognosis of Hepatocellular Carcinoma. *PLoS ONE* **2016**, *11*, e0168907. [CrossRef]
33. Schroeder, B.; Vander Steen, T.; Espinoza, I.; Venkatapoorna, C.M.K.; Hu, Z.; Silva, F.M.; Regan, K.; Cuyàs, E.; Meng, X.W.; Verdura, S.; et al. Fatty acid synthase (FASN) regulates the mitochondrial priming of cancer cells. *Cell Death Dis.* **2021**, *12*, 977. [CrossRef]
34. Kowalik, M.A.; Columbano, A.; Perra, A. Emerging Role of the Pentose Phosphate Pathway in Hepatocellular Carcinoma. *Front. Oncol.* **2017**, *7*, 87. [CrossRef]
35. Yang, H.-C.; Stern, A.; Chiu, D.T.-Y. G6PD: A hub for metabolic reprogramming and redox signaling in cancer. *Biomed. J.* **2021**, *44*, 285–292. [CrossRef] [PubMed]
36. Payen, V.L.; Mina, E.; Van Hée, V.F.; Porporato, P.E.; Sonveaux, P. Monocarboxylate transporters in cancer. *Mol. Metab.* **2020**, *33*, 48–66. [CrossRef]
37. Bungay, P.; Bagal, S. Drug Efflux Transporters: P-gp and BCRP. In *The Medicinal Chemist's Guide to Solving ADMET Challenges*; Schnider, P., Ed.; The Royal Society of Chemistry: London, UK, 2021; pp. 109–127.
38. Bose, S.; Zhang, C.; Le, A. Glucose Metabolism in Cancer. *Adv. Exp. Med. Biol.* **2021**, *1311*, 3–15. [PubMed]
39. Shuvalov, O.; Petukhov, A.; Daks, A.; Fedorova, O.; Vasileva, E.; Barlev, N.A. One-carbon metabolism and nucleotide biosynthesis as attractive targets for anticancer therapy. *Oncotarget* **2017**, *8*, 23955–23977. [CrossRef]
40. Patra, K.C.; Hay, N. The pentose phosphate pathway and cancer. *Trends Biochem. Sci.* **2014**, *39*, 347–354. [CrossRef]
41. Jaiswara, P.K.; Gupta, V.K.; Rawat, S.G.; Tiwari, R.K.; Sonker, P.; Maurya, R.P.; Kumar, A. Targeting of Aerobic Glycolysis: An Emerging Therapeutic Approach Against Colon Cancer. In *Colon Cancer Diagnosis and Therapy*; Vishvakarma, N.K., Nagaraju, G.P., Shukla, D., Eds.; Springer International Publishing: Cham, Germany, 2021; Volume 2, pp. 225–244.

42. Claps, G.; Faouzi, S.; Quidville, V.; Chehade, F.; Shen, S.; Vagner, S.; Robert, C. The multiple roles of LDH in cancer. *Nat. Rev. Clin. Oncol.* **2022**, *19*, 749–762. [CrossRef] [PubMed]
43. Feng, Y.; Xiong, Y.; Qiao, T.; Li, X.; Jia, L.; Han, Y. Lactate dehydrogenase A: A key player in carcinogenesis and potential target in cancer therapy. *Cancer Med.* **2018**, *7*, 6124–6136. [CrossRef]
44. Wu, G.; Yuan, S.; Chen, Z.; Chen, G.; Fan, Q.; Dong, H.; Ye, F.; Li, J.; Zhu, X. The KLF14 Transcription Factor Regulates Glycolysis by Downregulating LDHB in Colorectal Cancer. *Int. J. Biol. Sci.* **2019**, *15*, 628–635. [CrossRef]
45. Lin, Y.; Wang, Y.; Li, P.-f. Mutual regulation of lactate dehydrogenase and redox robustness. *Front. Physiol.* **2022**, *13*, 1038421. [CrossRef] [PubMed]
46. Cui, J.; Quan, M.; Jiang, W.; Hu, H.; Jiao, F.; Li, N.; Jin, Z.; Wang, L.; Wang, Y.; Wang, L. Suppressed expression of LDHB promotes pancreatic cancer progression via inducing glycolytic phenotype. *Med. Oncol.* **2015**, *32*, 143. [CrossRef]
47. Chang, Y.C.; Yang, Y.C.; Tien, C.P.; Yang, C.J.; Hsiao, M. Roles of Aldolase Family Genes in Human Cancers and Diseases. *Trends Endocrinol. Metab.* **2018**, *29*, 549–559. [CrossRef]
48. Jiang, Z.; Wang, X.; Li, J.; Yang, H.; Lin, X. Aldolase A as a prognostic factor and mediator of progression via inducing epithelial-mesenchymal transition in gastric cancer. *J. Cell. Mol. Med.* **2018**, *22*, 4377–4386. [CrossRef]
49. Saito, Y.; Takasawa, A.; Takasawa, K.; Aoyama, T.; Akimoto, T.; Ota, M.; Magara, K.; Murata, M.; Hirohashi, Y.; Hasegawa, T.; et al. Aldolase A promotes epithelial-mesenchymal transition to increase malignant potentials of cervical adenocarcinoma. *Cancer Sci.* **2020**, *111*, 3071–3081. [CrossRef]
50. Du, S.; Guan, Z.; Hao, L.; Song, Y.; Wang, L.; Gong, L.; Liu, L.; Qi, X.; Hou, Z.; Shao, S. Fructose-Bisphosphate Aldolase A Is a Potential Metastasis-Associated Marker of Lung Squamous Cell Carcinoma and Promotes Lung Cell Tumorigenesis and Migration. *PLoS ONE* **2014**, *9*, e85804. [CrossRef] [PubMed]
51. Tang, Y.; Yang, X.; Feng, K.; Hu, C.; Li, S. High expression of aldolase A is associated with tumor progression and poor prognosis in hepatocellular carcinoma. *J. Gastrointest. Oncol.* **2021**, *12*, 174–183. [CrossRef]
52. Xu, M.; Xi, S.; Li, H.; Xia, Y.; Mei, G.; Cheng, Z. Prognosis significance and potential association between ALDOA and AKT expression in colorectal cancer. *Sci. Rep.* **2024**, *14*, 6488. [CrossRef]
53. Lu, G.; Shi, W.; Zhang, Y. Prognostic Implications and Immune Infiltration Analysis of ALDOA in Lung Adenocarcinoma. *Front. Genet.* **2021**, *12*, 721021. [CrossRef] [PubMed]
54. Blanco, A.; Blanco, G. Carbohydrate Metabolism. In *Medical Biochemistry*; Blanco, A., Blanco, G., Eds.; Academic Press: Cambridge, MA, USA, 2017; pp. 283–323.
55. Principe, M.; Borgoni, S.; Cascione, M.; Chattaragada, M.S.; Ferri-Borgogno, S.; Capello, M.; Bulfamante, S.; Chapelle, J.; Di Modugno, F.; Defilippi, P.; et al. Alpha-enolase (ENO1) controls alpha v/beta 3 integrin expression and regulates pancreatic cancer adhesion, invasion, and metastasis. *J. Hematol. Oncol.* **2017**, *10*, 16. [CrossRef]
56. Huang, C.K.; Sun, Y.; Lv, L.; Ping, Y. ENO1 and Cancer. *Mol. Ther. Oncolytics* **2022**, *24*, 288–298. [CrossRef]
57. Sun, L.; Lu, T.; Tian, K.; Zhou, D.; Yuan, J.; Wang, X.; Zhu, Z.; Wan, D.; Yao, Y.; Zhu, X.; et al. Alpha-enolase promotes gastric cancer cell proliferation and metastasis via regulating AKT signaling pathway. *Eur. J. Pharmacol.* **2019**, *845*, 8–15. [CrossRef] [PubMed]
58. Cheng, Z.; Shao, X.; Xu, M.; Zhou, C.; Wang, J. ENO1 Acts as a Prognostic Biomarker Candidate and Promotes Tumor Growth and Migration Ability Through the Regulation of Rab1A in Colorectal Cancer. *Cancer Manag. Res.* **2019**, *11*, 9969–9978. [CrossRef]
59. Yang, T.; Shu, X.; Zhang, H.-W.; Sun, L.-X.; Yu, L.; Liu, J.; Sun, L.-C.; Yang, Z.-H.; Ran, Y.-L. Enolase 1 regulates stem cell-like properties in gastric cancer cells by stimulating glycolysis. *Cell Death Dis.* **2020**, *11*, 870. [CrossRef] [PubMed]
60. Qiao, H.; Wang, Y.; Zhu, B.; Jiang, L.; Yuan, W.; Zhou, Y.; Guan, Q. Enolase1 overexpression regulates the growth of gastric cancer cells and predicts poor survival. *J. Cell. Biochem.* **2019**, *120*, 18714–18723. [CrossRef] [PubMed]
61. Fahimeh, G.; Tahereh, F.; Saeed, S.; Alireza, G.; Mehdi, S. Oncogenic Alterations of Metabolism Associated with Resistance to Chemotherapy. *Curr. Mol. Med.* **2024**, *24*, 856–866. [CrossRef]
62. Ciscato, F.; Filadi, R.; Masgras, I.; Pizzi, M.; Marin, O.; Damiano, N.; Pizzo, P.; Gori, A.; Frezzato, F.; Chiara, F.; et al. Hexokinase 2 displacement from mitochondria-associated membranes prompts Ca^{2+}-dependent death of cancer cells. *EMBO Rep.* **2020**, *21*, e49117. [CrossRef]
63. Ho, N.; Coomber, B.L. Hexokinase II expression is correlated with colorectal cancer prognosis. *Cancer Treat. Commun.* **2016**, *6*, 11–16. [CrossRef]
64. Sato-Tadano, A.; Suzuki, T.; Amari, M.; Takagi, K.; Miki, Y.; Tamaki, K.; Watanabe, M.; Ishida, T.; Sasano, H.; Ohuchi, N. Hexokinase II in breast carcinoma: A potent prognostic factor associated with hypoxia-inducible factor-1alpha and Ki-67. *Cancer Sci.* **2013**, *104*, 1380–1388. [CrossRef]
65. Roberts, D.J.; Miyamoto, S. Hexokinase II integrates energy metabolism and cellular protection: Akting on mitochondria and TORCing to autophagy. *Cell Death Differ.* **2015**, *22*, 248–257. [CrossRef]

66. DeWaal, D.; Nogueira, V.; Terry, A.R.; Patra, K.C.; Jeon, S.-M.; Guzman, G.; Au, J.; Long, C.P.; Antoniewicz, M.R.; Hay, N. Hexokinase-2 depletion inhibits glycolysis and induces oxidative phosphorylation in hepatocellular carcinoma and sensitizes to metformin. *Nat. Commun.* **2018**, *9*, 446. [CrossRef]
67. Tian, R.F.; Li, X.F.; Xu, C.; Wu, H.; Liu, L.; Wang, L.H.; He, D.; Cao, K.; Cao, P.G.; Ma, J.K.; et al. SiRNA targeting PFK1 inhibits proliferation and migration and enhances radiosensitivity by suppressing glycolysis in colorectal cancer. *Am. J. Transl. Res.* **2020**, *12*, 4923–4940.
68. Maddalena, F.; Condelli, V.; Matassa, D.S.; Pacelli, C.; Scrima, R.; Lettini, G.; Li Bergolis, V.; Pietrafesa, M.; Crispo, F.; Piscazzi, A. TRAP1 enhances Warburg metabolism through modulation of PFK1 expression/activity and favors resistance to EGFR inhibitors in human colorectal carcinomas. *Mol. Oncol.* **2020**, *14*, 3030–3047. [CrossRef] [PubMed]
69. Chang, Y.C.; Chan, M.H.; Li, C.H.; Yang, C.J.; Tseng, Y.W.; Tsai, H.F.; Chiou, J.; Hsiao, M. Metabolic protein phosphoglycerate kinase 1 confers lung cancer migration by directly binding HIV Tat specific factor 1. *Cell Death Discov.* **2021**, *7*, 135. [CrossRef]
70. Li, X.; Jiang, Y.; Meisenhelder, J.; Yang, W.; Hawke, D.H.; Zheng, Y.; Xia, Y.; Aldape, K.; He, J.; Hunter, T.; et al. Mitochondria-Translocated PGK1 Functions as a Protein Kinase to Coordinate Glycolysis and the TCA Cycle in Tumorigenesis. *Mol. Cell* **2016**, *61*, 705–719. [CrossRef] [PubMed]
71. Ahmad, S.S.; Glatzle, J.; Bajaeifer, K.; Buhler, S.; Lehmann, T.; Konigsrainer, I.; Vollmer, J.P.; Sipos, B.; Ahmad, S.S.; Northoff, H.; et al. Phosphoglycerate kinase 1 as a promoter of metastasis in colon cancer. *Int. J. Oncol.* **2013**, *43*, 586–590. [CrossRef] [PubMed]
72. Alquraishi, M.; Puckett, D.L.; Alani, D.S.; Humidat, A.S.; Frankel, V.D.; Donohoe, D.R.; Whelan, J.; Bettaieb, A. Pyruvate kinase M2: A simple molecule with complex functions. *Free Radic. Biol. Med.* **2019**, *143*, 176–192. [CrossRef] [PubMed]
73. Li, X.; Kim, W.; Arif, M.; Gao, C.; Hober, A.; Kotol, D.; Strandberg, L.; Forsström, B.; Sivertsson, Å.; Oksvold, P.; et al. Discovery of Functional Alternatively Spliced PKM Transcripts in Human Cancers. *Cancers* **2021**, *13*, 348. [CrossRef]
74. Shiroki, T.; Yokoyama, M.; Tanuma, N.; Maejima, R.; Tamai, K.; Yamaguchi, K.; Oikawa, T.; Noguchi, T.; Miura, K.; Fujiya, T.; et al. Enhanced expression of the M2 isoform of pyruvate kinase is involved in gastric cancer development by regulating cancer-specific metabolism. *Cancer Sci.* **2017**, *108*, 931–940. [CrossRef]
75. Yang, W.; Zheng, Y.; Xia, Y.; Ji, H.; Chen, X.; Guo, F.; Lyssiotis, C.A.; Aldape, K.; Cantley, L.C.; Lu, Z. ERK1/2-dependent phosphorylation and nuclear translocation of PKM2 promotes the Warburg effect. *Nat. Cell Biol.* **2012**, *14*, 1295–1304. [CrossRef]
76. Pekel, G.; Ari, F. Therapeutic Targeting of Cancer Metabolism with Triosephosphate Isomerase. *Chem. Biodivers.* **2020**, *17*, e2000012. [CrossRef] [PubMed]
77. Chen, T.; Huang, Z.; Tian, Y.; Lin, B.; He, R.; Wang, H.; Ouyang, P.; Chen, H.; Wu, L. Clinical significance and prognostic value of Triosephosphate isomerase expression in gastric cancer. *Medicine* **2017**, *96*, e6865. [CrossRef]
78. Oliveira, N.C.; Gomig, T.H.; Milioli, H.H.; Cordeiro, F.; Costa, G.G.; Urban, C.A.; Lima, R.S.; Cavalli, I.J.; Ribeiro, E.M. Comparative proteomic analysis of ductal and lobular invasive breast carcinoma. *Genet. Mol. Res.* **2016**, *15*. [CrossRef]
79. Lone, S.N.; Maqbool, R.; Parray, F.Q.; Ul Hussain, M. Triose-phosphate isomerase is a novel target of miR-22 and miR-28, with implications in tumorigenesis. *J. Cell. Physiol.* **2018**, *233*, 8919–8929. [CrossRef]
80. Jiang, H.; Ma, N.; Shang, Y.; Zhou, W.; Chen, T.; Guan, D.; Li, J.; Wang, J.; Zhang, E.; Feng, Y.; et al. Triosephosphate isomerase 1 suppresses growth, migration and invasion of hepatocellular carcinoma cells. *Biochem. Biophys. Res. Commun.* **2017**, *482*, 1048–1053. [CrossRef] [PubMed]
81. Zhang, W.; Tong, D.; Liu, F.; Li, D.; Li, J.; Cheng, X.; Wang, Z. RPS7 inhibits colorectal cancer growth via decreasing HIF-1α-mediated glycolysis. *Oncotarget* **2016**, *7*, 5800–5814. [CrossRef]
82. Doyen, J.; Trastour, C.; Ettore, F.; Peyrottes, I.; Toussant, N.; Gal, J.; Ilc, K.; Roux, D.; Parks, S.K.; Ferrero, J.M.; et al. Expression of the hypoxia-inducible monocarboxylate transporter MCT4 is increased in triple negative breast cancer and correlates independently with clinical outcome. *Biochem. Biophys. Res. Commun.* **2014**, *451*, 54–61. [CrossRef]
83. Pelizzari, G.; Basile, D.; Zago, S.; Lisanti, C.; Bartoletti, M.; Bortot, L.; Vitale, M.G.; Fanotto, V.; Barban, S.; Cinausero, M.; et al. Lactate Dehydrogenase (LDH) Response to First-Line Treatment Predicts Survival in Metastatic Breast Cancer: First Clues for A Cost-Effective and Dynamic Biomarker. *Cancers* **2019**, *11*, 1243. [CrossRef]
84. Wang, S.; Ma, L.; Wang, Z.; He, H.; Chen, H.; Duan, Z.; Li, Y.; Si, Q.; Chuang, T.-H.; Chen, C.; et al. Lactate Dehydrogenase-A (LDH-A) Preserves Cancer Stemness and Recruitment of Tumor-Associated Macrophages to Promote Breast Cancer Progression. *Front. Oncol.* **2021**, *11*, 654452. [CrossRef] [PubMed]
85. Huang, X.; Li, X.; Xie, X.; Ye, F.; Chen, B.; Song, C.; Tang, H.; Xie, X. High expressions of LDHA and AMPK as prognostic biomarkers for breast cancer. *Breast* **2016**, *30*, 39–46. [CrossRef]
86. Jia, Z.; Zhang, J.; Wang, Z.; Wang, B.; Wang, L.; Cao, J.; Tao, Z.; Hu, X. An explorative analysis of the prognostic value of lactate dehydrogenase for survival and the chemotherapeutic response in patients with advanced triple-negative breast cancer. *Oncotarget* **2018**, *9*, 10714–10722. [CrossRef]
87. Dennison, J.B.; Molina, J.R.; Mitra, S.; González-Angulo, A.M.; Balko, J.M.; Kuba, M.G.; Sanders, M.E.; Pinto, J.A.; Gómez, H.L.; Arteaga, C.L.; et al. Lactate Dehydrogenase B: A Metabolic Marker of Response to Neoadjuvant Chemotherapy in Breast Cancer. *Clin. Cancer Res.* **2013**, *19*, 3703–3713. [CrossRef]

88. Dong, T.; Liu, Z.; Xuan, Q.; Wang, Z.; Ma, W.; Zhang, Q. Tumor LDH-A expression and serum LDH status are two metabolic predictors for triple negative breast cancer brain metastasis. *Sci. Rep.* **2017**, *7*, 6069. [CrossRef] [PubMed]
89. Xiao, H.; Zhang, L.; Chen, Y.; Zhou, C.; Wang, X.; Wang, D.; Liu, Z. PKM2 Promotes Breast Cancer Progression by Regulating Epithelial Mesenchymal Transition. *Anal. Cell. Pathol.* **2020**, *2020*, 8396023. [CrossRef] [PubMed]
90. Lin, Y.; Lv, F.; Liu, F.; Guo, X.; Fan, Y.; Gu, F.; Gu, J.; Fu, L. High Expression of Pyruvate Kinase M2 is Associated with Chemosensitivity to Epirubicin and 5-Fluorouracil in Breast Cancer. *J. Cancer* **2015**, *6*, 1130–1139. [CrossRef]
91. Wang, G.; Xu, Z.; Wang, C.; Yao, F.; Li, J.; Chen, C.; Sun, S. Differential phosphofructokinase-1 isoenzyme patterns associated with glycolytic efficiency in human breast cancer and paracancer tissues. *Oncol. Lett.* **2013**, *6*, 1701–1706. [CrossRef]
92. Coelho, R.G.; Calaca, I.C.; Celestrini, D.M.; Correia-Carneiro, A.H.; Costa, M.M.; Zancan, P.; Sola-Penna, M. Hexokinase and phosphofructokinase activity and intracellular distribution correlate with aggressiveness and invasiveness of human breast carcinoma. *Oncotarget* **2015**, *6*, 29375–29387. [CrossRef] [PubMed]
93. Zhang, D.; Jin, N.; Sun, W.; Li, X.; Liu, B.; Xie, Z.; Qu, J.; Xu, J.; Yang, X.; Su, Y.; et al. Phosphoglycerate mutase 1 promotes cancer cell migration independent of its metabolic activity. *Oncogene* **2017**, *36*, 2900–2909. [CrossRef]
94. Jang, S.M.; Han, H.; Jang, K.S.; Jun, Y.J.; Jang, S.H.; Min, K.W.; Chung, M.S.; Paik, S.S. The Glycolytic Phenotype is Correlated with Aggressiveness and Poor Prognosis in Invasive Ductal Carcinomas. *J. Breast Cancer* **2012**, *15*, 172–180. [CrossRef]
95. Kim, S.; Kim, D.H.; Jung, W.-H.; Koo, J.S. Metabolic phenotypes in triple-negative breast cancer. *Tumor Biol.* **2013**, *34*, 1699–1712. [CrossRef]
96. Hussein, Y.R.; Bandyopadhyay, S.; Semaan, A.; Ahmed, Q.; Albashiti, B.; Jazaerly, T.; Nahleh, Z.; Ali-Fehmi, R. Glut-1 Expression Correlates with Basal-like Breast Cancer. *Transl. Oncol.* **2011**, *4*, 321–327. [CrossRef] [PubMed]
97. Sun, X.F.; Shao, Y.B.; Liu, M.G.; Chen, Q.; Liu, Z.J.; Xu, B.; Luo, S.X.; Liu, H. High-concentration glucose enhances invasion in invasive ductal breast carcinoma by promoting Glut1/MMP2/MMP9 axis expression. *Oncol. Lett.* **2017**, *13*, 2989–2995. [CrossRef] [PubMed]
98. Pinheiro, C.; Sousa, B.; Albergaria, A.; Paredes, J.; Dufloth, R.; Vieira, D.; Schmitt, F.; Baltazar, F. GLUT1 and CAIX expression profiles in breast cancer correlate with adverse prognostic factors and MCT1 overexpression. *Histol. Histopathol.* **2011**, *26*, 1279–1286. [CrossRef] [PubMed]
99. Ahn, S.; Woo, J.W.; Lee, K.; Park, S.Y. HER2 status in breast cancer: Changes in guidelines and complicating factors for interpretation. *J. Pathol. Transl. Med.* **2020**, *54*, 34–44. [CrossRef]
100. Wang, J.; Ji, H.; Niu, X.; Yin, L.; Wang, Y.; Gu, Y.; Li, D.; Zhang, H.; Lu, M.; Zhang, F.; et al. Sodium-Dependent Glucose Transporter 1 (SGLT1) Stabled by HER2 Promotes Breast Cancer Cell Proliferation by Activation of the PI3K/Akt/mTOR Signaling Pathway in HER2+ Breast Cancer. *Dis. Markers* **2020**, *2020*, 6103542. [CrossRef]
101. Han, L.; Qu, Q.; Aydin, D.; Panova, O.; Robertson, M.J.; Xu, Y.; Dror, R.O.; Skiniotis, G.; Feng, L. Structure and mechanism of the SGLT family of glucose transporters. *Nature* **2022**, *601*, 274–279. [CrossRef]
102. Krzeslak, A.; Wojcik-Krowiranda, K.; Forma, E.; Jozwiak, P.; Romanowicz, H.; Bienkiewicz, A.; Brys, M. Expression of GLUT1 and GLUT3 Glucose Transporters in Endometrial and Breast Cancers. *Pathol. Oncol. Res.* **2012**, *18*, 721–728. [CrossRef]
103. Hong, C.S.; Graham, N.A.; Gu, W.; Espindola Camacho, C.; Mah, V.; Maresh, E.L.; Alavi, M.; Bagryanova, L.; Krotee, P.A.L.; Gardner, B.K.; et al. MCT1 Modulates Cancer Cell Pyruvate Export and Growth of Tumors that Co-express MCT1 and MCT4. *Cell Rep.* **2016**, *14*, 1590–1601. [CrossRef]
104. Johnson, J.M.; Cotzia, P.; Fratamico, R.; Mikkilineni, L.; Chen, J.; Colombo, D.; Mollaee, M.; Whitaker-Menezes, D.; Domingo-Vidal, M.; Lin, Z.; et al. MCT1 in Invasive Ductal Carcinoma: Monocarboxylate Metabolism and Aggressive Breast Cancer. *Front. Cell Dev. Biol.* **2017**, *5*, 27. [CrossRef] [PubMed]
105. Huang, C.-K.; Chang, P.-H.; Kuo, W.-H.; Chen, C.-L.; Jeng, Y.-M.; Chang, K.-J.; Shew, J.-Y.; Hu, C.-M.; Lee, W.-H. Adipocytes promote malignant growth of breast tumours with monocarboxylate transporter 2 expression via β-hydroxybutyrate. *Nat. Commun.* **2017**, *8*, 14706. [CrossRef] [PubMed]
106. Yang, Y.; Chong, Y.; Chen, M.; Dai, W.; Zhou, X.; Ji, Y.; Qiu, G.; Du, X. Targeting lactate dehydrogenase a improves radiotherapy efficacy in non-small cell lung cancer: From bedside to bench. *J. Transl. Med.* **2021**, *19*, 170. [CrossRef]
107. McCleland, M.L.; Adler, A.S.; Deming, L.; Cosino, E.; Lee, L.; Blackwood, E.M.; Solon, M.; Tao, J.; Li, L.; Shames, D.; et al. Lactate dehydrogenase B is required for the growth of KRAS-dependent lung adenocarcinomas. *Clin. Cancer Res.* **2013**, *19*, 773–784. [CrossRef]
108. Chang, Y.C.; Chan, Y.C.; Chang, W.M.; Lin, Y.F.; Yang, C.J.; Su, C.Y.; Huang, M.S.; Wu, A.T.H.; Hsiao, M. Feedback regulation of ALDOA activates the HIF-1alpha/MMP9 axis to promote lung cancer progression. *Cancer Lett.* **2017**, *403*, 28–36. [CrossRef]
109. Chen, J.M.; Chiu, S.C.; Chen, K.C.; Huang, Y.J.; Liao, Y.A.; Yu, C.R. Enolase 1 differentially contributes to cell transformation in lung cancer but not in esophageal cancer. *Oncol. Lett.* **2020**, *19*, 3189–3196. [CrossRef]
110. Li, H.-J.; Ke, F.-Y.; Lin, C.-C.; Lu, M.-Y.; Kuo, Y.-H.; Wang, Y.-P.; Liang, K.-H.; Lin, S.-C.; Chang, Y.-H.; Chen, H.-Y.; et al. ENO1 Promotes Lung Cancer Metastasis via HGFR and WNT Signaling–Driven Epithelial-to-Mesenchymal Transition. *Cancer Res.* **2021**, *81*, 4094–4109. [CrossRef]

111. Fu, Q.F.; Liu, Y.; Fan, Y.; Hua, S.N.; Qu, H.Y.; Dong, S.W.; Li, R.L.; Zhao, M.Y.; Zhen, Y.; Yu, X.L.; et al. Alpha-enolase promotes cell glycolysis, growth, migration, and invasion in non-small cell lung cancer through FAK-mediated PI3K/AKT pathway. *J. Hematol. Oncol.* **2015**, *8*, 22. [CrossRef]
112. Shen, J.; Jin, Z.; Lv, H.; Jin, K.; Jonas, K.; Zhu, C.; Chen, B. PFKP is highly expressed in lung cancer and regulates glucose metabolism. *Cell. Oncol.* **2020**, *43*, 617–629. [CrossRef]
113. Sun, Q.; Li, S.; Wang, Y.; Peng, H.; Zhang, X.; Zheng, Y.; Li, C.; Li, L.; Chen, R.; Chen, X.; et al. Phosphoglyceric acid mutase-1 contributes to oncogenic mTOR-mediated tumor growth and confers non-small cell lung cancer patients with poor prognosis. *Cell Death Differ.* **2018**, *25*, 1160–1173. [CrossRef] [PubMed]
114. Li, F.; Yang, H.; Kong, T.; Chen, S.; Li, P.; Chen, L.; Cheng, J.; Cui, G.; Zhang, G. PGAM1, regulated by miR-3614–5p, functions as an oncogene by activating transforming growth factor-β (TGF-β) signaling in the progression of non-small cell lung carcinoma. *Cell Death Dis.* **2020**, *11*, 710. [CrossRef] [PubMed]
115. Ito, R.; Yashiro, M.; Tsukioka, T.; Izumi, N.; Komatsu, H.; Inoue, H.; Yamamoto, Y.; Nishiyama, N. GLUT1 and PKM2 may be useful prognostic predictors in patients with non-small cell lung cancer following curative R0 resection. *Oncol Lett.* **2023**, *25*, 129. [CrossRef]
116. Eilertsen, M.; Andersen, S.; Al-Saad, S.; Kiselev, Y.; Donnem, T.; Stenvold, H.; Pettersen, I.; Al-Shibli, K.; Richardsen, E.; Busund, L.T.; et al. Monocarboxylate transporters 1–4 in NSCLC: MCT1 is an independent prognostic marker for survival. *PLoS ONE* **2014**, *9*, e105038. [CrossRef]
117. Kuo, T.C.; Huang, K.Y.; Yang, S.C.; Wu, S.; Chung, W.C.; Chang, Y.L.; Hong, T.M.; Wang, S.P.; Chen, H.Y.; Hsiao, T.H.; et al. Monocarboxylate Transporter 4 Is a Therapeutic Target in Non-small Cell Lung Cancer with Aerobic Glycolysis Preference. *Mol. Ther. Oncolytics* **2020**, *18*, 189–201. [CrossRef] [PubMed]
118. Katagiri, M.; Karasawa, H.; Takagi, K.; Nakayama, S.; Yabuuchi, S.; Fujishima, F.; Naitoh, T.; Watanabe, M.; Suzuki, T.; Unno, M.; et al. Hexokinase 2 in colorectal cancer: A potent prognostic factor associated with glycolysis, proliferation and migration. *Histol. Histopathol.* **2017**, *32*, 351–360. [CrossRef] [PubMed]
119. Hamabe, A.; Yamamoto, H.; Konno, M.; Uemura, M.; Nishimura, J.; Hata, T.; Takemasa, I.; Mizushima, T.; Nishida, N.; Kawamoto, K.; et al. Combined evaluation of hexokinase 2 and phosphorylated pyruvate dehydrogenase-E1alpha in invasive front lesions of colorectal tumors predicts cancer metabolism and patient prognosis. *Cancer Sci.* **2014**, *105*, 1100–1108. [CrossRef] [PubMed]
120. He, X.; Lin, X.; Cai, M.; Zheng, X.; Lian, L.; Fan, D.; Wu, X.; Lan, P.; Wang, J. Overexpression of Hexokinase 1 as a poor prognosticator in human colorectal cancer. *Tumor Biol.* **2016**, *37*, 3887–3895. [CrossRef]
121. Zhan, P.; Zhao, S.; Yan, H.; Yin, C.; Xiao, Y.; Wang, Y.; Ni, R.; Chen, W.; Wei, G.; Zhang, P. α-enolase promotes tumorigenesis and metastasis via regulating AMPK/mTOR pathway in colorectal cancer. *Mol. Carcinog.* **2017**, *56*, 1427–1437. [CrossRef]
122. Gu, J.; Zhong, K.; Wang, L.; Ni, H.; Zhao, Y.; Wang, X.; Yao, Y.; Jiang, L.; Wang, B.; Zhu, X. ENO1 contributes to 5-fluorouracil resistance in colorectal cancer cells via EMT pathway. *Front. Oncol.* **2022**, *12*, 1013035. [CrossRef]
123. Ma, L.; Zhang, X.; Liu, Y.; Jin, H.; Li, D.; Zhang, H.; Feng, L.; Zuo, J.; Wang, Y.; Liu, J.; et al. The ratio of PKM1/PKM2 is the key factor affecting the glucose metabolism and biological function of colorectal cancer cells. *Transl. Cancer Res.* **2024**, *13*, 3522–3535. [CrossRef]
124. Zhou, C.-F.; Li, X.-B.; Sun, H.; Zhang, B.; Han, Y.-S.; Jiang, Y.; Zhuang, Q.-L.; Fang, J.; Wu, G.-H. Pyruvate kinase type M2 is upregulated in colorectal cancer and promotes proliferation and migration of colon cancer cells. *IUBMB Life* **2012**, *64*, 775–782. [CrossRef]
125. Yang, P.; Li, Z.; Fu, R.; Wu, H.; Li, Z. Pyruvate kinase M2 facilitates colon cancer cell migration via the modulation of STAT3 signalling. *Cell. Signal.* **2014**, *26*, 1853–1862. [CrossRef] [PubMed]
126. Taniguchi, K.; Sugito, N.; Kumazaki, M.; Shinohara, H.; Yamada, N.; Matsuhashi, N.; Futamura, M.; Ito, Y.; Otsuki, Y.; Yoshida, K.; et al. Positive feedback of DDX6/c-Myc/PTB1 regulated by miR-124 contributes to maintenance of the Warburg effect in colon cancer cells. *Biochim. Biophys. Acta* **2015**, *1852*, 1971–1980. [CrossRef]
127. Cui, R.; Shi, X.Y. Expression of pyruvate kinase M2 in human colorectal cancer and its prognostic value. *Int. J. Clin. Exp. Pathol.* **2015**, *8*, 11393–11399. [PubMed]
128. Li, Y.; Wang, Y.; Liu, Z.; Guo, X.; Miao, Z.; Ma, S. Atractylenolide I Induces Apoptosis and Suppresses Glycolysis by Blocking the JAK2/STAT3 Signaling Pathway in Colorectal Cancer Cells. *Front. Pharmacol.* **2020**, *11*, 273. [CrossRef]
129. Cha, P.H.; Hwang, J.H.; Kwak, D.K.; Koh, E.; Kim, K.S.; Choi, K.Y. APC loss induces Warburg effect via increased PKM2 transcription in colorectal cancer. *Br. J. Cancer* **2021**, *124*, 634–644. [CrossRef]
130. Meng, W.; Zhu, H.H.; Xu, Z.F.; Cai, S.R.; Dong, Q.; Pan, Q.R.; Zheng, S.; Zhang, S.Z. Serum M2-pyruvate kinase: A promising non-invasive biomarker for colorectal cancer mass screening. *World J. Gastrointest. Oncol.* **2012**, *4*, 145–151. [CrossRef]
131. Rigi, F.; Jannatabad, A.; Izanloo, A.; Roshanravan, R.; Hashemian, H.R.; Kerachian, M.A. Expression of tumor pyruvate kinase M2 isoform in plasma and stool of patients with colorectal cancer or adenomatous polyps. *BMC Gastroenterol.* **2020**, *20*, 241. [CrossRef]

132. Wahib, A.A.; Seif El-Nasr, M.S.; Ismail, M.B.; Eltiby, D.M.; Sayed, S.A.-E.; Soliman, M.Y. Diagnostic Value of Plasma M2-Pyruvate Kinase in Egyptian Patients with Colorectal Cancer. *Egypt. J. Hosp. Med.* **2018**, *73*, 7997–8006. [CrossRef]
133. Liu, H.; Chen, X.; Wang, P.; Chen, M.; Deng, C.; Qian, X.; Bai, J.; Li, Z.; Yu, X. PRMT1-mediated PGK1 arginine methylation promotes colorectal cancer glycolysis and tumorigenesis. *Cell Death Dis.* **2024**, *15*, 170. [CrossRef] [PubMed]
134. Jun, Y.J.; Jang, S.M.; Han, H.L.; Lee, K.H.; Jang, K.S.; Paik, S.S. Clinicopathologic significance of GLUT1 expression and its correlation with Apaf-1 in colorectal adenocarcinomas. *World J. Gastroenterol.* **2011**, *17*, 1866–1873. [CrossRef]
135. Martins, S.F.; Amorim, R.; Viana-Pereira, M.; Pinheiro, C.; Costa, R.F.A.; Silva, P.; Couto, C.; Alves, S.; Fernandes, S.; Vilaça, S.; et al. Significance of glycolytic metabolism-related protein expression in colorectal cancer, lymph node and hepatic metastasis. *BMC Cancer* **2016**, *16*, 535. [CrossRef]
136. Gao, H.; Hao, Y.; Zhou, X.; Li, H.; Liu, F.; Zhu, H.; Song, X.; Niu, Z.; Ni, Q.; Chen, M.S.; et al. Prognostic value of glucose transporter 3 expression in hepatocellular carcinoma. *Oncol. Lett.* **2020**, *19*, 691–699. [CrossRef]
137. Lei, Y.; Hu, Q.; Gu, J. Expressions of Carbohydrate Response Element Binding Protein and Glucose Transporters in Liver Cancer and Clinical Significance. *Pathol. Oncol. Res.* **2020**, *26*, 1331–1340. [CrossRef] [PubMed]
138. Kwee, S.A.; Hernandez, B.; Chan, O.; Wong, L. Choline kinase alpha and hexokinase-2 protein expression in hepatocellular carcinoma: Association with survival. *PLoS ONE* **2012**, *7*, e46591. [CrossRef]
139. Guzman, G.; Chennuri, R.; Chan, A.; Rea, B.; Quintana, A.; Patel, R.; Xu, P.Z.; Xie, H.; Hay, N. Evidence for heightened hexokinase II immunoexpression in hepatocyte dysplasia and hepatocellular carcinoma. *Dig. Dis. Sci.* **2015**, *60*, 420–426. [CrossRef]
140. Cui, Z.; Li, Y.; Gao, Y.; Kong, L.; Lin, Y.; Chen, Y. Cancer-testis antigen lactate dehydrogenase C4 in hepatocellular carcinoma: A promising biomarker for early diagnosis, efficacy evaluation and prognosis prediction. *Aging* **2020**, *12*, 19455–19467. [CrossRef]
141. Zhang, J.P.; Wang, H.B.; Lin, Y.H.; Xu, J.; Wang, J.; Wang, K.; Liu, W.L. Lactate Dehydrogenase Is an Important Prognostic Indicator for Hepatocellular Carcinoma after Partial Hepatectomy. *Transl. Oncol.* **2015**, *8*, 497–503. [CrossRef]
142. Li, M.X.; Zhao, H.; Bi, X.Y.; Li, Z.Y.; Yao, X.S.; Li, H.; Huang, Z.; Han, Y.; Zhou, J.G.; Zhao, J.J.; et al. Lactate dehydrogenase is a prognostic indicator in patients with hepatocellular carcinoma treated by sorafenib: Results from the real life practice in HBV endemic area. *Oncotarget* **2016**, *7*, 86630–86647. [CrossRef]
143. Zhao, R.; Li, L.; Yang, J.; Niu, Q.; Wang, H.; Qin, X.; Zhu, N.; Shi, A. Overexpression of Pyruvate Kinase M2 in Tumor Tissues Is Associated with Poor Prognosis in Patients with Hepatocellular Carcinoma. *Pathol. Oncol. Res.* **2020**, *26*, 853–860. [CrossRef]
144. Ye, G.; Qin, Y.; Wang, S.; Pan, D.; Xu, S.; Wu, C.; Wang, X.; Wang, J.; Ye, H.; Shen, H. Lamc1 promotes the Warburg effect in hepatocellular carcinoma cells by regulating PKM2 expression through AKT pathway. *Cancer Biol. Ther.* **2019**, *20*, 711–719. [CrossRef]
145. Chen, Z.; Lu, X.; Wang, Z.; Jin, G.; Wang, Q.; Chen, D.; Chen, T.; Li, J.; Fan, J.; Cong, W.; et al. Co-expression of PKM2 and TRIM35 predicts survival and recurrence in hepatocellular carcinoma. *Oncotarget* **2015**, *6*, 2538–2548. [CrossRef]
146. Liu, Y.; Wu, H.; Mei, Y.; Ding, X.; Yang, X.; Li, C.; Deng, M.; Gong, J. Clinicopathological and prognostic significance of PKM2 protein expression in cirrhotic hepatocellular carcinoma and non-cirrhotic hepatocellular carcinoma. *Sci. Rep.* **2017**, *7*, 15294. [CrossRef]
147. Jeon, J.Y.; Lee, M.; Whang, S.H.; Kim, J.W.; Cho, A.; Yun, M. Regulation of Acetate Utilization by Monocarboxylate Transporter 1 (MCT1) in Hepatocellular Carcinoma (HCC). *Oncol. Res.* **2018**, *26*, 71–81. [CrossRef]
148. Alves, V.A.; Pinheiro, C.; Morais-Santos, F.; Felipe-Silva, A.; Longatto-Filho, A.; Baltazar, F. Characterization of monocarboxylate transporter activity in hepatocellular carcinoma. *World J. Gastroenterol.* **2014**, *20*, 11780–11787. [CrossRef]
149. Gu, M.; Jiang, B.; Li, H.; Zhu, D.; Jiang, Y.; Xu, W. Aldolase A promotes cell proliferation and cisplatin resistance via the EGFR pathway in gastric cancer. *Am. J. Transl. Res.* **2022**, *14*, 6586–6595.
150. Chen, L.; Wu, Z.; Guo, J.; Wang, X.; Zhao, Z.; Liang, H.; Zhang, R.; Deng, J. Initial clinical and experimental analyses of ALDOA in gastric cancer, as a novel prognostic biomarker and potential therapeutic target. *Clin. Exp. Med.* **2023**, *23*, 2443–2456. [CrossRef]
151. Morisaki, T.; Yashiro, M.; Kakehashi, A.; Inagaki, A.; Kinoshita, H.; Fukuoka, T.; Kasashima, H.; Masuda, G.; Sakurai, K.; Kubo, N.; et al. Comparative Proteomics Analysis of Gastric Cancer Stem Cells. *PLoS ONE* **2014**, *9*, e110736. [CrossRef]
152. Zhou, D.; Jiang, L.; Jin, L.; Yao, Y.; Wang, P.; Zhu, X. Glucose Transporter-1 Cooperating with AKT Signaling Promote Gastric Cancer Progression. *Cancer Manag. Res.* **2020**, *12*, 4151–4160. [CrossRef]
153. Yin, C.; Gao, B.; Yang, J.; Wu, J. Glucose Transporter-1 (GLUT-1) Expression is Associated with Tumor Size and Poor Prognosis in Locally Advanced Gastric Cancer. *Med. Sci. Monit Basic Res.* **2020**, *26*, e920778. [CrossRef]
154. Yan, S.; Wang, Y.; Chen, M.; Li, G.; Fan, J. Deregulated SLC2A1 Promotes Tumor Cell Proliferation and Metastasis in Gastric Cancer. *Int. J. Mol. Sci.* **2015**, *16*, 16144–16157. [CrossRef]
155. Schlößer, H.A.; Drebber, U.; Urbanski, A.; Haase, S.; Baltin, C.; Berlth, F.; Neiß, S.; von Bergwelt-Baildon, M.; Fetzner, U.K.; Warnecke-Eberz, U.; et al. Glucose transporters 1, 3, 6, and 10 are expressed in gastric cancer and glucose transporter 3 is associated with UICC stage and survival. *Gastric Cancer* **2017**, *20*, 83–91. [CrossRef]
156. Yang, H.; Yang, S.; He, J.; Li, W.; Zhang, A.; Li, N.; Zhou, G.; Sun, B. Glucose transporter 3 (GLUT3) promotes lactylation modifications by regulating lactate dehydrogenase A (LDHA) in gastric cancer. *Cancer Cell Int.* **2023**, *23*, 303. [CrossRef]

157. Sun, X.; Sun, Z.; Zhu, Z.; Li, C.; Zhang, J.; Xu, H.; Sun, M. Expression of SIP1 is strongly correlated with LDHA and shows a significantly poor outcome in gastric cancer. *Tumor Biol.* **2015**, *36*, 7521–7530. [CrossRef]
158. Jiang, W.; Zhou, F.; Li, N.; Li, Q.; Wang, L. FOXM1-LDHA signaling promoted gastric cancer glycolytic phenotype and progression. *Int. J. Clin. Exp. Pathol.* **2015**, *8*, 6756–6763.
159. Wang, X.H.; Jiang, Z.H.; Yang, H.M.; Zhang, Y.; Xu, L.H. Hypoxia-induced FOXO4/LDHA axis modulates gastric cancer cell glycolysis and progression. *Clin. Transl. Med.* **2021**, *11*, e279. [CrossRef]
160. Zhu, W.; Ma, L.; Qian, J.; Xu, J.; Xu, T.; Pang, L.; Zhou, H.; Shu, Y.; Zhou, J. The molecular mechanism and clinical significance of LDHA in HER2-mediated progression of gastric cancer. *Am. J. Transl. Res.* **2018**, *10*, 2055–2067.
161. Kwon, O.H.; Kang, T.W.; Kim, J.H.; Kim, M.; Noh, S.M.; Song, K.S.; Yoo, H.S.; Kim, W.H.; Xie, Z.; Pocalyko, D.; et al. Pyruvate kinase M2 promotes the growth of gastric cancer cells via regulation of Bcl-xL expression at transcriptional level. *Biochem. Biophys. Res. Commun.* **2012**, *423*, 38–44. [CrossRef]
162. Lim, J.Y.; Yoon, S.O.; Seol, S.Y.; Hong, S.W.; Kim, J.W.; Choi, S.H.; Cho, J.Y. Overexpression of the M2 isoform of pyruvate kinase is an adverse prognostic factor for signet ring cell gastric cancer. *World J. Gastroenterol.* **2012**, *18*, 4037–4043. [CrossRef]
163. Chen, J.; Cui, L.; Lu, S.; Xu, S. Amino acid metabolism in tumor biology and therapy. *Cell Death Dis.* **2024**, *15*, 42. [CrossRef]
164. Li, T.; Copeland, C.; Le, A. Glutamine Metabolism in Cancer. In *The Heterogeneity of Cancer Metabolism*; Le, A., Ed.; Springer International Publishing: Cham, Germany, 2021; pp. 17–38.
165. Pan, S.; Fan, M.; Liu, Z.; Li, X.; Wang, H. Serine, glycine and one-carbon metabolism in cancer (Review). *Int. J. Oncol.* **2021**, *58*, 158–170. [CrossRef]
166. Sivanand, S.; Vander Heiden, M.G. Emerging Roles for Branched-Chain Amino Acid Metabolism in Cancer. *Cancer Cell* **2020**, *37*, 147–156. [CrossRef]
167. Tanner, J.J.; Fendt, S.M.; Becker, D.F. The Proline Cycle As a Potential Cancer Therapy Target. *Biochemistry* **2018**, *57*, 3433–3444. [CrossRef]
168. Chen, C.L.; Hsu, S.C.; Ann, D.K.; Yen, Y.; Kung, H.J. Arginine Signaling and Cancer Metabolism. *Cancers* **2021**, *13*, 3541. [CrossRef]
169. Zhang, Z.; Liu, R.; Shuai, Y.; Huang, Y.; Jin, R.; Wang, X.; Luo, J. ASCT2 (SLC1A5)-dependent glutamine uptake is involved in the progression of head and neck squamous cell carcinoma. *Br. J. Cancer* **2020**, *122*, 82–93. [CrossRef]
170. Yu, W.; Yang, X.; Zhang, Q.; Sun, L.; Yuan, S.; Xin, Y. Targeting GLS1 to cancer therapy through glutamine metabolism. *Clin. Transl. Oncol.* **2021**, *23*, 2253–2268. [CrossRef]
171. Song, Z.; Wei, B.; Lu, C.; Li, P.; Chen, L. Glutaminase sustains cell survival via the regulation of glycolysis and glutaminolysis in colorectal cancer. *Oncol. Lett.* **2017**, *14*, 3117–3123. [CrossRef]
172. Altman, B.J.; Stine, Z.E.; Dang, C.V. From Krebs to clinic: Glutamine metabolism to cancer therapy. *Nat. Rev. Cancer* **2016**, *16*, 619–634. [CrossRef]
173. Scalise, M.; Galluccio, M.; Console, L.; Pochini, L.; Indiveri, C. The Human SLC7A5 (LAT1): The Intriguing Histidine/Large Neutral Amino Acid Transporter and Its Relevance to Human Health. *Front. Chem.* **2018**, *6*, 243. [CrossRef]
174. Ming, T.; Lei, J.; Peng, Y.; Wang, M.; Liang, Y.; Tang, S.; Tao, Q.; Wang, M.; Tang, X.; He, Z.; et al. Curcumin suppresses colorectal cancer by induction of ferroptosis via regulation of p53 and solute carrier family 7 member 11/glutathione/glutathione peroxidase 4 signaling axis. *Phytother. Res.* **2024**, *38*, 3954–3972. [CrossRef]
175. Poettler, M.; Unseld, M.; Braemswig, K.; Haitel, A.; Zielinski, C.C.; Prager, G.W. CD98hc (SLC3A2) drives integrin-dependent renal cancer cell behavior. *Mol. Cancer* **2013**, *12*, 169. [CrossRef]
176. Digomann, D.; Kurth, I.; Tyutyunnykova, A.; Chen, O.; Löck, S.; Gorodetska, I.; Peitzsch, C.; Skvortsova, I.I.; Negro, G.; Aschenbrenner, B.; et al. The CD98 Heavy Chain Is a Marker and Regulator of Head and Neck Squamous Cell Carcinoma Radiosensitivity. *Clin. Cancer Res.* **2019**, *25*, 3152–3163. [CrossRef]
177. Jin, M.; Lee, W.K.; You, M.H.; Jang, A.; Cheng, S.Y.; Kim, W.G.; Jeon, M.J.; Lee, Y.M. SHMT2 expression as a diagnostic and prognostic marker for thyroid cancer. *Endocr. Connect.* **2021**, *10*, 630–636. [CrossRef]
178. Bernhardt, S.; Bayerlova, M.; Vetter, M.; Wachter, A.; Mitra, D.; Hanf, V.; Lantzsch, T.; Uleer, C.; Peschel, S.; John, J.; et al. Proteomic profiling of breast cancer metabolism identifies SHMT2 and ASCT2 as prognostic factors. *Breast Cancer Res.* **2017**, *19*, 112. [CrossRef]
179. Kim, S.; Kim, D.H.; Jung, W.-H.; Koo, J.S. Expression of glutamine metabolism-related proteins according to molecular subtype of breast cancer. *Endocr. Relat. Cancer* **2013**, *20*, 339–348. [CrossRef]
180. Ahn, S.G.; Jeong, J.; Hong, S.; Jung, W.H. Current Issues and Clinical Evidence in Tumor-Infiltrating Lymphocytes in Breast Cancer. *J. Pathol. Transl. Med.* **2015**, *49*, 355–363. [CrossRef]
181. Chang, C.H.; Qiu, J.; O'Sullivan, D.; Buck, M.D.; Noguchi, T.; Curtis, J.D.; Chen, Q.; Gindin, M.; Gubin, M.M.; van der Windt, G.J.; et al. Metabolic Competition in the Tumor Microenvironment Is a Driver of Cancer Progression. *Cell* **2015**, *162*, 1229–1241. [CrossRef]

182. Kim, J.Y.; Heo, S.-H.; Choi, S.K.; Song, I.H.; Park, I.A.; Kim, Y.-A.; Park, H.S.; Park, S.Y.; Bang, W.S.; Gong, G.; et al. Glutaminase expression is a poor prognostic factor in node-positive triple-negative breast cancer patients with a high level of tumor-infiltrating lymphocytes. *Virchows Arch.* **2017**, *470*, 381–389. [CrossRef]
183. Zhou, M.; Chen, Q.-x.; Yang, Y.-z.; Liang, Z.-z.; Li, Y.-l.; Huang, Z.-y.; Weng, Z.-j.; Zhang, X.-f.; Guan, J.-x.; Tang, L.-y.; et al. Prognostic value of glutaminase 1 in breast cancer depends on H3K27me3 expression and menopausal status. *Virchows Arch.* **2022**, *480*, 259–267. [CrossRef]
184. Alfarsi, L.H.; El Ansari, R.; Craze, M.L.; Mohammed, O.J.; Masisi, B.K.; Ellis, I.O.; Rakha, E.A.; Green, A.R. SLC1A5 co-expression with TALDO1 associates with endocrine therapy failure in estrogen receptor-positive breast cancer. *Breast Cancer Res. Treat.* **2021**, *189*, 317–331. [CrossRef]
185. Alfarsi, L.H.; El-Ansari, R.; Craze, M.L.; Masisi, B.K.; Mohammed, O.J.; Ellis, I.O.; Rakha, E.A.; Green, A.R. Co-Expression Effect of SLC7A5/SLC3A2 to Predict Response to Endocrine Therapy in Oestrogen-Receptor-Positive Breast Cancer. *Int. J. Mol. Sci.* **2020**, *21*, 1407. [CrossRef]
186. El Ansari, R.; Craze, M.L.; Miligy, I.; Diez-Rodriguez, M.; Nolan, C.C.; Ellis, I.O.; Rakha, E.A.; Green, A.R. The amino acid transporter SLC7A5 confers a poor prognosis in the highly proliferative breast cancer subtypes and is a key therapeutic target in luminal B tumours. *Breast Cancer Res.* **2018**, *20*, 21. [CrossRef]
187. El Ansari, R.; Craze, M.L.; Diez-Rodriguez, M.; Nolan, C.C.; Ellis, I.O.; Rakha, E.A.; Green, A.R. The multifunctional solute carrier 3A2 (SLC3A2) confers a poor prognosis in the highly proliferative breast cancer subtypes. *Br. J. Cancer* **2018**, *118*, 1115–1122. [CrossRef]
188. Yang, C.; Zhang, J.; Liao, M.; Yang, Y.; Wang, Y.; Yuan, Y.; Ouyang, L. Folate-mediated one-carbon metabolism: A targeting strategy in cancer therapy. *Drug Discov. Today* **2021**, *26*, 817–825. [CrossRef]
189. Yin, K. Positive correlation between expression level of mitochondrial serine hydroxymethyltransferase and breast cancer grade. *OncoTargets Ther.* **2015**, *8*, 1069–1074. [CrossRef]
190. Zhang, L.; Chen, Z.; Xue, D.; Zhang, Q.; Liu, X.; Luh, F.; Hong, L.; Zhang, H.; Pan, F.; Liu, Y.; et al. Prognostic and therapeutic value of mitochondrial serine hydroxyl-methyltransferase 2 as a breast cancer biomarker. *Oncol. Rep.* **2016**, *36*, 2489–2500. [CrossRef]
191. Xie, S.Y.; Shi, D.B.; Ouyang, Y.; Lin, F.; Chen, X.Y.; Jiang, T.C.; Xia, W.; Guo, L.; Lin, H.X. SHMT2 promotes tumor growth through VEGF and MAPK signaling pathway in breast cancer. *Am. J. Cancer Res.* **2022**, *12*, 3405–3421.
192. Giatromanolaki, A.; Harris, A.L.; Koukourakis, M.I. The prognostic and therapeutic implications of distinct patterns of argininosuccinate synthase 1 (ASS1) and arginase-2 (ARG2) expression by cancer cells and tumor stroma in non-small-cell lung cancer. *Cancer Metab.* **2021**, *9*, 28. [CrossRef]
193. Hassanein, M.; Hoeksema, M.D.; Shiota, M.; Qian, J.; Harris, B.K.; Chen, H.; Clark, J.E.; Alborn, W.E.; Eisenberg, R.; Massion, P.P. SLC1A5 Mediates Glutamine Transport Required for Lung Cancer Cell Growth and Survival. *Clin. Cancer Res.* **2013**, *19*, 560–570. [CrossRef]
194. Hassanein, M.; Qian, J.; Hoeksema, M.D.; Wang, J.; Jacobovitz, M.; Ji, X.; Harris, F.T.; Harris, B.K.; Boyd, K.L.; Chen, H.; et al. Targeting SLC1a5-mediated glutamine dependence in non-small cell lung cancer. *Int. J. Cancer* **2015**, *137*, 1587–1597. [CrossRef]
195. Shimizu, K.; Kaira, K.; Tomizawa, Y.; Sunaga, N.; Kawashima, O.; Oriuchi, N.; Tominaga, H.; Nagamori, S.; Kanai, Y.; Yamada, M.; et al. ASC amino-acid transporter 2 (ASCT2) as a novel prognostic marker in non-small cell lung cancer. *Br. J. Cancer* **2014**, *110*, 2030–2039. [CrossRef]
196. Yazawa, T.; Shimizu, K.; Kaira, K.; Nagashima, T.; Ohtaki, Y.; Atsumi, J.; Obayashi, K.; Nagamori, S.; Kanai, Y.; Oyama, T.; et al. Clinical significance of coexpression of L-type amino acid transporter 1 (LAT1) and ASC amino acid transporter 2 (ASCT2) in lung adenocarcinoma. *Am. J. Transl. Res.* **2015**, *7*, 1126–1139.
197. Takeuchi, K.; Ogata, S.; Nakanishi, K.; Ozeki, Y.; Hiroi, S.; Tominaga, S.; Aida, S.; Matsuo, H.; Sakata, T.; Kawai, T. LAT1 expression in non-small-cell lung carcinomas: Analyses by semiquantitative reverse transcription-PCR (237 cases) and immunohistochemistry (295 cases). *Lung Cancer* **2010**, *68*, 58–65. [CrossRef]
198. Imai, H.; Kaira, K.; Oriuchi, N.; Yanagitani, N.; Sunaga, N.; Ishizuka, T.; Kanai, Y.; Endou, H.; Nakajima, T.; Mori, M. L-type amino acid transporter 1 expression is a prognostic marker in patients with surgically resected stage I non-small cell lung cancer. *Histopathology* **2009**, *54*, 804–813. [CrossRef]
199. Liu, D.; An, M.; Wen, G.; Xing, Y.; Xia, P. Both In Situ and Circulating SLC3A2 Could Be Used as Prognostic Markers for Human Lung Squamous Cell Carcinoma and Lung Adenocarcinoma. *Cancers* **2022**, *14*, 5191. [CrossRef]
200. Li, Z.; Chen, S.; He, X.; Gong, S.; Sun, L.; Weng, L. SLC3A2 promotes tumor-associated macrophage polarization through metabolic reprogramming in lung cancer. *Cancer Sci.* **2023**, *114*, 2306–2317. [CrossRef]
201. Zhou, X.; Tian, C.; Cao, Y.; Zhao, M.; Wang, K. The role of serine metabolism in lung cancer: From oncogenesis to tumor treatment. *Front. Genet.* **2023**, *13*, 1084609. [CrossRef]
202. Zhu, J.; Ma, J.; Wang, X.; Ma, T.; Zhang, S.; Wang, W.; Zhou, X.; Shi, J. High Expression of PHGDH Predicts Poor Prognosis in Non–Small Cell Lung Cancer. *Transl. Oncol.* **2016**, *9*, 592–599. [CrossRef]

203. Zhang, B.; Zheng, A.; Hydbring, P.; Ambroise, G.; Ouchida, A.T.; Goiny, M.; Vakifahmetoglu-Norberg, H.; Norberg, E. PHGDH Defines a Metabolic Subtype in Lung Adenocarcinomas with Poor Prognosis. *Cell Rep.* **2017**, *19*, 2289–2303. [CrossRef]
204. Paone, A.; Marani, M.; Fiascarelli, A.; Rinaldo, S.; Giardina, G.; Contestabile, R.; Paiardini, A.; Cutruzzolà, F. SHMT1 knockdown induces apoptosis in lung cancer cells by causing uracil misincorporation. *Cell Death Dis.* **2014**, *5*, e1525. [CrossRef]
205. Li, L.; Wei, Y.; To, C.; Zhu, C.-Q.; Tong, J.; Pham, N.-A.; Taylor, P.; Ignatchenko, V.; Ignatchenko, A.; Zhang, W.; et al. Integrated Omic analysis of lung cancer reveals metabolism proteome signatures with prognostic impact. *Nat. Commun.* **2014**, *5*, 5469. [CrossRef]
206. Xiang, L.; Mou, J.; Shao, B.; Wei, Y.; Liang, H.; Takano, N.; Semenza, G.L.; Xie, G. Glutaminase 1 expression in colorectal cancer cells is induced by hypoxia and required for tumor growth, invasion, and metastatic colonization. *Cell Death Dis.* **2019**, *10*, 40. [CrossRef]
207. Mizuno, Y.; Hattori, K.; Taniguchi, K.; Tanaka, K.; Uchiyama, K.; Hirose, Y. Intratumoral heterogeneity of glutaminase and lactate dehydrogenase A protein expression in colorectal cancer. *Oncol. Lett.* **2020**, *19*, 2934–2942. [CrossRef]
208. Du, F.; Chen, J.; Liu, H.; Cai, Y.; Cao, T.; Han, W.; Yi, X.; Qian, M.; Tian, D.; Nie, Y.; et al. SOX12 promotes colorectal cancer cell proliferation and metastasis by regulating asparagine synthesis. *Cell Death Dis.* **2019**, *10*, 239. [CrossRef]
209. Huang, F.; Zhang, Q.; Ma, H.; Lv, Q.; Zhang, T. Expression of glutaminase is upregulated in colorectal cancer and of clinical significance. *Int. J. Clin. Exp. Pathol.* **2014**, *7*, 1093–1100.
210. Liu, G.; Zhu, J.; Yu, M.; Cai, C.; Zhou, Y.; Yu, M.; Fu, Z.; Gong, Y.; Yang, B.; Li, Y.; et al. Glutamate dehydrogenase is a novel prognostic marker and predicts metastases in colorectal cancer patients. *J. Transl. Med.* **2015**, *13*, 144. [CrossRef]
211. Huang, G.; Ma, L.; Shen, L.; Lei, Y.; Guo, L.; Deng, Y.; Ding, Y. MIF/SCL3A2 depletion inhibits the proliferation and metastasis of colorectal cancer cells via the AKT/GSK-3β pathway and cell iron death. *J. Cell. Mol. Med.* **2022**, *26*, 3410–3422. [CrossRef]
212. Toda, K.; Nishikawa, G.; Iwamoto, M.; Itatani, Y.; Takahashi, R.; Sakai, Y.; Kawada, K. Clinical Role of ASCT2 (SLC1A5) in KRAS-Mutated Colorectal Cancer. *Int. J. Mol. Sci.* **2017**, *18*, 1632. [CrossRef]
213. Huang, F.; Zhao, Y.; Zhao, J.; Wu, S.; Jiang, Y.; Ma, H.; Zhang, T. Upregulated SLC1A5 promotes cell growth and survival in colorectal cancer. *Int. J. Clin. Exp. Pathol.* **2014**, *7*, 6006–6014.
214. Ma, H.; Wu, Z.; Peng, J.; Li, Y.; Huang, H.; Liao, Y.; Zhou, M.; Sun, L.; Huang, N.; Shi, M.; et al. Inhibition of SLC1A5 sensitizes colorectal cancer to cetuximab. *Int. J. Cancer* **2018**, *142*, 2578–2588. [CrossRef]
215. Ogawa, M.; Tanaka, A.; Maekawa, M.; Namba, K.; Otani, Y.; Shia, J.; Wang, J.Y.; Roehrl, M.H. Protein expression of the amino acid transporter SLC7A5 in tumor tissue is prognostic in early-stage colorectal cancer. *PLoS ONE* **2024**, *19*, e0298362. [CrossRef]
216. Najumudeen, A.K.; Ceteci, F.; Fey, S.K.; Hamm, G.; Steven, R.T.; Hall, H.; Nikula, C.J.; Dexter, A.; Murta, T.; Race, A.M.; et al. The amino acid transporter SLC7A5 is required for efficient growth of KRAS-mutant colorectal cancer. *Nat. Genet.* **2021**, *53*, 16–26. [CrossRef]
217. Hu, S.S.; Han, Y.; Tan, T.Y.; Chen, H.; Gao, J.W.; Wang, L.; Yang, M.H.; Zhao, L.; Wang, Y.Q.; Ding, Y.Q.; et al. SLC25A21 downregulation promotes KRAS-mutant colorectal cancer progression by increasing glutamine anaplerosis. *JCI Insight* **2023**, *8*, e167874. [CrossRef]
218. Ahmed, A.; Iaconisi, G.N.; Di Molfetta, D.; Coppola, V.; Caponio, A.; Singh, A.; Bibi, A.; Capobianco, L.; Palmieri, L.; Dolce, V.; et al. The Role of Mitochondrial Solute Carriers SLC25 in Cancer Metabolic Reprogramming: Current Insights and Future Perspectives. *Int. J. Mol. Sci.* **2025**, *26*, 92. [CrossRef]
219. Wu, J.B.; Li, X.J.; Liu, H.; Liu, Y.J.; Liu, X.P. Association of KRAS, NRAS, BRAF and PIK3CA gene mutations with clinicopathological features, prognosis and ring finger protein 215 expression in patients with colorectal cancer. *Biomed. Rep.* **2023**, *19*, 104. [CrossRef]
220. Midthun, L.; Shaheen, S.; Deisch, J.; Senthil, M.; Tsai, J.; Hsueh, C.T. Concomitant KRAS and BRAF mutations in colorectal cancer. *J. Gastrointest. Oncol.* **2019**, *10*, 577–581. [CrossRef]
221. Zhang, L.; Su, K.; Liu, Q.; Li, B.; Wang, Y.; Cheng, C.; Li, Y.; Xu, C.; Chen, J.; Wu, H.; et al. Kidney-type glutaminase is a biomarker for the diagnosis and prognosis of hepatocellular carcinoma: A prospective study. *BMC Cancer* **2023**, *23*, 1081. [CrossRef]
222. Yu, D.; Shi, X.; Meng, G.; Chen, J.; Yan, C.; Jiang, Y.; Wei, J.; Ding, Y. Kidney-type glutaminase (GLS1) is a biomarker for pathologic diagnosis and prognosis of hepatocellular carcinoma. *Oncotarget* **2015**, *6*, 7619–7631. [CrossRef]
223. Liu, J.; Zhang, C.; Lin, M.; Zhu, W.; Liang, Y.; Hong, X.; Zhao, Y.; Young, K.H.; Hu, W.; Feng, Z. Glutaminase 2 negatively regulates the PI3K/AKT signaling and shows tumor suppression activity in human hepatocellular carcinoma. *Oncotarget* **2014**, *5*, 2635–2647. [CrossRef]
224. Zhou, Y.; Yu, H.; Cheng, S.; Chen, Y.; He, L.; Ren, J.; He, X.; Chen, J.; Zheng, L.; Li, F. Glutamate dehydrogenase 1 mediated glutaminolysis sustains HCC cells survival under glucose deprivation. *J. Cancer* **2022**, *13*, 1061–1072. [CrossRef]
225. Jiang, J.; Hu, Y.; Fang, D.; Luo, J. Glutamine synthetase and hepatocellular carcinoma. *Clin. Res. Hepatol. Gastroenterol.* **2023**, *47*, 102248. [CrossRef]

226. Liu, P.; Lu, D.; Al-ameri, A.; Wei, X.; Ling, S.; Li, J.; Zhu, H.; Xie, H.; Zhu, L.; Zheng, S.; et al. Glutamine synthetase promotes tumor invasion in hepatocellular carcinoma through mediating epithelial–mesenchymal transition. *Hepatol. Res.* **2020**, *50*, 246–257. [CrossRef]
227. Zhang, B.; Dong, L.W.; Tan, Y.X.; Zhang, J.; Pan, Y.F.; Yang, C.; Li, M.H.; Ding, Z.W.; Liu, L.J.; Jiang, T.Y.; et al. Asparagine synthetase is an independent predictor of surgical survival and a potential therapeutic target in hepatocellular carcinoma. *Br. J. Cancer* **2013**, *109*, 14–23. [CrossRef]
228. Kerk, S.A.; Garcia-Bermudez, J.; Birsoy, K.; Sherman, M.H.; Shah, Y.M.; Lyssiotis, C.A. Spotlight on GOT2 in Cancer Metabolism. *OncoTargets Ther.* **2023**, *16*, 695–702. [CrossRef]
229. Li, Y.; Li, B.; Xu, Y.; Qian, L.; Xu, T.; Meng, G.; Li, H.; Wang, Y.; Zhang, L.; Jiang, X.; et al. GOT2 Silencing Promotes Reprogramming of Glutamine Metabolism and Sensitizes Hepatocellular Carcinoma to Glutaminase Inhibitors. *Cancer Res.* **2022**, *82*, 3223–3235. [CrossRef]
230. Li, J.; Qiang, J.; Chen, S.F.; Wang, X.; Fu, J.; Chen, Y. The impact of L-type amino acid transporter 1 (LAT1) in human hepatocellular carcinoma. *Tumour Biol.* **2013**, *34*, 2977–2981. [CrossRef]
231. Liang, Y.; Su, S.; Lun, Z.; Zhong, Z.; Yu, W.; He, G.; Wang, Q.; Wang, J.; Huang, S. Ferroptosis regulator SLC7A11 is a prognostic marker and correlated with PD-L1 and immune cell infiltration in liver hepatocellular carcinoma. *Front. Mol. Biosci.* **2022**, *9*, 1012505. [CrossRef]
232. He, Q.; Liu, M.; Huang, W.; Chen, X.; Zhang, B.; Zhang, T.; Wang, Y.; Liu, D.; Xie, M.; Ji, X.; et al. IL-1β-Induced Elevation of Solute Carrier Family 7 Member 11 Promotes Hepatocellular Carcinoma Metastasis Through Up-regulating Programmed Death Ligand 1 and Colony-Stimulating Factor 1. *Hepatology* **2021**, *74*, 3174–3193. [CrossRef]
233. Zhang, L.; Huang, Y.; Ling, J.; Zhuo, W.; Yu, Z.; Luo, Y.; Zhu, Y. Overexpression of SLC7A11: A novel oncogene and an indicator of unfavorable prognosis for liver carcinoma. *Future Oncol.* **2018**, *14*, 927–936. [CrossRef]
234. Xu, L.; Zhou, D.; Li, F.; Ji, L. Glutaminase 2 functions as a tumor suppressor gene in gastric cancer. *Transl. Cancer Res.* **2020**, *9*, 4906–4913. [CrossRef]
235. Wu, Y.-J.; Hu, Z.-L.; Hu, S.-D.; Li, Y.-X.; Xing, X.-W.; Yang, Y.; Du, X.-H. Glutamate dehydrogenase inhibits tumor growth in gastric cancer through the Notch signaling pathway. *Cancer Biomark.* **2019**, *26*, 303–312. [CrossRef]
236. Ma, H.; Wu, J.; Zhou, M.; Wu, J.; Wu, Z.; Lin, L.; Huang, N.; Liao, W.; Sun, L. Inhibition of Glutamine Uptake Improves the Efficacy of Cetuximab on Gastric Cancer. *Integr. Cancer Ther.* **2021**, *20*, 15347354211045349. [CrossRef]
237. Ye, J.; Huang, Q.; Xu, J.; Huang, J.; Wang, J.; Zhong, W.; Chen, W.; Lin, X.; Lin, X. Targeting of glutamine transporter ASCT2 and glutamine synthetase suppresses gastric cancer cell growth. *J. Cancer Res. Clin. Oncol.* **2018**, *144*, 821–833. [CrossRef]
238. Hong, L.; Tang, X.; Han, J.; Wang, J.; Xu, Q.; Zhu, X. Abnormal arginine synthesis confers worse prognosis in patients with middle third gastric cancer. *Cancer Cell Int.* **2024**, *24*, 6. [CrossRef]
239. Tsai, C.-Y.; Chi, H.-C.; Chi, L.-M.; Yang, H.-Y.; Tsai, M.-M.; Lee, K.-F.; Huang, H.-W.; Chou, L.-F.; Cheng, A.-J.; Yang, C.-W.; et al. Argininosuccinate synthetase 1 contributes to gastric cancer invasion and progression by modulating autophagy. *FASEB J.* **2018**, *32*, 2601–2614. [CrossRef]
240. Shi, L.; Luo, W.; Huang, W.; Huang, S.; Huang, G. Downregulation of L-type amino acid transporter 1 expression inhibits the growth, migration and invasion of gastric cancer cells. *Oncol. Lett.* **2013**, *6*, 106–112. [CrossRef]
241. Ichinoe, M.; Mikami, T.; Yoshida, T.; Igawa, I.; Tsuruta, T.; Nakada, N.; Anzai, N.; Suzuki, Y.; Endou, H.; Okayasu, I. High expression of L-type amino-acid transporter 1 (LAT1) in gastric carcinomas: Comparison with non-cancerous lesions. *Pathol. Int.* **2011**, *61*, 281–289. [CrossRef]
242. Jiang, Y.; Cui, J.; Cui, M.; Jing, R. SLC7A11 promotes the progression of gastric cancer and regulates ferroptosis through PI3K/AKT pathway. *Pathol. Res. Pract.* **2023**, *248*, 154646. [CrossRef]
243. Blanco, A.; Blanco, G. Lipid Metabolism. In *Medical Biochemistry*; Blanco, A., Blanco, G., Eds.; Academic Press: Cambridge, MA, USA, 2017; pp. 325–365.
244. Piccinin, E.; Cariello, M.; Moschetta, A. Lipid metabolism in colon cancer: Role of Liver X Receptor (LXR) and Stearoyl-CoA Desaturase 1 (SCD1). *Mol. Asp. Med.* **2021**, *78*, 100933. [CrossRef] [PubMed]
245. Nath, A.; Li, I.; Roberts, L.R.; Chan, C. Elevated free fatty acid uptake via CD36 promotes epithelial-mesenchymal transition in hepatocellular carcinoma. *Sci. Rep.* **2015**, *5*, 14752. [CrossRef]
246. Gaffar, S.; Aathirah, A.S. Fatty-Acid-Binding Proteins: From Lipid Transporters to Disease Biomarkers. *Biomolecules* **2023**, *13*, 1753. [CrossRef]
247. Kim, S.; Lee, Y.; Koo, J.S. Differential expression of lipid metabolism-related proteins in different breast cancer subtypes. *PLoS ONE* **2015**, *10*, e0119473. [CrossRef]
248. Jung, Y.Y.; Kim, H.M.; Koo, J.S. Expression of Lipid Metabolism-Related Proteins in Metastatic Breast Cancer. *PLoS ONE* **2015**, *10*, e0137204. [CrossRef]
249. Kim, Y.H.; Jung, W.H.; Koo, J.S. Expression of metabolism-related proteins in invasive lobular carcinoma: Comparison to invasive ductal carcinoma. *Tumor Biol.* **2014**, *35*, 10381–10393. [CrossRef]

250. Barroso-Sousa, R.; Metzger-Filho, O. Differences between invasive lobular and invasive ductal carcinoma of the breast: Results and therapeutic implications. *Ther. Adv. Med. Oncol.* **2016**, *8*, 261–266. [CrossRef]
251. Jung, Y.Y.; Lee, Y.K.; Koo, J.S. Expression of lipid metabolism-related proteins in breast phyllodes tumors. *Neoplasma* **2016**, *63*, 254–262. [CrossRef] [PubMed]
252. Ditsatham, C.; Chongruksut, W. Phyllodes tumor of the breast: Diagnosis, management and outcome during a 10-year experience. *Cancer Manag. Res.* **2019**, *11*, 7805–7811. [CrossRef] [PubMed]
253. Liu, L.; Yan, H.; Ruan, M.; Yang, H.; Wang, L.; Lei, B.; Sun, X.; Chang, C.; Huang, G.; Xie, W. An AKT/PRMT5/SREBP1 axis in lung adenocarcinoma regulates de novo lipogenesis and tumor growth. *Cancer Sci.* **2021**, *112*, 3083–3098. [CrossRef]
254. Pernemalm, M.; De Petris, L.; Branca, R.M.; Forshed, J.; Kanter, L.; Soria, J.-C.; Girard, P.; Validire, P.; Pawitan, Y.; van den Oord, J.; et al. Quantitative Proteomics Profiling of Primary Lung Adenocarcinoma Tumors Reveals Functional Perturbations in Tumor Metabolism. *J. Proteome Res.* **2013**, *12*, 3934–3943. [CrossRef] [PubMed]
255. Ma, Y.; Nenkov, M.; Berndt, A.; Abubrig, M.; Schmidt, M.; Sandhaus, T.; Huber, O.; Clement, J.H.; Lang, S.M.; Chen, Y.; et al. The Diagnostic Value of ACSL1, ACSL4, and ACSL5 and the Clinical Potential of an ACSL Inhibitor in Non-Small-Cell Lung Cancer. *Cancers* **2024**, *16*, 1170. [CrossRef]
256. Zhu, H.; Shi, H.; Lu, J.; Zhu, K.; Yang, L.; Guo, L.; Tang, L.; Shi, Y.; Hu, X. Proteomic profiling reveals the significance of lipid metabolism in small cell lung cancer recurrence and metastasis. *J. Transl. Med.* **2024**, *22*, 1117. [CrossRef]
257. Wen, J.; Min, X.; Shen, M.; Hua, Q.; Han, Y.; Zhao, L.; Liu, L.; Huang, G.; Liu, J.; Zhao, X. ACLY facilitates colon cancer cell metastasis by CTNNB1. *J. Exp. Clin. Cancer Res.* **2019**, *38*, 401. [CrossRef]
258. Qiao, C.; Huang, W.; Chen, J.; Feng, W.; Zhang, T.; Wang, Y.; Liu, D.; Ji, X.; Xie, M.; Sun, M.; et al. IGF1-mediated HOXA13 overexpression promotes colorectal cancer metastasis through upregulating ACLY and IGF1R. *Cell Death Dis.* **2021**, *12*, 564. [CrossRef]
259. Garcia, J.A.; Chen, R.; Xu, M.; Comerford, S.A.; Hammer, R.E.; Melton, S.D.; Feagins, L.A. Acss2/HIF-2 signaling facilitates colon cancer growth and metastasis. *PLoS ONE* **2023**, *18*, e0282223. [CrossRef]
260. Lu, T.; Sun, L.; Wang, Z.; Zhang, Y.; He, Z.; Xu, C. Fatty acid synthase enhances colorectal cancer cell proliferation and metastasis via regulating AMPK/mTOR pathway. *OncoTargets Ther.* **2019**, *12*, 3339–3347. [CrossRef]
261. Wang, H.; Xi, Q.; Wu, G. Fatty acid synthase regulates invasion and metastasis of colorectal cancer via Wnt signaling pathway. *Cancer Med.* **2016**, *5*, 1599–1606. [CrossRef]
262. Zaytseva, Y.Y.; Harris, J.W.; Mitov, M.I.; Kim, J.T.; Butterfield, D.A.; Lee, E.Y.; Weiss, H.L.; Gao, T.; Evers, B.M. Increased expression of fatty acid synthase provides a survival advantage to colorectal cancer cells via upregulation of cellular respiration. *Oncotarget* **2015**, *6*, 18891–18904. [CrossRef]
263. Ventura, R.; Mordec, K.; Waszczuk, J.; Wang, Z.; Lai, J.; Fridlib, M.; Buckley, D.; Kemble, G.; Heuer, T.S. Inhibition of de novo Palmitate Synthesis by Fatty Acid Synthase Induces Apoptosis in Tumor Cells by Remodeling Cell Membranes, Inhibiting Signaling Pathways, and Reprogramming Gene Expression. *EBioMedicine* **2015**, *2*, 808–824. [CrossRef]
264. Zaytseva, Y.Y.; Rychahou, P.G.; Gulhati, P.; Elliott, V.A.; Mustain, W.C.; O'Connor, K.; Morris, A.J.; Sunkara, M.; Weiss, H.L.; Lee, E.Y.; et al. Inhibition of fatty acid synthase attenuates CD44-associated signaling and reduces metastasis in colorectal cancer. *Cancer Res.* **2012**, *72*, 1504–1517. [CrossRef] [PubMed]
265. Long, Q.-Q.; Yi, Y.-X.; Qiu, J.; Xu, C.-J.; Huang, P.-L. Fatty acid synthase (FASN) levels in serum of colorectal cancer patients: Correlation with clinical outcomes. *Tumor Biol.* **2014**, *35*, 3855–3859. [CrossRef] [PubMed]
266. Sánchez-Martínez, R.; Cruz-Gil, S.; Gómez de Cedrón, M.; Álvarez-Fernández, M.; Vargas, T.; Molina, S.; García, B.; Herranz, J.; Moreno-Rubio, J.; Reglero, G.; et al. A link between lipid metabolism and epithelial-mesenchymal transition provides a target for colon cancer therapy. *Oncotarget* **2015**, *6*, 38719–38736. [CrossRef] [PubMed]
267. Hartmann, F.; Sparla, D.; Tute, E.; Tamm, M.; Schneider, U.; Jeon, M.K.; Kasperk, R.; Gassler, N.; Kaemmerer, E. Low acyl-CoA synthetase 5 expression in colorectal carcinomas is prognostic for early tumour recurrence. *Pathol. Res. Pract.* **2017**, *213*, 261–266. [CrossRef]
268. Zhang, Y.; Zhang, W.; Xia, M.; Xie, Z.; An, F.; Zhan, Q.; Tian, W.; Zhu, T. High expression of FABP4 in colorectal cancer and its clinical significance. *J. Zhejiang Univ. Sci. B* **2021**, *22*, 136–145. [CrossRef]
269. Zhang, Y.; Zhao, X.; Deng, L.; Li, X.; Wang, G.; Li, Y.; Chen, M. High expression of FABP4 and FABP6 in patients with colorectal cancer. *World J. Surg. Oncol.* **2019**, *17*, 171. [CrossRef]
270. Hao, Q.; Li, T.; Zhang, X.; Gao, P.; Qiao, P.; Li, S.; Geng, Z. Expression and roles of fatty acid synthase in hepatocellular carcinoma. *Oncol. Rep.* **2014**, *32*, 2471–2476. [CrossRef]
271. Han, Q.; Chen, C.-A.; Yang, W.; Liang, D.; Lv, H.-W.; Lv, G.-S.; Zong, Q.-N.; Wang, H.-Y. ATP-citrate lyase regulates stemness and metastasis in hepatocellular carcinoma via the Wnt/β-catenin signaling pathway. *Hepatobiliary Pancreat. Dis. Int.* **2021**, *20*, 251–261. [CrossRef]

272. Zhong, C.-Q.; Zhang, X.-P.; Ma, N.; Zhang, E.-B.; Li, J.-J.; Jiang, Y.-B.; Gao, Y.-Z.; Yuan, Y.-M.; Lan, S.-Q.; Xie, D.; et al. FABP4 suppresses proliferation and invasion of hepatocellular carcinoma cells and predicts a poor prognosis for hepatocellular carcinoma. *Cancer Med.* **2018**, *7*, 2629–2640. [CrossRef] [PubMed]
273. Ohira, M.; Yokoo, H.; Ogawa, K.; Fukai, M.; Kamiyama, T.; Sakamoto, N.; Taketomi, A. Serum fatty acid-binding protein 5 is a significant factor in hepatocellular carcinoma progression independent of tissue expression level. *Carcinogenesis* **2021**, *42*, 794–803. [CrossRef]
274. Lu, J.B.; Cai, S.H.; Pan, Y.H.; Yun, J.P. Altered epidermal fatty acid-binding protein expression in hepatocellular carcinoma predicts unfavorable outcomes. *Cancer Manag. Res.* **2018**, *10*, 6275–6284. [CrossRef]
275. Ndiaye, H.; Liu, J.Y.; Hall, A.; Minogue, S.; Morgan, M.Y.; Waugh, M.G. Immunohistochemical staining reveals differential expression of ACSL3 and ACSL4 in hepatocellular carcinoma and hepatic gastrointestinal metastases. *Biosci. Rep.* **2020**, *40*, BSR20200219. [CrossRef]
276. Chen, J.; Ding, C.; Chen, Y.; Hu, W.; Yu, C.; Peng, C.; Feng, X.; Cheng, Q.; Wu, W.; Lu, Y.; et al. ACSL4 reprograms fatty acid metabolism in hepatocellular carcinoma via c-Myc/SREBP1 pathway. *Cancer Lett.* **2021**, *502*, 154–165. [CrossRef]
277. Wu, D.; Zuo, Z.; Sun, X.; Li, X.; Yin, F.; Yin, W. ACSL4 promotes malignant progression of Hepatocellular carcinoma by targeting PAK2 transcription. *Biochem. Pharmacol.* **2024**, *224*, 116206. [CrossRef]
278. Feng, J.; Bin, J.-L.; Liao, X.-W.; Wu, Y.; Tang, Y.; Lu, P.-Z.; Zhu, G.-Z.; Cui, Q.-R.; Dan, Y.Y.; Yang, G.-H.; et al. The prognostic role of ACSL4 in postoperative adjuvant TACE-treated HCC: Implications for therapeutic response and mechanistic insights. *J. Exp. Clin. Cancer Res.* **2024**, *43*, 306. [CrossRef]
279. Chen, J.; Ding, C.; Chen, Y.; Hu, W.; Lu, Y.; Wu, W.; Zhang, Y.; Yang, B.; Wu, H.; Peng, C.; et al. ACSL4 promotes hepatocellular carcinoma progression via c-Myc stability mediated by ERK/FBW7/c-Myc axis. *Oncogenesis* **2020**, *9*, 42. [CrossRef] [PubMed]
280. Feng, J.; Lu, P.-z.; Zhu, G.-z.; Hooi, S.C.; Wu, Y.; Huang, X.-w.; Dai, H.-q.; Chen, P.-h.; Li, Z.-j.; Su, W.-j.; et al. ACSL4 is a predictive biomarker of sorafenib sensitivity in hepatocellular carcinoma. *Acta Pharmacol. Sin.* **2021**, *42*, 160–170. [CrossRef]
281. Li, C.; Yang, W.; Zhang, J.; Zheng, X.; Yao, Y.; Tu, K.; Liu, Q. SREBP-1 has a prognostic role and contributes to invasion and metastasis in human hepatocellular carcinoma. *Int. J. Mol. Sci.* **2014**, *15*, 7124–7138. [CrossRef]
282. Chen, Y.; Zhong, Z.; Ruan, X.; Zhan, X.; Ding, Y.; Wei, F.; Qin, X.; Yu, H.; Lu, Y. Novel biomarker in hepatocellular carcinoma: Stearoyl-CoA desaturase 1. *Dig. Liver Dis.* **2025**, *57*, 770–781. [CrossRef]
283. Gao, Q.; Zhang, G.; Zheng, Y.; Yang, Y.; Chen, C.; Xia, J.; Liang, L.; Lei, C.; Hu, Y.; Cai, X.; et al. SLC27A5 deficiency activates NRF2/TXNRD1 pathway by increased lipid peroxidation in HCC. *Cell Death Differ.* **2020**, *27*, 1086–1104. [CrossRef]
284. Wang, J.; Qiao, Y.; Sun, H.; Chang, H.; Zhao, H.; Zhang, S.; Shan, C. Decreased SLC27A5 Suppresses Lipid Synthesis and Tyrosine Metabolism to Activate the Cell Cycle in Hepatocellular Carcinoma. *Biomedicines* **2022**, *10*, 234. [CrossRef]
285. Xiang, H.G.; Hao, J.; Zhang, W.J.; Lu, W.J.; Dong, P.; Liu, Y.B.; Chen, L. Expression of Fatty Acid Synthase Negatively Correlates with PTEN and Predicts Peritoneal Dissemination of Human Gastric Cancer. *Asian Pac. J. Cancer Prev.* **2015**, *16*, 6851–6855. [CrossRef]
286. Li, H.E.; Wang, X.; Tang, Z.; Liu, F.; Chen, W.; Fang, Y.; Wang, C.; Shen, K.; Qin, J.; Shen, Z.; et al. A concordant expression pattern of fatty acid synthase and membranous human epidermal growth factor receptor 2 exists in gastric cancer and is associated with a poor prognosis in gastric adenocarcinoma patients. *Oncol. Lett.* **2015**, *10*, 2107–2117. [CrossRef]
287. Duan, J.; Sun, L.; Huang, H.; Wu, Z.; Wang, L.; Liao, W. Overexpression of fatty acid synthase predicts a poor prognosis for human gastric cancer. *Mol. Med. Rep.* **2016**, *13*, 3027–3035. [CrossRef]
288. Hou, W.; Fei, M.; Qin, X.; Zhu, X.; Greshock, J.; Liu, P.; Zhou, Y.; Wang, H.; Ye, B.C.; Qin, C.Y. High overexpression of fatty acid synthase is associated with poor survival in Chinese patients with gastric carcinoma. *Exp. Ther. Med.* **2012**, *4*, 999–1004. [CrossRef]
289. Ito, T.; Sato, K.; Maekawa, H.; Sakurada, M.; Orita, H.; Shimada, K.; Daida, H.; Wada, R.; Abe, M.; Hino, O.; et al. Elevated levels of serum fatty acid synthase in patients with gastric carcinoma. *Oncol. Lett.* **2014**, *7*, 616–620. [CrossRef]
290. Yan, Z.; Liu, K.; Xu, P.; Chen, Z.; Zhang, P.; Pei, S.; Cheng, Q.; Huang, S.; Li, B.; Lv, J.; et al. ACLY promotes gastric tumorigenesis and accelerates peritoneal metastasis of gastric cancer regulated by HIF-1A. *Cell Cycle* **2023**, *22*, 2288–2301. [CrossRef]
291. Qian, X.; Hu, J.; Zhao, J.; Chen, H. ATP citrate lyase expression is associated with advanced stage and prognosis in gastric adenocarcinoma. *Int. J. Clin. Exp. Med.* **2015**, *8*, 7855–7860. [PubMed]
292. Li, C.; Liu, W.; Liu, Y.; Wang, W.; Deng, W. Role of ATP citrate lyase and its complementary partner on fatty acid synthesis in gastric cancer. *Sci. Rep.* **2024**, *14*, 30043. [CrossRef] [PubMed]
293. Pan, J.; Fan, Z.; Wang, Z.; Dai, Q.; Xiang, Z.; Yuan, F.; Yan, M.; Zhu, Z.; Liu, B.; Li, C. CD36 mediates palmitate acid-induced metastasis of gastric cancer via AKT/GSK-3beta/beta-catenin pathway. *J. Exp. Clin. Cancer Res.* **2019**, *38*, 52. [CrossRef] [PubMed]
294. Aoki, T.; Kinoshita, J.; Munesue, S.; Hamabe-Horiike, T.; Yamaguchi, T.; Nakamura, Y.; Okamoto, K.; Moriyama, H.; Nakamura, K.; Harada, S.; et al. Hypoxia-Induced CD36 Expression in Gastric Cancer Cells Promotes Peritoneal Metastasis via Fatty Acid Uptake. *Ann. Surg. Oncol.* **2023**, *30*, 3125–3136. [CrossRef]

295. Wang, C.; Shi, M.; Ji, J.; Cai, Q.; Zhao, Q.; Jiang, J.; Liu, J.; Zhang, H.; Zhu, Z.; Zhang, J. Stearoyl-CoA desaturase 1 (SCD1) facilitates the growth and anti-ferroptosis of gastric cancer cells and predicts poor prognosis of gastric cancer. *Aging* **2020**, *12*, 15374–15391. [CrossRef]
296. Li, X.; Luo, J.; Mou, K.; Peng, L.; Zhou, H.; Lei, Y.; Wang, H.; Zhao, Z.; Wang, J.; Wu, J.; et al. SDPR Inhibits TGF-β Induced Cancer Metastasis Through Fatty Acid Oxidation Regulation in Gastric Cancer. *Int. J. Biol. Sci.* **2023**, *19*, 2999–3014. [CrossRef]
297. Huang, J.; Wang, X.; Wen, J.; Zhao, X.; Wu, C.; Wang, L.; Cao, X.; Dong, H.; Xu, X.; Huang, F.; et al. Gastric cancer cell-originated small extracellular vesicle induces metabolic reprogramming of BM-MSCs through ERK-PPARγ-CPT1A signaling to potentiate lymphatic metastasis. *Cancer Cell Int.* **2023**, *23*, 87. [CrossRef]
298. Fang, W.; Cui, H.; Yu, D.; Chen, Y.; Wang, J.; Yu, G. Increased expression of phospho-acetyl-CoA carboxylase protein is an independent prognostic factor for human gastric cancer without lymph node metastasis. *Med. Oncol.* **2014**, *31*, 15. [CrossRef] [PubMed]
299. Xiao, J.; Cao, S.; Wang, J.; Li, P.; Cheng, Q.; Zhou, X.; Dong, J.; Li, Y.; Zhao, X.; Xu, Z.; et al. Leptin-mediated suppression of lipoprotein lipase cleavage enhances lipid uptake and facilitates lymph node metastasis in gastric cancer. *Cancer Commun.* **2024**, *44*, 855–878. [CrossRef]
300. Pih, G.Y.; Gong, E.J.; Choi, J.Y.; Kim, M.J.; Ahn, J.Y.; Choe, J.; Bae, S.E.; Chang, H.S.; Na, H.K.; Lee, J.H.; et al. Associations of Serum Lipid Level with Gastric Cancer Risk, Pathology, and Prognosis. *Cancer Res. Treat.* **2021**, *53*, 445–456. [CrossRef]
301. Song, S.; Wang, Z.; Li, Y.; Ma, L.; Jin, J.; Scott, A.W.; Xu, Y.; Estrella, J.S.; Song, Y.; Liu, B.; et al. PPARδ Interacts with the Hippo Coactivator YAP1 to Promote SOX9 Expression and Gastric Cancer Progression. *Mol. Cancer Res.* **2020**, *18*, 390–402. [CrossRef]
302. Ma, X.M.; Yu, H.; Huai, N. Peroxisome proliferator-activated receptor-gamma is essential in the pathogenesis of gastric carcinoma. *World J. Gastroenterol.* **2009**, *15*, 3874–3883. [CrossRef]
303. Scheerens, H.; Malong, A.; Bassett, K.; Boyd, Z.; Gupta, V.; Harris, J.; Mesick, C.; Simnett, S.; Stevens, H.; Gilbert, H.; et al. Current Status of Companion and Complementary Diagnostics: Strategic Considerations for Development and Launch. *Clin. Transl. Sci.* **2017**, *10*, 84–92. [CrossRef]
304. Kang, S.L.; Woo, J.H.; Kim, N.H.; Kwon, J.Y.; Kim, S.M. Necessity of strengthening the current clinical regulatory for companion diagnostics: An institutional comparison of the FDA, EMA, and MFDS. *Mol. Ther. Methods Clin. Dev.* **2023**, *30*, 447–458. [CrossRef]
305. Xu, C.; Liu, W.; Li, L.; Wang, Y.; Yuan, Q. Serum tumour M2-pyruvate kinase as a biomarker for diagnosis and prognosis of early-stage non-small cell lung cancer. *J. Cell. Mol. Med.* **2021**, *25*, 7335–7341. [CrossRef]
306. El-Abd, E.; El-Sheikh, M.; Zaky, S.; Fayed, W.; El-Zoghby, S. Plasma TuM2-PK correlates with tumor size, CRP and CA 15-3 in metastatic breast carcinomas; short versus long term follow up study of the Egyptian breast cancer patients. *Cancer Biomark.* **2017**, *20*, 123–133. [CrossRef] [PubMed]
307. Hathurusinghe, H.R.; Goonetilleke, K.S.; Siriwardena, A.K. Current status of tumor M2 pyruvate kinase (tumor M2-PK) as a biomarker of gastrointestinal malignancy. *Ann. Surg. Oncol.* **2007**, *14*, 2714–2720. [CrossRef]
308. Bandara, I.A.; Baltatzis, M.; Sanyal, S.; Siriwardena, A.K. Evaluation of tumor M2-pyruvate kinase (Tumor M2-PK) as a biomarker for pancreatic cancer. *World J. Surg. Oncol.* **2018**, *16*, 56. [CrossRef]
309. Che Alhadi, S.; Wan Zain, W.Z.; Zahari, Z.; Md Hashim, M.N.; Syed Abd Aziz, S.H.; Zakaria, Z.; Wong, M.P.; Zakaria, A.D. The Use of M2-Pyruvate Kinase as a Stool Biomarker for Detection of Colorectal Cancer in Tertiary Teaching Hospital: A Comparative Study. *Ann. Coloproctol.* **2020**, *36*, 409–414. [CrossRef]
310. Hardt, P.D.; Mazurek, S.; Toepler, M.; Schlierbach, P.; Bretzel, R.G.; Eigenbrodt, E.; Kloer, H.U. Faecal tumour M2 pyruvate kinase: A new, sensitive screening tool for colorectal cancer. *Br. J. Cancer* **2004**, *91*, 980–984. [CrossRef]
311. Gray, M.R.; Martin del Campo, S.; Zhang, X.; Zhang, H.; Souza, F.F.; Carson, W.E.; Smith, A.D. Metastatic Melanoma: Lactate Dehydrogenase Levels and CT Imaging Findings of Tumor Devascularization Allow Accurate Prediction of Survival in Patients Treated with Bevacizumab. *Radiology* **2014**, *270*, 425–434. [CrossRef]
312. Shamoon, R.; Polus, R. Serum LDH activity in lymphomas: Prognostic Significance and Relationship to Presentation, Stage and Histologic Type. *Zanco J. Med. Sci.* **2010**, *14*, 85.
313. William, B.M.; Bongu, N.R.; Bast, M.; Bociek, R.G.; Bierman, P.J.; Vose, J.M.; Armitage, J.O. The utility of lactate dehydrogenase in the follow up of patients with diffuse large B-cell lymphoma. *Rev. Bras. Hematol. Hemoter.* **2013**, *35*, 189–191. [CrossRef]
314. Bar, J.; Spencer, S.; Morgan, S.; Brooks, L.; Cunningham, D.; Robertson, J.; Jurgensmeier, J.M.; Goss, G.D. Correlation of lactate dehydrogenase isoenzyme profile with outcome in patients with advanced colorectal cancer treated with chemotherapy and bevacizumab or cediranib: Retrospective analysis of the HORIZON I study. *Clin. Color. Cancer* **2014**, *13*, 46–53. [CrossRef]
315. Anami, S.; Doi, H.; Nakamatsu, K.; Uehara, T.; Wada, Y.; Fukuda, K.; Inada, M.; Ishikawa, K.; Kanamori, S.; Nishimura, Y. Serum lactate dehydrogenase predicts survival in small-cell lung cancer patients with brain metastases that were treated with whole-brain radiotherapy. *J. Radiat. Res.* **2019**, *60*, 257–263. [CrossRef]
316. Zhang, Z.; Li, Y.; Yan, X.; Song, Q.; Wang, G.; Hu, Y.; Jiao, S.; Wang, J. Pretreatment lactate dehydrogenase may predict outcome of advanced non small-cell lung cancer patients treated with immune checkpoint inhibitors: A meta-analysis. *Cancer Med.* **2019**, *8*, 1467–1473. [CrossRef]

317. Oya, Y.; Yoshida, T.; Kuroda, H.; Mikubo, M.; Kondo, C.; Shimizu, J.; Horio, Y.; Sakao, Y.; Hida, T.; Yatabe, Y. Predictive clinical parameters for the response of nivolumab in pretreated advanced non-small-cell lung cancer. *Oncotarget* **2017**, *8*, 103117–103128. [CrossRef]
318. Internò, V.; Melaccio, A.; Vitale, P.; Spedaliere, R.; Buonfantino, M.; Messina, R.; Lippolis, A.L.; Signorelli, F.; Addeo, R.; Giuliani, F. Therapeutic inhibition of isocitrate dehydrogenase mutations in glioma and cholangiocarcinoma: New insights and promises-a narrative review. *Chin. Clin. Oncol.* **2024**, *13*, 69. [CrossRef]
319. Lee, C.L.; O'Kane, G.M.; Mason, W.P.; Zhang, W.J.; Spiliopoulou, P.; Hansen, A.R.; Grant, R.C.; Knox, J.J.; Stockley, T.L.; Zadeh, G.; et al. Circulating Oncometabolite 2-hydroxyglutarate as a Potential Biomarker for Isocitrate Dehydrogenase (IDH1/2) Mutant Cholangiocarcinoma. *Mol. Cancer Ther.* **2024**, *23*, 394–399. [CrossRef]
320. Nakagawa, M.; Yamaguchi, M.; Endo, M.; Machida, Y.; Hattori, A.; Tanzawa, F.; Tsutsumi, S.; Kitabayashi, I.; Kawai, A.; Nakatani, F. Clinical usefulness of 2-hydroxyglutarate as a biomarker in IDH-mutant chondrosarcoma. *J. Bone Oncol.* **2022**, *34*, 100430. [CrossRef]
321. Tong, S.; Wu, J.; Song, Y.; Fu, W.; Yuan, Y.; Zhong, P.; Liu, Y.; Wang, B. IDH1-mutant metabolite D-2-hydroxyglutarate inhibits proliferation and sensitizes glioma to temozolomide via down-regulating ITGB4/PI3K/AKT. *Cell Death Discov.* **2024**, *10*, 317. [CrossRef]
322. Wang, Y.-P.; Li, J.-T.; Qu, J.; Yin, M.; Lei, Q.-Y. Metabolite sensing and signaling in cancer. *J. Biol. Chem.* **2020**, *295*, 11938–11946. [CrossRef]
323. Abou-Alfa, G.K.; Macarulla, T.; Javle, M.M.; Kelley, R.K.; Lubner, S.J.; Adeva, J.; Cleary, J.M.; Catenacci, D.V.; Borad, M.J.; Bridgewater, J.; et al. Ivosidenib in IDH1-mutant, chemotherapy-refractory cholangiocarcinoma (ClarIDHy): A multicentre, randomised, double-blind, placebo-controlled, phase 3 study. *Lancet Oncol.* **2020**, *21*, 796–807. [CrossRef]
324. Corrigan, L.; Lowery, M. Ivosidenib for the treatment of isocitrate dehydrogenase-1 mutant cholangiocarcinoma. *Expert Rev. Gastroenterol. Hepatol.* **2021**, *15*, 475–481. [CrossRef]
325. Wu, Z.; Yang, H.; Li, X.; Ji, X.; Mo, C.; Zheng, Z.; Xu, Y.; Xiong, D. Circulating tumor DNA laboratory processes and clinical applications in nasopharyngeal carcinoma. *Front. Oncol.* **2025**, *15*, 1520733. [CrossRef]
326. Peng, H.; Pan, M.; Zhou, Z.; Chen, C.; Xing, X.; Cheng, S.; Zhang, S.; Zheng, H.; Qian, K. The impact of preanalytical variables on the analysis of cell-free DNA from blood and urine samples. *Front. Cell Dev. Biol.* **2024**, *12*, 1385041. [CrossRef] [PubMed]
327. Lee, J.-S.; Cho, E.H.; Kim, B.; Hong, J.; Kim, Y.-G.; Kim, Y.; Jang, J.-H.; Lee, S.-T.; Kong, S.-Y.; Lee, W.; et al. Clinical Practice Guideline for Blood-based Circulating Tumor DNA Assays. *Ann. Lab. Med.* **2024**, *44*, 195–209. [CrossRef]
328. Mavroudakis, L.; Golubova, A.; Lanekoff, I. Spatial metabolomics platform combining mass spectrometry imaging and in-depth chemical characterization with capillary electrophoresis. *Talanta* **2025**, *286*, 127460. [CrossRef]
329. Saunders, K.D.G.; Lewis, H.-M.; Beste, D.J.V.; Cexus, O.; Bailey, M.J. Spatial single cell metabolomics: Current challenges and future developments. *Curr. Opin. Chem. Biol.* **2023**, *75*, 102327. [CrossRef] [PubMed]
330. Petrova, B.; Guler, A.T. Recent Developments in Single-Cell Metabolomics by Mass Spectrometry–A Perspective. *J. Proteome Res.* **2025**, *24*, 1493–1518. [CrossRef] [PubMed]
331. Chen, X.; Peng, Z.; Yang, Z. Metabolomics studies of cell-cell interactions using single cell mass spectrometry combined with fluorescence microscopy. *Chem. Sci.* **2022**, *13*, 6687–6695. [CrossRef] [PubMed]
332. Danzi, F.; Pacchiana, R.; Mafficini, A.; Scupoli, M.T.; Scarpa, A.; Donadelli, M.; Fiore, A. To metabolomics and beyond: A technological portfolio to investigate cancer metabolism. *Signal Transduct. Target. Ther.* **2023**, *8*, 137. [CrossRef]
333. Guo, L.; Wang, Y.; Yang, W.; Wang, C.; Guo, T.A.; Yang, J.; Shao, Z.; Cai, G.; Cai, S.; Zhang, L.; et al. Molecular Profiling Provides Clinical Insights Into Targeted and Immunotherapies as Well as Colorectal Cancer Prognosis. *Gastroenterology* **2023**, *165*, 414–428.e7. [CrossRef]

MDPI AG
Grosspeteranlage 5
4052 Basel
Switzerland
Tel.: +41 61 683 77 34

Biomolecules Editorial Office
E-mail: biomolecules@mdpi.com
www.mdpi.com/journal/biomolecules

www.ingramcontent.com/pod-product-compliance
Lightning Source LLC
Chambersburg PA
CBHW040811250926
51850CB00007BA/1136
9783725875061